PRINCIPLES OF SNOW H

Snow hydrology is a specialized field of hydrology that is of particular importance for high latitudes and mountainous terrain. In many parts of the world, river and groundwater supplies for domestic, irrigation, industrial, and ecosystem needs are generated from snowmelt, and an in-depth understanding of snow hydrology is of clear importance. Study of the impacts of global warming has also stimulated interest in snow hydrology because increased air temperatures are projected to have major impacts on the snow hydrology of cold regions.

Principles of Snow Hydrology describes the factors that control the accumulation, melting, and runoff of water from seasonal snowpacks over the surface of the earth. The book addresses not only the basic principles governing snow in the hydrologic cycle, but also the latest applications of remote sensing, and principles applicable to modelling streamflow from snowmelt across large, mixed land-use river basins. Individual chapters are devoted to climatology and distribution of snow, ground-based measurements and remote sensing of snowpack characteristics, snowpack energy exchange, snow chemistry, modelling snowmelt runoff (including the SRM model developed by Rango and others), and principles of snowpack management on urban, agricultural, forest, and range lands. There are lists of terms, review questions, and problems with solutions for many chapters available online at www.cambridge.org/9780521823623.

This book is invaluable for all those needing an in-depth knowledge of snow hydrology. It is a reference book for practising water resources managers and a textbook for advanced hydrology and water resources courses which span fields such as engineering, Earth sciences, meteorology, biogeochemistry, forestry and range management, and water resources planning.

DAVID R. DEWALLE is a Professor of Forest Hydrology with the School of Forest Resources at the Pennsylvania State University, and is also Director of the Pennsylvania Water Resources Research Center. He received his BS and MS degrees in forestry from the University of Missouri, and his PhD in watershed management from Colorado State University. DeWalle has conducted research on the impacts of atmospheric deposition, urbanization, forest harvesting, and climate change on the hydrology and health of watersheds in Pennsylvania. He regularly teaches courses in watershed management, snow hydrology, and forest microclimatology. In addition to holding numerous administrative positions at Penn State, such as Associate Director of the Institutes of the Environment and Forest Science Program Chair, DeWalle has been major advisor to over 50 MS and PhD students since coming to Penn State in 1969. DeWalle has also been a visiting scientist with the University of Canterbury in New Zealand, University of East Anglia in England, and most recently the USDA, Agricultural Research Service in Las Cruces, New Mexico. He has served as President and is a fellow of the American Water Resources Association.

ALBERT RANGO is a Research Hydrologist with the USDA Agricultural Research Service, Jornada Experimental Range, Las Cruces, New Mexico, USA. He received his BS and MS in meteorology from the Pennsylvania State University and his PhD in watershed management from Colorado State University. Rango has conducted research on snow hydrology, hydrological modelling, effects of climate change, rangeland health and remediation, and applications of remote sensing. He has been President of the International Commission on Remote Sensing, the Western Snow Conference, and the American Water Resources Association. He is a fellow of the Western Snow Conference and the American Water Resources Association. He received the NASA Exceptional Service Medal (1974), the Agricultural Research Service Scientist of the Year Award (1999), and the Presidential Rank Award – Meritorious Senior Professional (2005). He has published over 350 professional papers.

PRINCIPLES OF
SNOW HYDROLOGY

DAVID R. DEWALLE
Pennsylvania State University, USA

ALBERT RANGO
United States Department of Agriculture

CAMBRIDGE
UNIVERSITY PRESS

CAMBRIDGE UNIVERSITY PRESS
Cambridge, New York, Melbourne, Madrid, Cape Town,
Singapore, São Paulo, Delhi, Tokyo, Mexico City

Cambridge University Press
The Edinburgh Building, Cambridge CB2 8RU, UK

Published in the United States of America by Cambridge University Press, New York

www.cambridge.org
Information on this title: www.cambridge.org/9780521290326

First published 2008
First paperback edition 2011

A catalogue record for this publication is available from the British Library

ISBN 978-0-521-82362-3 Hardback
ISBN 978-0-521-29032-6 Paperback

Additional resources for this publication at www.cambridge.org/9780521290326

Contents

The color plates are situated between pages 234 and 235.*

* These plates are available in color for download from www.cambridge.org/9780521290326

Preface

This book is the culmination of several years of effort to create an up-to-date text book and reference book on snow hydrology. Our interest in snow hydrology was initiated while we were both taking an early snow hydrology class taught by Dr. James R. Meiman at Colorado State University. The book is an outgrowth of our interest in snow hydrology borne out of that class and later experiences while teaching snow hydrology courses of our own and conducting snow-related research. The book includes the basics of snow hydrology and updated information about remote sensing, blowing snow, soil frost, melt prediction, climate change, snow avalanches, and distributed modelling of snowmelt runoff, especially considering the effects of topography and forests. A separate chapter is devoted to the SRM or Snowmelt Runoff Model that Rango helped to develop, which includes the use of satellite snow-cover data. We have also added a chapter on management of snowpacks in rangeland, cropland, forest, alpine, and urban settings. Topics related to glaciology and glacial hydrology were largely avoided. The chapters are sequenced so that students with a basic understanding of hydrology and physics can progressively learn the principles of snow hydrology in a semester-long class. The book can serve as a text in an upper-level undergraduate or graduate-level class. We have included example computations in some of the chapters to enhance understanding. A website has been created for students with example problems and discussion questions related to specific chapters.

The authors are indebted to employers, colleagues, students, and family for making this book possible. Work began on the book during spring 2001 while DeWalle was on sabbatical leave from Penn State at USDA, Agricultural Research Service at Beltsville, MD. During summer 2001 the USDA ARS kindly provided support for Dr. DeWalle in Las Cruces, NM to continue this work. Over the intervening years many colleagues offered advice and comments on the various chapters, notably John Pomeroy (Chapter 2), Takeshi Ohta (Chapter 7), Martyn Tranter (Chapter 8), Doug Kane (Chapter 9), George Leavesley and Kevin Dressler (Chapter 10), and

James Meiman (Chapters 2, 3, and 12). Students in several of DeWalle's snow hydrology classes at Penn State also provided useful feedback. We appreciate the many helpful comments we have received, but in the final analysis we are responsible for any errors and mistakes. Penn State students Anthony Buda helped with literature searches, Brian Younkin helped prepare figures, and Sarah MacDougall helped with formatting the text. Staff at Penn State Institutes of the Environment, especially Sandy Beck, Chris Pfeiffer, and Patty Craig also provided valuable help. We are also very appreciative of the many people who kindly gave permissions to use figures, data, photos, etc. for the book. Finally, we are grateful to our families, especially spouses Nancy and Josie, for their support while this book writing project was underway.

Book Cover

Cover photo courtesy Dr. Randy Julander (United States Department of Agriculture, Natural Resources Conservation Service) showing Upper Stillwater Dam on Rock Creek, tributary to the Duchesne River and then Green River, on the south slope of the Uintah Mountains near Tabiona, Utah.

1

Introduction

1.1 Perceptions of snow

There is a general lack of appreciation by society of the importance of snow to everyday life. One good example of this is found in the Rio Grande Basin in the southwestern United States and Mexico. The Rio Grande, the third longest river in the United States, is sustained by snow accumulation and melt in the mountain rim regions which provide a major contribution to the total streamflow, despite its flowing right through the heart of North America's largest desert (Chihuahuan). Because the majority of the population in the basin resides in a few large cities in the Rio Grande Valley, which are all located in the desert (see Figure 1.1), there is little realization on the part of the urban residents that snowmelt far to the north is an important factor in their lives. This same situation is true in many arid mountainous regions around the globe. Where agricultural water use predominates, however, the importance of snow for the water supply and food production is more widely known, at least by farmers and ranchers and the rural populace.

The importance of snow during and in the aftermath of a snowstorm is immediately evident because of its significant effect on transportation (see Figure 1.2). The effects of snowstorms on wagon trains (in the past), railroads, and motorized transport are widely documented by Mergen (1997). Except for very small countries, the effect of a snowstorm on transportation is localized and does not affect the entire country. Such local effects are evident in the United States as in the leeward areas of the Great Lakes where persistent and heavy annual snowfalls occur, such as over 370 in (940 cm) on the Tughill Plateau area of upstate New York (Macierowski, 1979). Areas just a short distance away, however, have annual averages of only about 65 in (165 cm) a year. Storm tracks or snowbands are relatively narrow, so that the majority of the population is seldom impacted but small regions can be severely impacted (see Figure 1.3). One exception to this in winter in the United States is when a low-pressure area tracks its way from the Gulf of Mexico all the

1

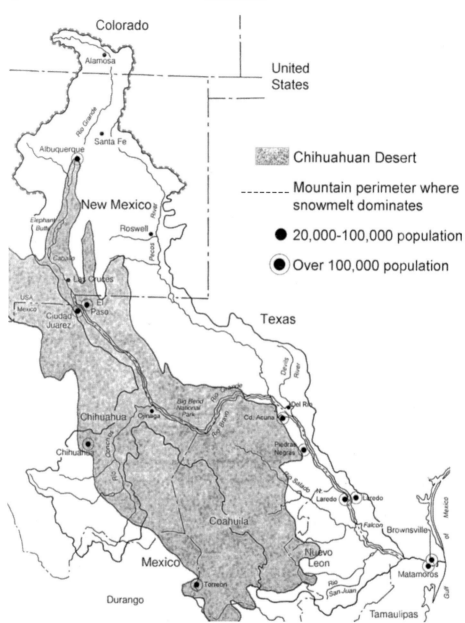

Figure 1.1 Location of the Rio Grande Basin in the United States and Mexico with large cities identified in relation to mountain snowpack areas and desert areas.

Figure 1.2 The effect of snow on transportation as documented on the front page of the Providence, Rhode Island, "The Evening Bulletin," on Wednesday, February 8, 1978 (from www.quahog.org/include/image.php?id=133).

way up the eastern US coast where the majority of the US population is concentrated. Figure 1.4 shows the snow depth map compiled from ground measurements associated with the "Blizzard of 1996" which was a "Nor'easter."

If the storm is particularly severe, normal transport can be shut down for extended periods. Figure 1.5 shows the clearing of mountain highways in the

Figure 1.3 Red Cross workers search for possible victims after a lake effect storm near Buffalo, New York, on January 28–29, 1977 (note that snow depth is up to the top of an automobile). (Courtesy NOAA Photo Library, wea00952, Historic NWS Collection and the American Red Cross.)

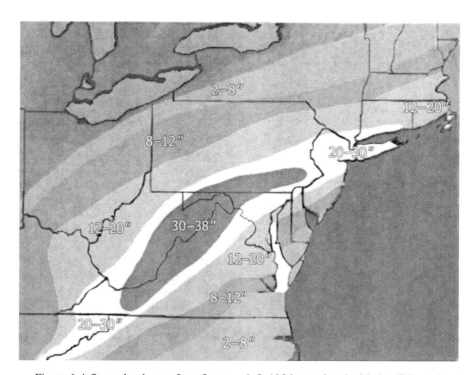

Figure 1.4 Snow depth map from January 6–8, 1996 associated with the "Blizzard of 1996," centered over the middle Atlantic United States (after WRC-TV/NBC 4, Washington, DC analysis). See also color plate.

Figure 1.5 Transportation department snow blower at work clearing Highway 143 between Cedar Breaks and Panguitch, Utah in 2005. (Courtesy R. Julander.) See also color plate.

western United States. Additionally, in these severe snowstorms, communication is frequently disrupted by downed utility lines. Those affected for that period of time would certainly support the (in this case, negative) importance of snow.

Because of the disruptions to human activities caused by extreme snowfalls, Kocin and Uccellini (2004) developed a Northeast Snowfall Impact Scale (NESIS) to categorize snow storms in the northeast United States. The scale is based upon the amount of snowfall, the areal distribution of the snowfall and the human population density in the affected areas. Maximum category 5 "Extreme" events are those like the January 1996 event (NESIS rating = 11.54) depicted in Figure 1.4 with up to 75-cm snowfall depths which affected about 82 million people over nearly 0.81 million square kilometers in the Northeast. At the other extreme, category 1 "Notable" events were those like the February 2003 event (NESIS rating = 1.18) with up to 25-cm snowfall depths which affected 50 million people over 0.23 million square kilometers. The NESIS ratings of storms combined with information about wind speeds controlling drifting during and after snowfall should significantly improve our appreciation of the severity of impacts of snowfall.

When a fresh snowfall blankets the landscape, most people will agree that snow is aesthetically pleasing and a positive visual experience. Less widely known and more important, however, is that a snow cover radically changes the properties of the Earth's surface by increasing albedo and also insulating the surface. Extremely cold air temperatures may exist right above the surface of the snow, but the insulating effect can protect the underlying soil and keep it relatively warm and unfrozen. In General Circulation Models (GCM), it is important to precisely locate the areas of the Earth's surface covered by snow because of the great differences in energy and water fluxes between the atmosphere and snow-covered and snow-free portions of the landscape. Because such differences are important in the reliability of simulations produced by GCMs, providing correct snow cover inputs to GCMs is critical.

In many areas of the world like the western United States, snow accumulation and subsequent melt are the most critical determining factors for producing an adequate water supply. To quantitatively estimate this water supply, including volume, timing, and quality, it is important to have a detailed understanding of snow hydrology processes, the goal of this book. As water demand outstrips the water supply, which is happening in most of the world today, this knowledge of snow hydrology becomes increasingly important.

The technology of remote sensing has had a major impact on data collection for measuring snow accumulation and snow ablation rates. The reasons for the ready application of remote sensing to snow hydrology are multifaceted. Significant snowpacks accumulate and deplete in remote, inaccessible areas that are easily imaged with remote-sensing platforms. These snowpack processes in mountain regions are active generally during the most inhospitable time of year which makes considerations of human safety important. The use of remote-sensing approaches is much safer than employing ground access during these times. In most cases, the appearance of snow in various types of remote-sensing data products is strikingly different from snow-free surfaces, allowing snow mapping in different spectral bands.

A surprising amount of biological activity occurs within and beneath the snow cover, especially so for deep snowpacks, because of warm soil conditions promoted by the insulating effect of the snow (see Jones *et al.*, 2001). This insulating effect protects many types of vegetation from the low air temperatures just above the snow surface. In many cases, plant survival is dependent upon the occurrence of a regular winter snowpack. In agriculture, this property is relied upon for the survival of the winter wheat crop which requires a snowpack of 10 cm or more (Steppuhn, 1981). For the world's winter wheat crop, the Food and Agriculture Organization (1978) has estimated that each centimeter of snow from 5–10 cm in depth would produce a crop survival benefit of $297\,000\,000\ \text{cm}^{-1}$ (Steppuhn, 1981).

Small animals survive beneath the snowpack for the same reasons as certain plants. About 20 cm of snow depth seems to be the breakpoint to allow such activity

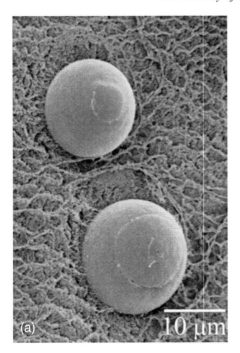 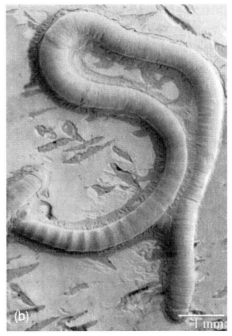

Figure 1.6 (a) Sample of red snow (containing green algae cells, *Chlamydomonas nivalis*) taken from a snowfield surface near the South Cascade Glacier and imaged with the Low-Temperature Scanning Electron Microscope (LTSEM) showing enlarged view of two individual cells. (b) LTSEM image of an ice worm (a species of oligochaetes) which was collected 1 m below the top of the seasonal snowpack near South Cascade Glacier. (Courtesy USDA, Agricultural Research Service, Beltsville, MD.)

(Jones *et al.*, 2001). The snow itself is the habitat for various micro-organisms like snow worms and algae which are shown in Figures 1.6(a) and (b) taken with a Low-Temperature Scanning Electron Microscope. See Jones, *et al.* (2001) for further details on snow ecology.

1.2 History of snow hydrology

Although history suggests that technical understanding of snow hydrology was a relatively recent phenomenon, some evidence exists that the role of snow was understood by some very early in our study of the physical world. References to the philosophy of the ancient Greek, Anaxagoras (500–428 BCE), indicate a rather surprising early understanding of the relationships between river flows and freezing and thawing of water, for example (Franks 1898): *"The Nile comes from the snow*

in Ethiopia which melts in summer and freezes in winter" (*Aet. Plac.* iv 1;385);
*"And the Nile increases in summer because waters flow down into it from snows
at the north"* (*Hipp. Phil.* 8; Dox. 561). Passages from the Bible also show an
early general understanding of the role of snow in the natural world, most notably
perhaps: "*For as the rain and the snow come down from heaven, And do not return
there without watering the earth, And making it bear and sprout, And furnishing
seed to the sower and bread to the eater.*" (Isaiah 55:10, New American Standard
Bible®, Copyright © 1995 by the Lockman Foundation, used by permission). These
early references show that some basic concepts underpinning snow hydrology have
existed for millennia.

Much later, literature from the writings of naturalist/geologist Antonio Vallisnieri
(1661–1730) in Italy showed specific recognition of the role of snow in hydrology.
He correctly theorized that rivers arising from springs in the Italian Alps came from
rain and snowmelt seeping into underground channels (attributed to Lupi, F. W.,
www.killerplants.com/whats-in-a-name/20030725.asp).

In the United States during World War II, the US Army Corps of Engineers and
the US Weather Bureau initiated the Cooperative Snow Investigations in 1944 (US
Army Corps of Engineers, 1956). The snow investigations were organized to address
specific snow hydrology problems that were being encountered by both agencies. In
order to meet snow hydrology objectives of both agencies, it was deemed necessary
to establish fundamental research in the physics of snow. An extensive laboratory
program across the western United States was established and observations were
gathered starting in 1945. Analysis of these data formed the basis for developing
the basic relationships and methods of application derived to develop solutions to
the key snow hydrology problems (US Army Corps of Engineers, 1956).

Three snow laboratories were established: the Central Sierra Snow Laboratory
(CSSL), Soda Springs, CA (see Figure 1.7); the Upper Columbia Snow Laboratory
(UCSL), Marias Pass, MT; and the Willamette Basin Snow Laboratory (WBSL)
Blue River, OR. The drainage areas where the research was concentrated had areas
as follows: CSSL, 10.26 km^2; UCSL, 53.61 km^2; and WBSL, 29.81 km^2. Although
the major report coming from these studies was written in 1956 (US Army Corps
of Engineers, 1956), this book, *Snow Hydrology*, is still an excellent reference book
for students and forms the basis for much of the information on snow hydrology in
basic hydrology texts.

Both the CSSL and WBSL received snow indicative of maritime-influenced climate conditions. The UCSL snowfall was influenced by both maritime and continental climate conditions. A fourth snow laboratory was established for cooperative
snow investigations by the US Bureau of Reclamation and the US Forest Service at
the Fraser Experimental Forest, Fraser, CO. Continuous measurements were made
there starting in 1947. The climate conditions influencing snowfall at Fraser are

Figure 1.7 Instrumentation at the Central Sierra Snow Laboratory in Soda Springs, California. (Courtesy R. Osterhuber.) See also color plate.

more of a true continental origin than the UCSL. The emphasis at Fraser was on evaluation of various snow and runoff measurements, development of snowmelt-runoff forecasting techniques, and the effect of forest management on water yield from snow-fed basins. The final report of the project, *Factors Affecting Snowmelt and Streamflow* (Garstka *et al.*, 1958), has also been identified as a significant contribution to understanding snow hydrology. Snow hydrology research has continued to the current day at the CSSL and the Fraser Experimental Forest.

Outside the United States, a number of snow laboratories and research watersheds were established. The Marmot Creek Basin (9.4 km^2), about 80 km west of Calgary, Alberta, Canada, was instrumented in 1962 to study the water balance in a typical subalpine spruce-fir forested watershed (Storr, 1967). A network of 40 precipitation gauges and 20 snow courses cover the basin. The overriding reason for establishing this research basin was to determine the effects of forest clearing practices on snow accumulation and streamflow. Treatments were performed in 1974 which involved clear-cutting five separate blocks ranging from 8–13 ha (Forsythe, 1997).

Work in Russia on snow hydrology began in the 1930s, but, as in the United States, specific field research sites were set up in the mid- to late 1940s. The primary field sites were the Valdai Hydrological Research Laboratory and the Dubovskoye

Hydrological Laboratory. Much Russian work on heat and water balance of snow, snow cover observations, and snow metamorphism is reported by Kuz′min (1961). In the same work, the arrangements of permanent snow stakes and instrumentation, like gamma ray snow water equivalent detection systems, are reported.

In 1959, the Chinese Academy of Sciences established the Tianshan Glaciological Station at the source of the Urumqi River in the Tianshan Mountains at 3600 m above sea level (a.s.l.) to provide comprehensive observations and studies of snowmelt and glacier hydrology. The station has full research facilities and living accommodation for both permanent staff and visiting scholars (Liu *et al.*, 1991). Studies are conducted on the total Urumqi River basin (4684 km^2) and also on the part of the basin only in the mountains (Ying Qiongqia hydrometric station, 924 km^2). Recent studies at the Tianshan Glaciological Station have emphasized effects of global climate change, ice and snow physical processes, energy and mass balances of glaciers and snow, water balances of the upper mountain snow zone, and application of advanced observation technology, including remote sensing. Applications of outside investigators to participate in research at the Tianskan Glaciological Station are encouraged.

Several notable texts on snow hydrology also exist from early and recent literature. Kuz′min's (1961) summary book on *Melting of Snow*, which has been translated into English, is a rich source of information about early Russian studies. *The Handbook of Snow* edited by Gray and Male (1981) dealt with a wide variety of snow topics, including hydrology. Singh and Singh (2001) have written a book, *Snow and Glacier Hydrology*, that covers topics in snow hydrology and glaciology and some fundamentals of hydrology. Seidel and Martinec (2004) concentrate on remote-sensing applications in *Remote Sensing and Snow Hydrology*.

1.3 Snow hydrology research basins

Experimental basins established for snow hydrology research have had two major functions. First would be for the purpose of collecting all types of snow information for better understanding the physics of snow hydrology. An example of such a basin would be the Reynolds Creek Watershed in southwestern Idaho. The second purpose is for evaluating the effectiveness of snow management treatments to manipulate the quantity, quality, and timing of streamflow. The first example of this in the United States was the Wagon Wheel Gap experiment in the Upper Rio Grande Basin of Colorado (Bates and Henry, 1922; 1928). Both the data collection and snow management functions can be satisfied in the same research area. A good example of this would be at the previously mentioned Fraser Experimental Forest where long-term snow hydrology data have been collected and a classic, paired-watershed study was used to evaluate the effects of forest clear-cutting on snowmelt

Figure 1.8 Fool Creek clear-cut watershed near the center and the East St. Louis Creek control watershed to the right of center, both in the Fraser Experimental Forest, Colorado. (US Forest Service Photo, courtesy C. Leaf.)

runoff using the Fool Creek and East St. Louis Creek Basins (Goodell, 1958; 1959) (Figure 1.8).

Various basins are representative of different snow hydrology regions. Reynolds Creek (239 km^2) (Figure 1.9) is administered by the USDA Agricultural Research Service (ARS) and is indicative of high relief, semiarid rangelands where seasonal snow and frozen soil dominate the annual hydrologic cycle (Slaughter and Richardson, 2000). Data have been collected for 38 years. Another USDA/ARS original experimental watershed, the Sleepers River Basin (111 km^2) in Vermont, started operation in 1958 (Figure 1.10). It is indicative of snow hydrology processes in northeastern United States forested basins. After ARS, this basin was administered by the National Weather Service and now jointly by the US Geological Survey and the US Army Cold Regions Research and Engineering Laboratory. The US Forest Service administers another forested snow research basin in the northeastern United States – the Hubbard Brook Basin (31.4 km^2) near North Woodstock, New Hampshire. In the central Rocky Mountains, the USGS administers the Loch Vale Watershed in the Rocky Mountain National Park, Colorado, a small glacierized basin (6.6 km^2). Seasonal snow accumulation and melt are the major drivers of the

Figure 1.9 Site of the 239 km² Reynolds Creek Experimental Watershed in the Owyhee Mountains about 80 km southwest of Boise, Idaho, operated by the Agriculture Research Service. (Courtesy USDA, Agriculture Research Service; www.ars.usda.gov/is/graphics/photos/.) See also color plate.

hydrologic cycle in this basin. A similar basin (7.1 km²) in the Green Lakes region of Colorado is part of the Niwot Ridge LTER project and is run by the University of Colorado (Bowman and Seastedt, 2001). There are numerous additional basins across the snow zone of the United States that are administered by US Government agencies or universities where significant snow research is being conducted. Many times it is possible to become a collaborating investigator and thereby gain access to a valuable snow data resource.

Today, given Internet availability of data and services, key sources of information related to snow hydrology from Federal agencies within the United States can also be found at:

- USDA, Natural Resources Conservation Service – SNOTEL network of pressure pillows, snow survey data, forecasts of streamflow from snowmelt.
- USDC, NOAA, National Operational Hydrologic Remote-Sensing Center – satellite snow cover, gamma radiation surveys, maps and interactive products.
- USDI, Geological Survey – streamflow and groundwater data.
- USDC, NOAA, National Snow and Ice Data Center – polar and cyrospheric research center, snow cover, sea and ground ice, and more.

Figure 1.10 Outlet of the W-3 watershed in the Sleepers River Basin in Vermont during snowmelt. (Courtesy S. Sebestyen.)

- USDC, NOAA, National Climate Data Center – precipitation, snowfall, temperature, and other data.
- US Dept. of the Army, Corps of Engineers, Cold Regions Research and Engineering Laboratory – basic research on snow, ice, and frozen ground with application to the military.
- US National Atmospheric Deposition Program – precipitation chemistry.

1.4 Properties of water, ice, and snow

Snow hydrology is affected in important ways by the physical and chemical properties of water. Properties such as the density, specific heat, melting and freezing point, adhesion and cohesion, viscosity, and solubility of water are all affected by molecular structure. In this section a brief review is given of the molecular structure of water and ice and how it affects properties of water in all its phases, in preparation for more detailed discussions in later chapters. Physical properties of air, water, and ice useful to snow hydrologists are given in Appendix Tables A1 and A2.

(a) Oxygen – 8 protons +
8 electrons –
(2 inner shell and 6 valence)
Hydrogen –1 proton +
1 electron –

Oxygen of one water molecule can bind
with up to four hydrogen of adjacent water
molecules, which produces hexagonal lattice
structure in common ice.
Hydrogen bonds: ice > liquid > vapor.
Hydrogen bonds cause cohesion among molecules and
increase melting and boiling temperatures.

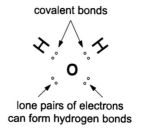

covalent bonds

lone pairs of electrons
can form hydrogen bonds

(b) Protons from oxygen attract
more (-) charge to oxygen side.
Leaves more (+) on hydrogen side.
Resulting dipolar molecules makes
water a good solvent and adhesive.

Angle: 104.5 degrees – liquid,
109 degrees – ice

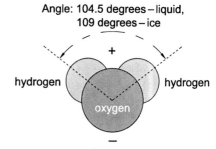

Figure 1.11 Water molecules (a) showing the pairing of electrons between an oxygen atom and two hydrogen atoms and (b) the general shape and dipolar nature of a water molecule with positive charge near the oxygen atom and negative charge near the hydrogen atoms and an angle between the axes of the H–O–H covalent bonds that varies between liquid water and ice.

1.4.1 Nature of liquid water and ice molecules

Water is a dipolar molecule formed from two hydrogen atoms and a single oxygen atom. The hydrogen atoms each have a nucleus of one positively charged proton around which a single negatively charged electron rapidly spins. Oxygen has a nucleus consisting of eight protons and eight uncharged neutrons that is surrounded by eight electrons. Together these ten electrons align to form five orbitals: one pair remains closely associated with the oxygen atom in the inner or 1s shell, two pairs of electrons form the covalent O–H bonds, and two pairs represent two so-called lone pairs of electrons. The structure of the water molecule with the eight electrons can be represented as shown in Figure 1.11(a) where the electron pairs between the O and H represent the covalent bonds and the other electron pairs are the lone pairs. These electrons form a cloud of electric charge moving about the molecule.

The shape of the water molecule with its cloud of rotating electrons can be better represented as a rather rotund molecule with an angle of about 104.5° between the H–O–H branches for liquid water and about 109° for normal hexagonal ice (Figure 1.11(b)). Since opposite charges attract, the eight protons in the oxygen nucleus

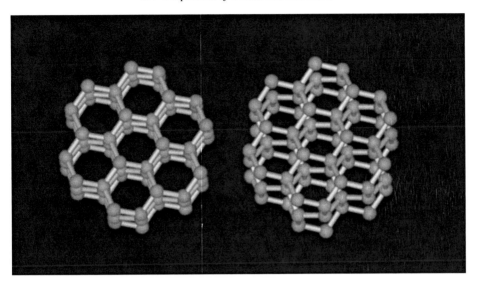

Figure 1.12 Hexagonal structure of normal or I_h ice showing the oxygen atoms (dark circles) and hydrogen atoms (gray bars) with four hydrogen bonds between oxygen atoms in one molecule and hydrogen atoms in adjacent molecules (with permission from K. G. Libbrecht, Caltech from the SnowCrystals.com website).

exert a stronger attraction for the electrons than the hydrogen nucleii with one proton each, leaving the hydrogen side of the molecule with slightly less electron charge than the oxygen side which has a higher density of negative charge. Thus, as shown in Figure 1.11(b), the oxygen side represents the negatively charged part and the hydrogen side represents the positively charged part of the electric field around the water molecule, giving the molecule its dipolar nature. The dipolar nature of the water molecule gives water its ability to act as a solvent and its adhesive nature.

Since there are two lone electron pairs in a water molecule, each oxygen can align with up to four hydrogen atoms from adjacent water molecules forming what are known as hydrogen bonds. These hydrogen bonds are much weaker than covalent bonds and in a liquid form near freezing the bonds are continually being formed and broken. Each oxygen in liquid water is associated with about 3.4 hydrogen atoms, while in ice the bonds are more permanent with about four hydrogen bonds per oxygen being formed. Even fewer hydrogen bonds exist for water in a vapor form. Regardless, hydrogen bonding among molecules of water leads to much higher intermolecular attractive forces which causes higher melting and boiling points, higher specific heat, and higher latent heats of fusion and vaporization for water than most other liquids.

The hydrogen bonds in ice produce a characteristic hexagonal lattice structure represented in Figure 1.12. Ice can exist in at least 10 other forms, but for

Introduction

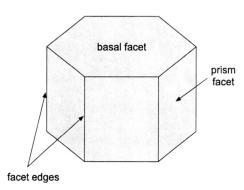

Figure 1.13 Typical hexagonal shape of tiny ice crystals that grow to produce snow crystals which shows basal facets on top and bottom and six side prism facets. Crystal growth often occurs with condensation/sublimation in clouds along the axes on the rougher edges between side facets producing characteristic six-sided snow flakes.

temperature and pressure conditions at the Earth's surface the hexagonal I_h form is the form encountered by snow hydrologists. The hexagonal lattice structure of ice is more open than for liquid water at about the same temperature, thus the density of ice is lower than the density of liquid water near freezing. As liquid water warms above $0\,°C$, molecules of water can become more closely packed giving a maximum density for water at about $4\,°C$. Thus ice floats on a cold liquid water surface. Ice density at $0\,°C$ is generally given as 917–920 kg m^{-3}, while water density at $4\,°C$ is about 1000 kg m^{-3}. Warming water above temperatures of $4\,°C$, does produce the expected decrease in density with increasing temperature common for other liquids, causing density to decrease to about 996 kg m^{-3} at $30\,°C$. Density of ice and water are given in Appendix Table A.2 as a function of temperature.

At subfreezing temperatures there is evidence that a quasi-liquid layer of water exists at the surface of a crystalline ice lattice. This layer may be only one or a few molecules thick or exist as islands of liquid on the ice surfaces. The presence of a liquid-like layer of water on the surface of ice may be responsible for some chemical reactions observed at subfreezing temperatures and gives ice surfaces their slippery nature.

The hexagonal I_h ice structure produces crystals with flat basal planes on top and bottom and six prism faces as shown in Figure 1.13. Growth of the crystal generally occurs more rapidly on the rougher edges between these crystal faces or facets, producing branches that lead to the normal hexagonal snow flakes encountered in nature. However, atmospheric conditions in clouds, temperature, and degree of supersaturation, also dictate how these crystals grow and various crystal types including needles and columnar crystals can develop as described later in Chapter 2.

Table 1.1 *Description of phase changes of water. (Based upon data in List, 1963.)*

PHASE CHANGE	PROCESS	ENERGY EXCHANGED[a]	MJ kg^{-1} @ 0 °C
Liquid → Vapor	Evaporation	Latent heat of vaporization (L_v)	−2.501
Vapor → Liquid	Condensation	Latent heat of vaporization (L_v)	+2.501
Solid → Liquid	Melting	Latent heat of fusion (L_f)	−0.334
Liquid → Solid	Freezing	Latent heat of fusion (L_f)	+0.334
Solid ↔ Vapor	Sublimation, either way	Latent heat of sublimation (L_s)	±2.835

[a] L_v in MJ kg^{-1} $= 3 \times 10^{-6} T^2 - 0.0025T + 2.4999, -50\,°C \leq T \leq 40\,°C$
L_f in MJ kg^{-1} $= -1 \times 10^{-5} T^2 + 0.0019T + 0.3332, -50\,°C \leq T \leq 0\,°C$
L_s in MJ kg^{-1} $= L_v + L_f, -50\,°C \leq T \leq 0\,°C$

1.4.2 Phase changes of water

An understanding of phase changes of water in the environment is fundamental to snow hydrology. Water commonly exists in all three phases – gas, liquid, and solid – at the same time in cold environments. Changes in phase involve energy transfer expressed as latent heats in MJ kg^{-1} (Table 1.1), which vary slightly with temperature. Latent heats can be adjusted to a specific temperature using equations given at the bottom of Table 1.1. The latent heat of sublimation at subfreezing temperatures is the sum of the respective latent heats of vaporization and fusion. At the freezing point, the latent heat of vaporization is approximately 7.5 times greater than the latent heat of fusion. Thus, condensation/evaporation of water involves 7.5 times more energy exchange per kg than melt/freeze conditions.

1.4.3 Snowpack water equivalent

One of the most common properties of snowpacks needed by snow hydrologists is snowpack water equivalent. The water equivalent of a snowpack represents the liquid water that would be released upon complete melting of the snowpack. Water equivalent is measured directly or computed from measurements of depth and density of the snowpack as:

$$\text{SWE} = d(\rho_s / \rho_w) \tag{1.1}$$

SWE = water equivalent, m
d = snowpack depth, m
ρ_s = snowpack density, kg m^{-3}
ρ_w = density of liquid water, approx. 1×10^3 kg m^{-3}

Given measurements of snowpack depth of 0.22 m and snowpack density of 256 kg m^{-3}, the snowpack water equivalent would be:

$$\text{SWE} = (0.22)(256/1000) = 0.0563 \text{ m or } 5.6 \text{ cm}$$

Snowpack water equivalent includes any liquid water that may be stored in the snowpack along with the ice crystals at the time of measurement. Snowpack water equivalent is treated as a primary input to the discussion of snow hydrology in the following chapters.

1.5 References

Bates, C. G. and Henry, A. J. (1928). Forest and streamflow experiment at Wagon Wheel Gap, CO.: final report on completion of second phase of the experiment. *Mon. Weather Rev.*, **53**(3), 79–85.

Bates, C. G. and Henry, A. J. (1922). Streamflow experiment at Wagon Wheel Gap, CO.: preliminary report on termination of first stage of the experiment. *Mon. Weather Rev.*, (Suppl. 17).

Bowman, W. D. and Seastedt, T. R. (eds.) (2001). *Structure and Function of an Alpine Ecosystem, Niwot Ridge, CO*. Oxford: Oxford University Press.

Food and Agriculture Organization (1978). *1977 Production Yearbook: FAO Statistics*, vol. 31, Ser. No. 15. Rome: United Nations Food and Agriculture Organization.

Forsythe, K. W. (1997). Stepwise multiple regression snow models: GIS application in the Marmot Creek Basin (Kananaskis Country, Alberta) Canada and the National Park Berchtesgaden, (Bayern) Germany. In *Proceedings of the 65th Annual Meeting of the Western Snow Conference*, Banff, Alberta, Canada, pp. 238–47.

Franks, A. (1898). (ed., trans.). *The First Philosophers of Greece*. London: Kegan Paul, Trench, Trubner, and Company.

Garstka, W. U., Love, L. D., Goodell, B. C., and Bertle, F. A. (1958). *Factors Affecting Snowmelt and Streamflow; a Report on the 1946–53 Cooperative Snow Investigations at the Fraser Experimental Forest, Fraser, CO*. Denver, CO: USDA Forest Service, Rocky Mountain Forest and Range Experiment Station and USDI Bureau of Reclamation, Division of Project Investigations.

Goodell, B. C. (1958). *A Preliminary Report on the First Year's Effects of Timber Harvesting on Water Yield from a Colorado Watershed*. No. 36. Fort Collins, CO: Rocky Mountain Forest and Range Experiment Station, USDA Forest Service.

Goodell, B. C. (1959). Management of forest stands in western United States to influence the flow of snow-fed streams. *International Association of Scientific Hydrology*, **48**, 49–58.

Gray, D. M. and Male, D. H. (1981). *Handbook of Snow, Principles, Processes, Management and Use*. Ontario: Pergamon Press.

Jones, H. G., Pomeroy, J. W., Walker, D. A., and Hohem, R. W. (eds.). (2001). *Snow Ecology*. Cambridge: Cambridge University Press.

Kocin, P. J. and Uccellini, L. W. (2004). A snowfall impact scale derived from Northeast storm snowfall distributions. *Bulletin Amer. Meteorol. Soc.*, **85**(Feb), 177–194.

Kuz'min, P. P. (1961). *Protsess tayaniya shezhnogo pokrova (Melting of Snow Cover)*. *Glavnoe Upravlenie Gidrometeorologicheskoi Sluzhby Pri Sovete Ministrov SSSR Gosudarstvennyi Gidrologicheskii Institut*. Main Admin. Hydrometeorol. Service,

USSR Council Ministers, State Hydrol. Institute. Translated by Israel Program for Scientific Translations. Avail from US Dept. Commerce, National Tech. Inform. Service, 1971, TT 71–50095.

List, R. J. (1963). Smithsonian meteorological tables. In *Revised Edited Smithsonian Miscellaneous Collections*. Vol. 114, 6th edn. Washington, DC: Smithsonian Institution.

Liu, C., Zhang, Y., Ren, B., Qui, G., Yang, D., Wang, Z., Huang, M., Kang, E., and Zhang, Z. (1991). *Handbook of Tianshan Glaciological Station*. Gansu: Science and Technology Press.

Macierowski, M. J. (1979). *Lake Effect Snows East of Lake Ontario*. Booneville, NY: Booneville Graphics.

Mergen, B. (1997). *Snow in America*. Washington, DC: Smithsonian Institution Press.

Seidel, K. and Martinec, J. (2004). *Remote Sensing in Snow Hydrology: Runoff Modeling, Effect of Climate Change*. Berlin: Springer-Praxis.

Singh, P. and Singh, V. P. (2001). *Snow and Glacier Hydrology*. Water Science and Technol. Library, vol. 37, Dordrecht: Kluwer Academic Publishers.

Slaughter, C. W. and Richardson, C. W. (2000). Long-term watershed research in USDA-Agricultural Research Service. *Water Resources Impact*, **2**(4), 28–31.

Steppuhn, H. (1981). Snow and Agriculture. In *The Handbook of Snow*, ed. D. M. Gray and D. H. Male, Willowdale, Ontario: Pergamon Press, pp. 60–125.

Storr, D. (1967). Precipitation variations in a small forested watershed. In *Proceedings of the 35th Annual Meeting of the Western Snow Conference*, Boise, ID, pp. 11–17.

US Army Corps of Engineers. (1956). *Snow Hydrology: Summary Report of the Snow Investigations*. Portland, OR: US Army Corps of Engineers, North Pacific Division.

2

Snow climatology and snow distribution

One of the most fundamental aspects of snow hydrology is an understanding of the processes that lead to snowfall and the eventual distribution of a snowpack on the landscape. The factors that lead to the formation of snowfall are generally discussed in the beginning sections of this chapter. Since snowfall once formed, unlike rain, is quite easily borne by the wind and redistributed across the landscape before finally coming to rest to form a snowpack, the basic principles controlling blowing snow are also reviewed. Finally, the interception of snow by vegetation, that can have a profound effect on the amount and timing of snow that accumulates into a snowpack beneath a plant canopy, is described in the last section of the chapter.

2.1 Snowfall formation

The occurrence of snowfall in a region is generally dependent upon several geographic and climatic factors: latitude, altitude, the distance from major water bodies, and the nature of regional air mass circulation (McKay and Gray, 1981). General discussions of the climatic factors affecting snowfall formation and precipitation are given by Sumner (1988) and Ahrens (1988). Latitude and altitude largely control the temperature regime of a region and dictate where it is cold enough for snowfall to occur. Virtually no snowfall occurs in low latitudes where the heat balance at the Earth's surface causes air temperatures to average well above freezing during winter. Since air temperature declines with altitude, higher altitudes generally mean lower air temperatures and greater opportunity for snowfall to occur. Proximity to major water bodies affects atmospheric moisture supplies needed for snowfall formation. However, atmospheric circulation is needed to transport water vapor and ultimately controls the nature of interactions between the land surface and various air masses in a region. Regions typically downwind from oceans or large lakes receive moist air that can lead to snowfall, providing temperatures are low enough at the surface. Circulation of moist air over mountain ranges or moist, warm air over cold, dry air masses can cause uplifting and cooling leading to snowfall. Given cold air and an adequate supply of moisture, it is the atmospheric circulation that ultimately explains the local and regional snowfall climatology. In this section, the meteorological factors affecting snowfall are described.

All precipitation ultimately depends upon air becoming saturated with water vapor. A primary mechanism for saturation to occur in the atmosphere is uplifting which cools the air below the dewpoint temperature. Uplifting of air can occur by uplifting over mountains and other terrain features, frontal activity, convection, and convergence in the atmosphere. Saturation can also be achieved by contact of cold air with relatively warm water surfaces. In either case, saturation levels are reached in the air mass and ice crystal and snowflake formation may begin.

2.1.1 Ice crystal and snowflake formation in clouds

The creation of snow and ice crystals within cold clouds is a complicated process that largely involves the interaction of super-cooled water droplets and tiny ice crystals (Schemenauer *et al.*, 1981; Sumner, 1988). Super-cooled water droplets form in clouds by condensation of water vapor on condensation nuclei from soil dust, pollution, forest fires, sea salt spray, and other sources. These droplets can exist at temperatures below freezing in clouds that are supersaturated with water vapor. Tiny ice crystals also form in such clouds due to spontaneous freezing of super-cooled water droplets, sublimation of vapor onto freezing nuclei, and freezing of droplets onto freezing nuclei. Freezing nuclei are thought to be certain types of clay minerals carried in the atmosphere that resemble ice crystals.

Whatever the cause, tiny water droplets and ice crystals coexist in cold clouds at the same temperature. This coexistence is a dynamic one due to the fact that the saturation vapor pressure over the droplets is slightly greater than that over the ice crystals. Thus, there is a transfer of water vapor from the droplets to the crystals and crystals grow at the expense of droplets. This process is referred to as the Bergeron process and is schematically depicted in Figure 2.1. By this process, crystals can grow in clouds as long there is a sufficient supply of vapor to the cloud. The type and shape of ice crystals formed by this process is primarily a function of temperature and secondarily a function of the degree of supersaturation in the clouds (Figure 2.2). Crystals also grow due to interactions with other crystals and contact with super-cooled water droplets to create snow flakes and other types of frozen precipitation. Crystal contact with super-cooled water droplets causes freezing onto the surface and a gradual rounding of the crystal by a process referred to as riming. Heavily rimed crystals are referred to as graupel. When crystals accrete to a size and mass that allows gravitational settling, they fall from the cloud. Some pictures of falling snow crystal types taken with a scanning electron microscope are shown in

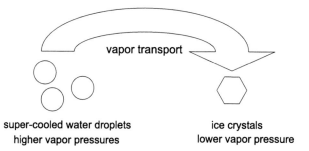

Figure 2.1 Schematic diagram of the Bergeron process for ice crystal growth within clouds.

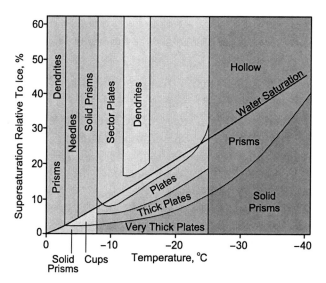

Figure 2.2 Interactions of cloud temperature and cloud super-saturation with water vapor upon the type of ice crystal formed. (Courtesy of D. Cline, after Kobayshi, 1961.)

Figure 2.3; methods for characterizing crystal types are described in Chapter 4. The international system for simple classification of solid precipitation crystal types or forms that ultimately reach the Earth's surface is portrayed in Figure 2.4.

2.1.2 Orographic lifting

In rugged terrain, snowfall formation commonly occurs due to orographic lifting when moving air masses are forced over mountain ranges (Figure 2.5). Many

Figure 2.3 Scanning electron microscope pictures of selected snow crystal types formed in the atmosphere; (a) steller crystals, (b) capped columns, (c) rimed plate, and (d) needles. (Courtesy USDA, Agricultural Research Service, Beltsville, MD.)

variables control the magnitude of the orographic effect such as the upper-level wind speeds, precipitable water in the air mass, ground slope, and the direction of the wind relative to the axis of the mountain range (Barry, 1992; Whiteman, 2000). Air laden with moisture moving at high velocity up steep slopes perpendicular to the long-axis of mountain ranges generally produces the most snowfall. Orographic lifting often gives rise to increasing snowfall with elevation on the windward side of the mountains. However, when wind directions vary, to become nearly parallel to mountain ranges, variation in snowfall with elevation may not occur. Each storm has somewhat unique characteristics and an entire winter's snowpack can be a mix of storm effects, so correlations of elevation with snowfall and snowpack water equivalents are often highly variable.

Orographic precipitation is responsible for snowfall generation in many parts of the world including the continental United States (Figure 2.6). In Figure 2.6, orographic effects that create bands of high snowfall at high elevations are clearly visible along the Pacific Coast Range, Intermountain and Rocky Mountain ranges in

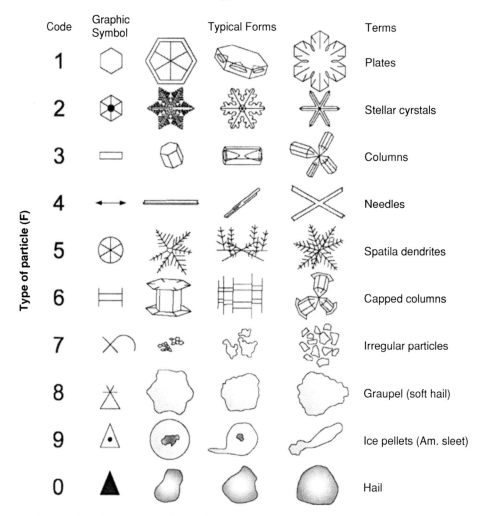

Figure 2.4 Snow crystal forms from the international classification system for solid precipitation (adapted from Mason, 1971).

the west and Appalachian Mountain range in the east. Bands of gradually increasing mean annual snowfall with increasing latitude are also visible in the central region of the country. High snowfall downwind from the Great Lakes due to lake-effect snow is also evident as discussed in the section on lake-effect snowfall.

2.1.3 Frontal activity

The uplifting of air masses associated with warm fronts (warm air moving horizontally and being uplifted over cool air) or cold fronts (cool air pushing under warm

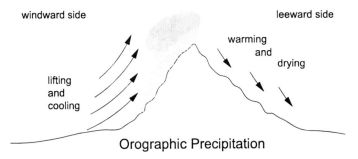

windward side

leeward side

warming
and
drying

lifting
and
cooling

Orographic Precipitation

Figure 2.5 Orographic snowfall generation.

air) also can lead to snowfall (Figure 2.7). The boundary between cool air and warm air is much steeper with cold fronts than with warm fronts. Thus, cold fronts give rise to more rapid uplifting and more intense snowfall, but generally over a smaller geographic area and for a shorter duration at a point than warm fronts. Snowfall associated with warm fronts is often spread over a broader area, but with lower intensity.

2.1.4 Convergence

Air converging towards regions of low pressure at the surface can result in uplift and cooling and widespread snowfall. Convergence at the surface is accompanied by divergence aloft. When divergence aloft is greater than convergence at the surface the low pressure and uplifting at the surface intensifies. Frontal activity often accompanies low-level convergence around a surface low-pressure center. Terrain features can also cause convergence of air flows around isolated mountains or through narrow valleys and cause localized uplift and snowfall.

2.1.5 Lake-effect snowfalls

Convection due to contact of cold air with warm surfaces such as large unfrozen water bodies in winter can cause uplifting and localized heavy snowfall in winter (Figure 2.8). Convection over mountains when exposed rock surfaces are heated by the Sun can also lead to localized convection and light snowfall/graupel events even in summer.

Lake-effect snowfalls influence snow accumulation over hundreds of kilometers downwind from the Great Lakes in North America (Figure 2.9) and can be responsible for some of the highest snowfall intensities recorded. Lake-effect snows can be attributed to uplifting caused by the addition of moisture and heat to cold, dry continental air in contact with a warm, unfrozen lake surface. The magnitude of the lake-effect snowfall effect is controlled by the temperature contrast between the

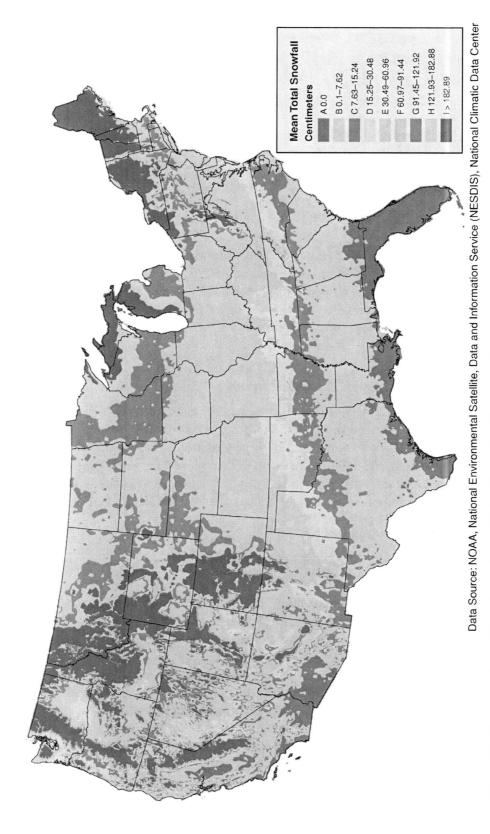

Mean Total Snowfall Centimeters

A 0.0
B 0.1–7.62
C 7.63–15.24
D 15.25–30.48
E 30.49–60.96
F 60.97–91.44
G 91.45–121.92
H 121.93–182.88
I > 182.89

Data Source: NOAA, National Environmental Satellite, Data and Information Service (NESDIS), National Climatic Data Center (modified from USDC, NOAA, National Climate Data Center, Climate Maps of the United States). See also color plate.

Figure 2.6 Mean annual snowfall in the continental United States

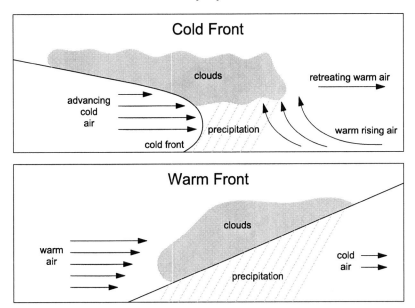

Figure 2.7 Schematic diagram showing relationships between air masses in cold fronts (upper) and warm fronts (lower).

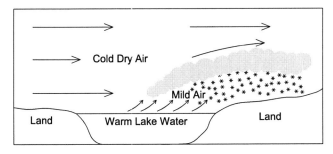

Figure 2.8 Diagram showing generation of lake-effect snowfall due to convection of warmed, moist air off the lake and downwind uplifting and convergence.

warm water and cold air and the distance or fetch that the cold air travels over the warm water, which in turn is controlled by wind direction and synoptic weather conditions. Air to water temperature contrasts and evaporation rates from unfrozen lakes are generally greatest in early winter leading to greater penetration of lake effect snowfall downwind from the lake in early winter. Slight increases in ground elevation downwind from the lake can help trigger cooling and snowfall when air masses laden with moisture experience even minor uplifting, greater turbulence due to rougher shoreline terrain and convergence with air aloft downwind from large water bodies.

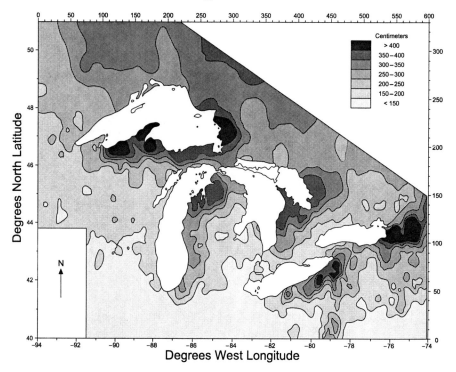

Figure 2.9 Distribution of annual snowfall around the Great Lakes in North America during 1951–80 showing enhanced downwind snowfall due to lake-effect snowfalls (adapted from Norton and Bolsenga, 1993, © 1993 American Meterological Society).

2.1.6 *El Niño/La Niña Southern Oscillation effects*

The Southern Oscillation which gives rise to El Niño conditions, or its counterpart La Niña conditions, is now known to have far-reaching effects on global climate and snowfall (Clark *et al.*, 2001; Smith and O'Brien, 2001; Kunkel and Angel, 1999). El Niño conditions are represented by warm southern Pacific Ocean surface temperatures off the west coast of Peru during winter, while La Niña conditions in contrast are represented by cool sea surface temperatures in that region. The occurrence of El Niño or La Niña conditions alternate within a period of about three to seven years, although the El Niño events have been much more frequent in recent years. The causes and forecasting of these oscillations are the topics for much research, but their effect on climate and snowfall is rapidly becoming appreciated.

These oscillations between ocean surface warming and cooling cause shifts in the position of the jet stream over North America that can influence regional and even global climate. El Niño events cause the jet stream to be deflected northward towards Alaska and are associated with warmer, dryer winter conditions and lower snowfall in the western United States. In contrast, La Niña conditions direct the

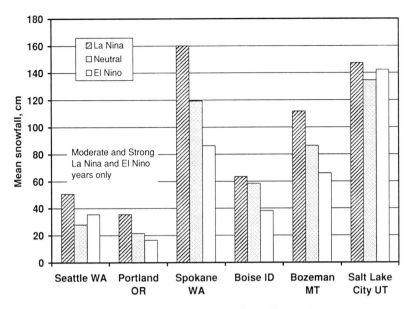

Figure 2.10 Mean winter snowfall for locations in the Pacific Northwest and western regions of the United States for La Niña, Neutral and El Niño years (adapted from USDC, NOAA, Climate Prediction Center, El Niño/LaNiña Snowfall for Selected US Cities).

jet stream toward the Pacific Northwest of the United States where it produces slightly cooler and relatively moist winters with greater snowfall. Figure 2.10 shows snowfall totals for November–March for El Niño, La Niña, and Neutral years at various locations in the Pacific Northwest and western regions of the United States. Snowfall totals in moderate to strong La Niña years averaged higher than in moderate to strong El Niño years, while neutral years generally had intermediate snowfall totals. Several investigators have found that snowpack and streamflow from snowmelt in the northwest United States can be related to indicies of the strength of the Southern Oscillation (for example see Beebee and Manga 2004, McCabe and Dettinger 2002).

2.2 Blowing snow

Snow is often transported by the wind for distances measured in km before it sublimates or comes to rest to form a snowpack. Blowing snow creates problems with direct measure of snowfall using standard precipitation gauges as described in Chapter 4 and also leads to very irregular snowpack distribution. Uneven snowpack distribution in turn causes serious problems with modelling snowmelt runoff due to uneven melt water delivery to soils across a watershed. Recent development of snowpack remote-sensing methods and theoretical models for computing fluxes of

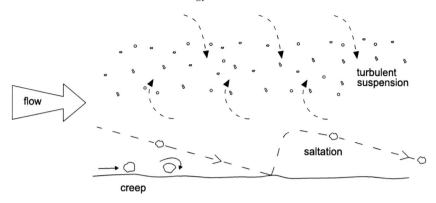

Figure 2.11 Modes of transport for blowing snow.

blowing snow have helped account for blowing-snow effects. Sublimation of blowing snow can also result in losses of as much as 50% of winter precipitation. Losses of this magnitude in regions with low precipitation and without supplemental irrigation can have significant impacts on spring soil moisture levels and agricultural productivity. Blowing snow also interacts with vegetation, especially trees, to influence the magnitude of canopy interception losses. Canopy interception losses are discussed in the following section of this chapter. For many reasons, an enhanced understanding of blowing-snow processes is important to snow hydrology.

2.2.1 Modes of blowing-snow transport

Three major modes of transport are commonly recognized for transport of blowing snow: turbulent suspension, saltation, and creep (Figure 2.11). Transport of snow in full turbulent suspension occurs when the uplift due to turbulent eddies in the air can completely support the ice grains against settling due to gravity. Turbulent suspension generally involves transport of smaller ice and snow particles and can involve an air layer up to several meters thick. Larger particles that are only partially supported by turbulent eddies are often transported in the saltation mode where the particles bounce and skip along the ground surface. Saltation occurs in a layer only a few centimeters thick over the snow surface, but blowing-snow episodes always begin with saltation and the subsequent breaking and loosening of surface snow particles. Saltating particles often loosen others upon impact with the surface. Under some conditions ice particles may also be transported by surface creep where large, loose particles simply slide or roll along the surface due to the force of the wind. The dominant mode and magnitude of blowing-snow transport depends upon the interaction of climate, snow surface conditions, and terrain features, but in general transport is dominated by suspension and secondarily by saltation.

2.2.2 Factors influencing blowing snow

Blowing-snow occurrence generally depends upon interactions among climatic, snow surface, and topographic conditions (Kind, 1981). Blowing snow over a smooth unobstructed snowpack surface begins whenever the force of the wind or shear stress on the surface exceeds the snowpack surface shear strength that resists movement. Shear stress of the wind depends upon the roughness of the surface and wind speed. Rougher surfaces generate greater turbulence and greater shear stress for a given wind speed. Wind shear also increases with wind speed and consequently the winter climatology of an area will play an important part in control of the occurrence of blowing snow. High amounts of snowfall, high wind speeds, and cold air temperatures that slow metamorphism and melt of the snowpack surface all contribute to greater masses of blowing snow.

Under neutral stability conditions in the atmosphere, the shear stress of wind over a uniform snow surface of infinite extent can be related to wind speed (u) and surface roughness (z_0) using the simple logarithmic profile Equation (2.1):

$$\tau = \rho k^2 u^2 [\ln(z/z_0)]^{-2} \tag{2.1}$$

where:

τ = shear stress, kg m^{-1} s^{-2}
ρ = air density, kg m^{-3}
k = von Karman's constant, = 0.4, dimensionless
u = wind speed, m s^{-1}
z_0 = roughness length, m
z = height of wind speed measurement, m

Shear stress is also often expressed as a friction velocity (u_*, m s^{-1}) where,

$$u_* = (\tau/\rho)^{1/2} = uk[\ln(z/z_0)]^{-1} \tag{2.2}$$

If a wind speed of 2 m s^{-1} is measured at a height of 2 m above a snow surface with a roughness length of 0.001 m and the air temperature is $-5\,°C$ giving an air density of 1.3 kg m^{-3}, then by Equation (2.1) the shear stress is:

$$\tau = (1.3)(0.4)^2(2)^2[\ln(2/0.001)]^{-2} = 0.0144 \text{ kg m}^{-1}\text{ s}^{-2}$$

The corresponding friction velocity by Equation (2.2) would be:

$$u_* = (0.0144/1.3)^{1/2} = 0.105 \text{ m s}^{-1}$$

The computed shear stress must be compared to the surface shear strength of the snow to determine if blowing snow will occur. The condition when the shear stress increases to the point when snow particles are first set in motion defines the threshold shear stress, wind speed or friction velocity. Alternatively, the threshold condition can also be defined as the shear stress, wind speed or friction velocity when particle motion ceases. Regardless, the threshold shear stress needed to initiate blowing snow is highly dependent on snow surface metamorphism and bonding of ice crystals. Fresh, light snow that has not been strongly bonded to surrounding crystals can exhibit relatively low surface shear strength and be set in motion at shear stresses of only about 0.01 kg m^{-1} s^{-2}, while hard, well-bonded or wet snow surfaces may require shear stresses of 1 kg m^{-1} s^{-2} to be scoured by the wind (see Kind, 1981). Thus in the above example, a wind shear stress of 0.0144 kg m^{-1} s^{-2}, corresponding to a 2 m s^{-1} wind speed, would be sufficient to transport fresh new snow, but this wind speed would be far too low to erode a hardened compacted snow surface. In fact, for the same snowpack roughness and air temperature used in the example, the wind speed would have to be nearly 12 m s^{-1} to initiate movement from a hard compacted snow surface. As a rough rule-of-thumb, friction velocities needed to initiate motion of 0.2 m s^{-1} can be used for fresh, light snow and 1 m s^{-1} for hard wind-packed or wet snow. Snowpack metamorphism processes that affect the shear strength of snowpack surfaces are discussed in Chapter 3 of this book. Although physical tests of the shear strength of the snow surface can be conducted, occurrence of blowing snow can be operationally predicted using wind speed, air temperature, and age of the snowpack surface (Li and Pomeroy, 1997a, 1997b).

During blowing-snow events part of the force of the wind is expended on the snow surface and part is used to transport blowing snow. Short vegetation like grasses, forbs, brush, and crop stubble, in areas with sufficient blowing snow, generally cause drifts to form to the tops of the plants; a process that increases the snow water stored on the landscape (McFadden *et al.*, 2001). Consequently, changes in vegetative cover over a watershed can markedly alter the snow storage. When the tops of plants protrude through the snowpack, part of the force of the wind will be expended on these objects and less energy will be available to transport snow from the surface. Intermittent melt events during a winter season allow vegetation to protrude through the snowpack surface and also increase the bonding of surface snow crystals, both of which reduce the total amount of blowing snow. Isolated larger objects protruding through the snow, like boulders and tree boles, can lead to localized scouring and drift formation, while numerous large protrusions such as groves of trees can completely eliminate the opportunity for blowing snow by absorbing momentum of the moving air and displacing the entire air stream upward above the snow surface. However in that case, the force of the wind can be expended, at least partially, in transport of snow intercepted on the plant canopy.

Patterns of snowpack accumulation over the landscape are related to regions with accelerating and decelerating winds due to surface obstructions and topography. The mass of blowing snow over flat terrain generally increases in the downwind direction and reaches a steady state over fetches of 300 m or more. Abrupt transitions such as fence lines and ridges and more gradual changes in watershed relief, in combination with prevailing wind directions, can give rise to repeating annual patterns of snow accumulation and scour. Large drifts often form downwind from obstacles and large open areas of terrain that represent source areas for blowing snow.

2.2.3 Sublimation of blowing snow

Static snowpack surfaces tend to lose mass by sublimation during winter (see Chapter 9), but sublimation losses are greatly increased when ice particles are transported by the wind due to the greater amount of exposed surface area and increased convection. Blowing-snow particles will lose mass due to sublimation whenever the vapor pressure at the ice-particle surface is greater than that in the air (Schmidt, 1972). The vapor pressure at the blowing ice-particle surface is the saturation vapor pressure with respect to ice and increases as air temperature increases to 0 °C. Sublimating particles will cool to the ice-bulb temperature and extract sensible heat from the air by slight cooling (0.5–1 °C) of the air layer in contact with the blowing snow. Solar radiation on clear days can also add energy to the blowing snow particles and enhance sublimation rates. Air will gradually increase in vapor content from the sublimating particles and the rise in humidity can limit the rate of sublimation downwind, unless the air is exchanged with dryer air from aloft. Despite the opportunity for humidity increases in the air layer near the ground, very significant losses of mass from sublimation of blowing snow have been reported.

Table 2.1 shows a wide range in estimates of sublimation losses due to blowing snow. Most of these results are based upon model calculations, augmented in some cases by comparisons with field measurements of snow accumulation on the ground or amounts of blowing snow. Estimated sublimation losses vary from over 40% of annual snowfall in warm windy climates to only about 10–15% of annual snowfall in colder and/or calmer sites. Modelling studies by Xiao et al. (2000) and Dery and Yau (1999) also show that sublimation rates from blowing snow ranging from about 2 to 3 mm d^{-1} are possible. In any location, existence of vegetative stubble protruding through the snow reduces the sublimation loss (Pomeroy and Gray, 1994). In regions with blowing snow and relatively low precipitation, it is obvious that sublimation can represent a significant loss to the water balance.

2.2.4 Modelling blowing snow

Advances have been made in recent years to improve our ability to model the impact of blowing snow on the spatial distribution of snowpack water equivalent.

Table 2.1 *Estimates of sublimation losses from blowing snow*

Reference	Location	Sublimation Loss from Blowing Snow (% of annual snowfall)
King *et al.* 2001	Antarctica	12.5%
Pomeroy *et al.* 1998	Arctic	19.5%
Pomeroy and Gray 1995	Canadian prairies	15–40%
Pomeroy *et al.* 1997	Western Canadian Arctic	28%
Pomeroy and Li 2000	Canadian prairie	29%
	Canadian arctic tundra	22%
Benson 1982	Alaska north slope	32%
Pomeroy *et al.* 1993	Saskatchewan, Canada	
	Cool, calm site	15–44%
	Warm, windy site	41–74%

One approach has been to solve the following mass balance for each landscape segment for months or an entire winter accumulation season (Pomeroy *et al.*, 1997):

$$Q_{net} = P - [dQ_t/dx](x) - Q_e \qquad (2.3)$$

where:

Q_{net} = net snow accumulation flux after sublimation, kg m^2 s^{-1}
P = snowfall flux, kg m^2 s^{-1}
Q_t = total blowing snow transport flux, kg m^2 s^{-1}
Q_e = blowing snow sublimation flux, kg m^2 s^{-1}

Q_{net} in Equation (2.3) can be either negative (surface is a source of blowing snow) or positive (surface is a sink for blowing snow) depending upon the balance between snowfall and the sum of sublimation and wind transport of snow at a specific point on the landscape. Assuming that P is accurately measured or modelled for a watershed, then computations of Q_e and Q_t as a function of downwind position (x) are needed to compute Q_{net}.

Tabler *et al.* (1990) reviewed the functions used to compute the mass flux and sublimation rates for blowing snow. By computing the transport of blowing snow and subtracting sublimation, they presented Equation (2.4) to predict the total seasonal transport of blowing snow that provides an "excellent first approximation" of the snow available for trapping using snow fencing as:

$$Q'_{net} = 500DP(1 - 0.14^{F/D}) \qquad (2.4)$$

where:

Q'_{net} = net mass of blowing snow for a winter season, kg m^{-1}
D = distance an average snow particle travels before completely sublimating, m
P = water equivalent of winter snowfall, m
F = fetch or length of upwind source area for blowing snow, m

In Equation (2.4) increases in winter snowfall (P) or fetch (F) that provides more source area for blowing snow can lead to increased blowing snow flux. The net flux of blowing snow after sublimation is measured as mass per unit width perpendicular to the prevailing wind direction. Average snow particle size, wind speed, and humidity conditions in an area are embodied in the D parameter in Equation (2.4), which for Wyoming conditions was computed as 3000 m. However, greater sublimation rates, which would lead to shorter distances for the average snow particle to sublimate (lower D), would reduce Q'_{net}.

To estimate the mass of blowing snow available downwind from a plain of 1000 m length (F) in a region with 0.3 m of annual snowfall water equivalent and an average distance that a snow particle travels before sublimating completely of 3500 m (D):

$$Q'_{net} = (500)(3500)(0.3)(1 - 0.14^{1000/3500}) = 225\,630\,\text{kg m}^{-1}$$

for the winter season.

More recently, Pomeroy *et al.* (1997) used a set of empirical relationships based upon theoretical modelling with the Prairie Blowing Snow Model (PBSM) to quantify the Q_t and Q_e fluxes in Equation (2.3) as a function of fetch distance, snowpack, and meteorological variables over level watershed surfaces. As shown in Figure 2.12 from their paper, the instantaneous flux of blowing snow (dQ_t/dx) increases rapidly and peaks within the first 300 m of fetch and then gradually declines with increasing distance (x). The total flux of blowing snow (Q_t) reaches a nearly steady rate as dQ_t/dx approaches zero. The total flux of sublimated water gradually increases with increasing fetch. Many other modelling approaches are being used to quantify the fluxes of blowing and sublimating snow (Xiao *et al.*, 2000; Dery and Yau, 1999) with some explicitly considering effects of varying alpine or mountain terrain and vegetation (Essery *et al.*, 1999, Gauer, 2001; Liston and Sturm, 1998; Prasad *et al.*, 2001; Liston *et al.*, 2002). Rapid advances are being made using these models for more realistic simulations of snowmelt hydrology for watersheds affected by blowing snow.

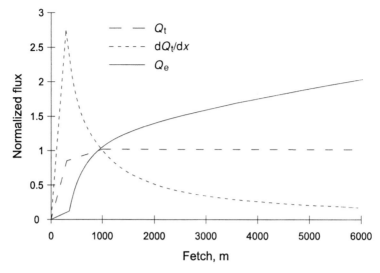

Figure 2.12 Estimated annual blowing-snow flux variations with fetch normalized to 1000 m values on level terrain; Q_t = blowing snow transport flux and Q_e = blowing snow sublimation flux (Pomeroy *et al.* 1997, © 1997, John Wiley & Sons, reproduced with permission).

2.3 Snow interception by vegetation

Snowfall often interacts with vegetation before it forms a snowpack over the landscape. Snow that is lodged within the canopies of plants is referred to as intercepted snow, while snow that falls or drips to the ground from or through the canopy is termed throughfall. Throughfall from intercepted snow can be in the form of ice particles dislodged by the wind that filter through the canopy during cold, windy conditions or in the form of large masses of intercepted snow that slide from the plant branches during warm, melting conditions or as branches bend under the weight of intercepted snow. Meltwater from intercepted snow that reaches the ground by flowing down the stems of plants is called stemflow. Stemflow is a minor pathway for intercepted water to reach the ground (Brooks *et al.*, 1997; Johnson, 1990), especially in winter when melt rates from intercepted snow are often much less than rainfall rates. Intercepted snow can be sublimated or evaporated from the canopy before reaching the ground and the mass lost to the atmosphere is referred to as interception loss. Interception loss can occur while the snow is stored within the canopy or as the intercepted snow is being redistributed within and over the landscape by the wind. Knowledge of the distribution and type of vegetation across a watershed is essential in successfully modelling effects of interception on snowmelt runoff.

Relationships among the various canopy components and pathways that snow can take to reach the ground can be simply stated as:

$$T + S_{st} = P - Ic \qquad (2.5)$$

where:

$T =$ throughfall water equivalent, m
$S_{st} =$ stemflow, m
$P =$ total snowfall water equivalent, m
$Ic =$ canopy interception loss, m

Many early studies to determine Ic as a residual in Equation (2.5) relied upon measurements of T beneath the plant canopy, S_{st} flowing down plant stems, and P in nearby open areas or above the plant canopy (Lundberg and Halldin, 2001). Although this approach is generally useful, errors can result due to canopy storage of snow that later falls or drips to the ground, wind redistribution of snow prior to measurement, unmeasured outflow from snowpacks prior to measurement, and undercatch of snowfall in precipitation gauges.

The problem for snow hydrologists is one of predicting the timing and amount of throughfall reaching the ground beneath vegetation. If reliable measurements of snowfall are available for a watershed, it cannot be assumed that all of the snowfall will reach the ground, a certain fraction will be lost to interception loss. The type of vegetation and density of the plants along with meteorological conditions during and after snowfall can affect the magnitude of interception loss. Neither can it be assumed that the throughfall accumulates into a snowpack during the actual snowfall event. The timing of throughfall after a snowfall event depends primarily upon climatic conditions, especially wind speed, solar radiation, and air temperature, which control melting or wind transport of intercepted snow. Factors controlling the magnitude of interception loss and delivery of snow to the ground are reviewed below.

2.3.1 Factors affecting snow interception

Vegetative and meteorological factors both affect the initial interception of snow in plant canopies. Vegetative factors can be divided into species-specific and plant community factors. Species-specific factors are those related to persistence of foliage (deciduous vs. evergreen), needle characteristics, branch angle, flexibility of branches (modulus of elasticity of wood), canopy form, etc. Foliage presence during winter adds to the initial snow interception due to the greater plant surface area. In conifers with needle-like foliage, snow particles can lodge at the base of needle bundles, bridge across the needles and gradually accumulate until the entire branch is covered with snow (Figure 2.13). Needle length, number of

Figure 2.13 Heavy accumulation of intercepted snow on pendulous conifer tree branches showing effects of bridging across needle bundles (top) and intercepted snow in a mixed hardwood/conifer forest after a wet, cohesive snowfall under calm conditions at 0 °C (bottom).

needles per bundle, and orientation of needle bundles and branches theoretically should influence snow catch in conifers, but most studies show minor effects attributed to species of conifers. Flexibility of branches then determines when branches bend and unload masses of snow (Schmidt and Gluns, 1991). Schmidt and Pomeroy (1990) showed that as air temperatures increased from -12 to $0\,^{\circ}$C branches became more flexible due to changes in elasticity which helped to explain the dumping of snow from conifer canopies as air temperatures increased. Scaling snow interception from individual branches and or an entire tree canopy to full plant community effects remains a challenge.

Plant community factors relate to the total biomass per unit land area available to intercept snow that is often indicated as a function of leaf-area index (LAI). Leaf-area index is defined as projected surface area of leaves and/or needles integrated from canopy top to ground per unit of ground area. In deciduous plant communities in winter the plant-area index is used which includes the total surface area of twigs, branches, and stems (Figure 2.13). Interception increases as leaf- and plant-area indices increase and species effects are difficult to separate from plant community effects.

Meteorological factors also affect canopy interception of snow. Conditions during snowfall that increase cohesiveness of snow particles, such as air temperatures approaching $0\,^{\circ}$C, lead to increased amounts of snow being held within the canopy (Figure 2.13). Cohesive snow combined with high winds can plaster intercepted snow deep within tree canopies and onto branches and plant stems. Snow particles falling at temperatures well below freezing are generally not as cohesive and either singly or in combination with high wind speeds during snowfall events can lead to reduced snow interception. Non-cohesive particles can also bounce off of or dislodge snow already intercepted on branches. Melting of intercepted snow is also thought to contribute to sliding of snow masses as well as dripping from canopies. High winds can dislodge intercepted snow long after snowfall events have occurred. Finally, rainfall occurring on intercepted snow in plant canopies can be initially stored by the snow, but with sufficient liquid-water inputs, due to added weight on branches, can eventually lead to liquid throughfall and delivery of large masses of snow to the ground. It is clear that many interactions between meteorological conditions and vegetation can occur and regional climates will help dictate the dominant types of snow interception processes that occur. Excellent reviews of early literature on snow interception processes were given by Miller (1964, 1966).

2.3.2 Modelling snow interception

Models of interactions between snow and more or less continuous plant canopies can vary from simple to complex. The simplest approach is to assume a fixed percentage of total snowfall is lost due to sublimation in the plant canopy. Many past mass balance studies show that Ic from snowfall can represent a loss ranging from about 45% of *P* for dense evergreens to about 10% of *P* for leafless deciduous forests (Lundberg and Halldin, 2001; Essery and Pomeroy, 2001). Storck *et al.* (2002) showed that in a coastal Douglas-fir forest in Oregon, USA interception efficiency remained a constant 60% of snowfall both within events and for total events. Assuming constant percentage interception loss did not account for variation in canopy density, Pomeroy *et al.* (2002) modified and validated a method of estimating snowpack accumulation beneath forest canopies developed by Kuz'min (1960) for cold climate forests of Canada with an equation in the form:

$$S_f = S_c\{1 - [0.144\ln(\mathrm{LAI'}) + 0.223]\} \tag{2.6}$$

and

$$\mathrm{LAI'} = \exp(C_c/0.29 - 1.9) \tag{2.7}$$

where:

S_f = snowpack accumulation beneath the forest canopy, mm
S_c = snowpack accumulation in a forest opening, mm
$\mathrm{LAI'}$ = effective winter forest leaf- and stem-area index, dimensionless
C_c = winter forest canopy density, dimensionless

These two equations can be used to estimate snow accumulation and indirectly canopy interception losses in cold regions, where sublimation of intercepted snow can be significant, using only measurements of snowfall or snowpack in openings and canopy density.

    More complex models involve computation of the amount of initial interception by the vegetation and the unloading of intercepted snow over time from branches due to melt, sublimation and dislodgement by gravity or the wind. Hedstrom and Pomeroy (1998) developed an exponential model that shows interception efficiency starting high for low snowfall amounts and gradually decreasing as snowfall increases, based upon measurements in boreal forest regions of Canada where sublimation losses are significant. This model also accounted for the gradual unloading of snow from the canopy and is described further below.

    According to Hedstrom and Pomeroy (1998) as modified in Pomeroy *et al.* (1998), an operational model to predict the amount of snow intercepted in the canopy for each time step could be written as:

$$I = c(S_m - I_0)[1 - \exp(-C_c P/S_m)] \tag{2.8}$$

where:

I = intercepted snow, mm
c = empirical snow canopy unloading coefficient, approx. 0.7 per hour

S_m = maximum canopy snow load, mm
I_0 = initial canopy snow load, mm
C_c = fractional canopy cover, dimensionless
P = snowfall, mm

They further computed the maximum canopy snow load (S) by scaling maximum snow storage loads per unit area of branches given by Schmidt and Gluns (1991) using the LAI for the plant stand as:

$$S_m = S_b(\text{LAI})(0.27 + 46/\rho_s) \qquad (2.9)$$

where:

S_m = maximum canopy snow storage, mm
S_b = maximum branch snow storage, 6.6 and 5.9 mm for pine and spruce
  forest respectively
LAI = winter leaf- and stem-area index corrected for clumping, dimensionless
ρ_s = fresh snow density, kg m^{-3}

Using this approach, first the maximum canopy snow storage was computed with knowledge of the LAI of the forest stand and the fresh snow density in Equation (2.9). The snow density was computed operationally from air temperatures. Then, using Equation (2.8), the new intercepted snowfall amount was computed knowing S_m, any intercepted snow from a previous time period, the canopy cover C_c, and the snowfall amount for the current time period. Further details are given in Hedstrom and Pomeroy (1998). If a snowfall of 1 mm with a density of 80 kg m^{-3} fell during an hour onto a dry pine forest ($S_b = 6.6$ mm, $I_0 = 0$) with canopy coverage of 82% and LAI = 2.2, the canopy interception during that hour would be computed as:

Equation (2.9): $S_m = (6.6)(2.2)(0.27 + 46/80) = 12.3$ mm
Equation (2.8): $I = (0.7)(12.3 - 0)(1 - \exp[(-0.82)(1)/12.3]) = 0.55$ mm

or 55% of snowfall would be intercepted at the end of the hour. This storage would then become the input snow storage (I_0) for the next hour's calculation.

In the above approach, the amount of intercepted snow was predicted over time and used to estimate sublimation losses using energy balance considerations. The difference between precipitation amounts and the sum of canopy storage and sublimation from the canopy, can be used to approximate throughfall and snowpack accumulation on the ground. The above relationships were developed for Canadian

Table 2.2 *Mean snow interception coefficients for the logistic growth curve model
Equation (2.10) on branches or individual trees*

Species	S_m mm	P_0 mm	k mm^{-1}	Reference
Douglas-fir (*Pseudotsuga menziesii* (Mirb.))	2.81	4.88	0.98	Satterlund and Haupt (1967), 2 weighed trees
Western white pine (*Pinus monticola* (Dougl.))	1.84	4.44	1.04	Satterlund and Haupt (1967), 2 weighed trees
Englemann spruce (*Picea Englemanii* (Parry))	5.22	4.14	0.6	Schmidt and Gluns (1991) branch interception
Subalpine fir (*Abies lasiocarpa* (Hook) Nutt.)	3.94	3.49	0.79	Schmidt and Gluns (1991) branch interception
Lodgepole pine (*Pinus contorta* Dougl.)	4.35	3.69	0.84	Schmidt and Gluns (1991) branch interception
Conifers	2.58	0.76	0.72	Nakai (1995), trees and branches

boreal forest conditions without frequent melting and rainfall during the winter, but
do allow for a more complete accounting of the fate of intercepted snow. Gelfan
et al. (2004) recently used this procedure to model accumulation, sublimation, and
melt in conifer forests in Russia. More detailed process studies are needed in the
future to extend such prediction schemes to a broader set of site conditions.

Observations of snow interception on isolated branches and individual trees suggest that cumulative snow interception follows a logistic growth curve. According
to Satterlund and Haupt (1967) snow interception on isolated conifer trees could
be modelled as Equation (2.10):

$$I = S_m/[1 + e^{-k(P-P_0)}]   \quad (2.10)$$

where:

I = water equivalent of intercepted snow storage for an event, mm
S_m = water equivalent of maximum snowfall canopy storage capacity, mm
k = rate of interception storage, mm^{-1}
P_0 = water equivalent of snowfall at inflection point on the sigmoid growth
    curve, mm
P = water equivalent of snowfall for an event, mm

The values for I, P_0, S_m are generally expressed over the projected horizontal area of
the crowns of trees or branches (Table 2.2). The general trend of snow interception
using this type of model is shown in Figure 2.14 using coefficients representing Douglas-fir trees from Satterlund and Haupt (1967). As snowfall increases the

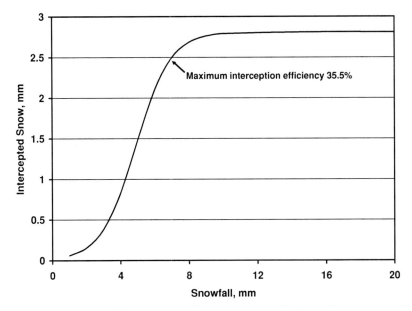

Figure 2.14 Predicted interception of snow on an isolated conifer canopy with varying snowfall amounts described using a logistic growth curve. Calculations are based upon average values for S_m, P_0 and k in Equation (2.10) of 2.81 mm, 4.88 mm and 0.98 mm^{-1}, respectively, for Douglas-fir (*Psuedotsuga menziesii* (Mirb.)) from Satterlund and Haupt (1967) given in Table 2.2.

interception of snow increases rapidly at a rate indexed by k to the point of inflection on the logistic growth curve (snowfall around P_0) due to cohesion and bridging and then gradually approaches an asymptote at the canopy or branch maximum canopy storage (S_m) and as the branches become laden with snow, bend and unload. The point of inflection of the curve, around a snowfall of 7 mm water equivalent for specific coefficients chosen in the example, represents a maximum interception efficiency expressed as a percentage of snowfall of about 35.5% of snowfall intercepted. Not all investigators have found that the logistic growth function applies to snow interception. Storck *et al.* (2002) also showed a variety of interception patterns from linear to logistic by studying weight gains of isolated conifer trees during snowfall.

It is important to realize that initial snow interception in a plant canopy described by the logistic function does not necessarily represent interception loss due to evaporation/sublimation. For example, Satterlund and Haupt (1970) estimated that 80% of snow initially intercepted on canopies in their isolated conifer tree experiments eventually fell to the ground as solid or liquid in a relatively humid environment. Storck *et al.* (2002) showed that under conditions favoring melting and events that ended as rain, evaporation or sublimation losses from intercepted snow in a Douglas-fir canopy in coastal Oregon, USA were essentially zero. Mass release of intercepted snow falling to the ground accounted for 28% and dripping from the canopy to the ground represented 72% of snow initially intercepted in the canopy. For non-melt conditions they found rates of canopy sublimation in excess of 0.5 mm d^{-1}, but true snow interception losses were only about 5% of winter precipitation in this humid coastal environment. In contrast, in drier climates where canopies can retain intercepted snow for days to months after snowfalls, sublimation losses can be much greater.

2.4 Applications in snow hydrology

Processes controlling snowfall amounts and timing are reasonably well understood, but snow hydrologists still must rely on precipitation and snowpack measurements to obtain basic liquid inputs to models. However, as is evident from details in this chapter, it is becoming possible to predict the distribution of snow across landscapes due to effects of blowing snow and vegetation interception. Blowing-snow adjustments are especially needed in cold locations with high wind speeds where entire slopes may be swept free of snow while others may be deeply drifted. Basins with a wide variety of land cover types and extensive dense tree and shrub vegetation will also require adjustment for canopy interception due to differences in types of vegetation as well as vegetation density. A simple reduction of snowfall for interception may suffice in some applications, but in others where time and space scales are expanded, accounting for storage of snow in the canopy and gradual sublimation losses and intermittent throughfall releases are needed. Much has been learned, but blowing snow and canopy interception remain two of the most important challenges for hydrologic simulation in the future.

2.5 References

Ahrens, C. D. (1988). *Meteorology Today*, 3rd edn. St. Paul, MN: West Publishing Co.
Barry, R. G. (1992). *Mountain Weather and Climate*, 2nd edn. New York: Routledge.
Beebee, R. A. and Manga, M. (2004). Variation in the relationship between snowmelt runoff in Oregon and ENSO and PDO. *J. Amer. Water Resour. Assoc.*, **40**, 1011–24.
Benson, C. S. (1982). *Reassessment of Winter Precipitation on Alaska's Arctic Slope and Measurements on the Flux of Windblown Snow, Report UAG R-288*. Univ. Alaska Fairbanks: Geophysical Instit.
Brooks, K. N., Ffolliott, P. F., Gregersen, H. M., and DeBano, L. F. (1997). *Hydrology and the Management of Watersheds*, 3rd edn. Ames, IA: Iowa State University Press.
Clark, M. P., Serreze, M. C., and McCabe, G. J. (2001). Historical effects of El Nino and La Nina events on the seasonal evolution of the montane snowpack in the Columbia and Colorado River Basins. *Water Resour. Res.*, **37**(3), 741–57.

Dery, S. J. and Yau, M. K. (1999). A bulk blowing snow model. *Boundary-Layer Meteorol.*, **93**, 237–51.

Essery, R. and Pomeroy, J. (2001). Sublimation of snow intercepted by coniferous forest canopies in a climate model. In *Proceedings Symposium Soil-Vegetation-Atmosphere Transfer Schemes and Large-Scale Hydrological Models*, IAHS Publ. No. 270, Sixth IAHS Scientific Assembly, Maastricht, The Netherlands July 2001, pp. 343–7.

Essery, R., Long, L., and Pomeroy, J. (1999). A distributed model of blowing snow over complex terrain. *Hydrol. Processes*, **13**, 2423–38.

Gauer, P. (2001). Numerical modelling of blowing and drifting snow in Alpine terrain. *J. Glaciol.*, **47**(156), 97–110.

Gelfan, A. N., Pomeroy, J. W., Kuchment, L. S. (2004). Modeling forest cover influences on snow accumulation, sublimation and melt. *J. Hydromet.*, **5**, 785–803.

Hedstrom, N. R. and Pomeroy, J. W. (1998). Measurements and modelling of snow interception in the boreal forest. *Hydrol. Processes*, **12**, 1611–25.

Johnson, R. C. (1990). The interception, throughfall and stemflow in a forest in Highland Scotland and the comparison with other upland forests in the U.K. *J. Hydrol.*, **118**, 281–7.

Kind, R. J. (1981). Chap. 8: Snow drifting. In *Handbook of Snow, Principles, Processes, Management and Use*, ed. D. M. Gray and D. H. Male. Toronto: Pergamon Press, pp. 338–58.

King, J. C., Anderson, P. S., and Mann, G. W. (2001). The seasonal cycle of sublimation at Halley, Antarctica. *J. Glaciol.*, **47**(156), 1–8.

Kobayashi, T. (1961). The growth of snow crystals at low supersaturations. *Philosophical Magazine B*, **6**, 1363–1370.

Kunkel, K. E. and Angel, J. R. (1999). Relationship of ENSO to snowfall and related cyclone activity in the contiguous United States. *J. Geophys. Res.*, **104**(D16), 19425–34.

Kuz'min, P. P. (1960). *Formirovanie Snezhnogo Pokrova i Metody Opredeleniya Snegozapasov*, Gidronmeteoizdat: Leningrad.[Published 1963 as *Snow Cover and Snow Reserves*. English translation by Israel Program for Scientific Translation, Jerusalem], National Science Foundation, Washington, DC.

Li, L. and Pomeroy, J. W. (1997a). Estimates of threshold wind speeds from snow transport using meteorological data. *J. Appl. Meteorol.*, **36**, 205–13.

Li, L. and Pomeroy, J. W. (1997b). Probability of occurrence of blowing snow. *J. Geophys. Res.*, **102**(D18), 21955–64.

Liston, G. E. and Sturm, M. (1998). A snow-transport model for complex terrain. *J. Glaciol.*, **44**(148), 498–516.

Liston, G. E., McFadden, J. P., Sturm, M., and Pielke Jr., R. A. (2002). Modelled changes in arctic tundra snow, energy and moisture fluxes due to increased shrubs. *Global Change Biol.*, **8**(1), 17–32.

Lundberg, A. and Halldin, S. (2001). Snow interception evaporation: review of measurement techniques, processes and models. *Theoretical and Applied Climatology*, **70**, 117–33.

Mason, B. J. (1971). *The Physics of Clouds*. Oxford: Clarendon Press.

McCabe, G. J., Dettinger, M. J. (2002). Primary modes and predictability of year-to-year snowpack variations in the western United States from teleconnections with Pacific Ocean Climate. *J. Hydromet.*, **3**, 13–25.

McFadden, J. P., Liston, G. E., Sturm, M., Pielke, R. A. Jr., and Chapin III, F. S. (2001). Interactions of shrubs and snow in arctic tundra measurements and models. In *Soil-Vegetation-Atmosphere Transfer Schemes and Large-Scale Hydrological*

Models, IAHS Publ. 270, Sixth IAHS Scientific Assembly, Masstricht, The Netherlands July 2001, pp. 317–25.

McKay, G. A. and Gray, D. M. (1981). Chapter 5: The distribution of snowcover. *In Handbook of Snow, Principles, Processes, Management and Use*, ed. D. M. Gray and D. H. Male. Toronto: Pergamon Press, pp. 153–90.

Miller, D. H. (1964). *Interception Processes during Snowstorms*. Res. Paper PSW-18, US Forest Service.

Miller, D. H. (1966). *Transport of Intercepted Snow from Trees during Snowstorms*. Res. Paper PSW-33, US Forest Service.

Nakai, Y. (1995). An observational study on evaporation from intercepted snow on forest canopies. Unpublished PhD thesis, Kyoto University Japan.

Norton, D. C. and Bolsenga, S. J. (1993). Spatiotemporal trends in lake effect and continental snowfall in the Laurentian Great Lakes, 1951–1980. *J. Climate*, **6**, 1943–56.

Pomeroy, J. W. and Gray, D. M. (1994). Sensitivity of snow relocation and sublimation to climate and surface vegetation. In *Snow and Ice Covers: Interactions with the Atmosphere and Ecosystems*, IAHS Publ. 223, Proceedings of Yokohama Symposia J2 and J5, July 1993, pp. 213–25.

Pomeroy, J. W. and Gray, D. M. (1995). *Snow Accumulation, Relocation and Management, NHRI Science Report 7*. Saskatoon: Environment Canada.

Pomeroy, J. W. and Li, L. (2000). Prairie and arctic areal snow cover mass balance using a blowing snow model. *J. Geophy. Res.*, **105**(D21), 26619–34.

Pomeroy, J. W., Gray, D. M., Hedstrom, N. R., and Janowicz, J. R. (2002). Prediction of seasonal snow accumulation in cold climate forests. *Hydrol. Process.*, **16**, 3543–58.

Pomeroy, J. W., Gray, D. M., and Landine, P. G. (1993). The prairie blowing snow model: characteristics, validation and operation. *J. Hydrol.*, **144**, 165–92.

Pomeroy, J. W., Marsh, P., and. Gray, D. M. (1997). Application of a distributed blowing snow model to the arctic. *Hydrol. Processes*, **11**, 1451–64.

Pomeroy, J. W., Parviainen, J. Hedstrom, N., and Gray, D. M. (1998). Coupled modelling of forest snow interception and sublimation. *Hydrol. Processes*, **12**, 2317–37.

Prasad, R., Tarboton, D. G., Liston, G. E., Luce, C. H., and Seyfried, M. S. (2001). Testing a blowing snow model against distributed snow measurements at Upper Sheep Creek Idaho, USA. *Water Resour. Res.*, **37**(5), 1341–50.

Satterlund, D. R. and Haupt, H. F. (1967). Snow catch by conifer crowns. *Water Resour. Res.*, **3**(4), 1035–9.

Satterlund, D. R. and Haupt, H. F. (1970). The disposition of snow caught by conifer crowns. *Water Resour. Res.*, **6**(2), 649–52.

Schemenauer, R. S., Berry, M. O. and Maxwell J. B. (1981). Chapter 4: Snowfall formation. In *Handbook of Snow, Principles, Processes, Management and Use*, ed. D. M. Gray and D. H. Male. Toronto: Pergamon Press, pp. 129–52.

Schmidt, R. A., Jr. (1972). *Sublimation of Wind-Transported Snow – a Model*, Research Paper RM-90. Rocky Mtn. For. and Range Exper. Station: USDA Forest Service.

Schmidt, R. A. and Gluns, D. R. (1991). Snowfall interception on branches of three conifer species. *Can. J. For. Res.*, **21**:1262–9.

Schmidt, R. A. and Pomeroy, J. W. (1990). Bending of a conifer branch at subfreezing temperatures: implications for snow interception. *Can. J. For. Res.* **20**, 1250–53.

Smith, S. R. and O'Brien, J. J. (2001). Regional snowfall distributions associated with ENSO: Implications for seasonal forecasting. *Bull. Amer. Meteorol. Soc.*, **82**(6), 1179–92.

Storck, P., Lettenmaier, D. P. and Bolton, S. M. (2002). Measurement of snow interception and canopy effects on snow accumulation and melt in a mountainous maritime climate, Oregon, U.S. *Water Resour. Res.*, **38**(11:1223), 5–1 to 5–16.

Sumner, G. (1988). *Precipitation, Processes and Analysis*. Chichester: John Wiley and Sons.

Tabler, R. D., Pomeroy, J. W., and Santana, B. W. (1990). Drifting snow. In *Cold Regions Hydrology and Hydraulics*, ed. W. L Ryan and R. D. Crissman. New York: American Society of Civil Engineers, pp. 95–145.

Whiteman, C. D. (2000). *Mountain Meteorology: Fundamentals and Applications*. New York: Oxford University Press.

Xiao, J., Bintanja, R., Dery, S. J., Mann, G. W., and Taylor, P. A. (2000). An intercomparison among four models of blowing snow. *Boundary-Layer Meteorol.*, **97**, 109–35.

3

Snowpack condition

Once snow has accumulated on the landscape, the individual ice grains can be rapidly transformed or metamorphosed into a structured snowpack. Metamorphism ultimately influences the thermal conductivity and liquid permeability of the snowpack that in turn influence the snowpack temperature regime and storage and release of liquid water. Snowpacks vary seasonally from low density, subfreezing snowpacks capable of refreezing any liquid-water inputs to isothermal, dense snowpacks that rapidly transmit liquid water to the ground below. In this chapter, important processes controlling snowpack metamorphism and the conduction of heat and liquid water in snow are described along with some methods hydrologists use to describe snowpack condition.

3.1 Snowpack metamorphism

Snow crystals that fall from the atmosphere undergo changes in size and shape over time as they become part of the snowpack. Owing to their large surface area to volume ratio, crystals are quite unstable in the snowpack and can transform to rounded or faceted ice grains. Processes causing metamorphism of ice grains and snowpacks vary between dry snow and wet snow. Metamorphism within the snowpack is important to hydrologists because it can ultimately lead to changes in thermal conductivity and liquid permeability that influences the snowpack energy budget and release of meltwater. The following discussion of snowpack metamorphism is largely based upon writings by Perla and Martinelli (1978) and McClung and Schaerer (1993). Major snowpack processes for snowpack metamorphism are summarized in Table 3.1.

3.1.1 Formation of rounded grains in dry snow

Ice crystals with irregular shapes can be transformed into rounded grains by migration of vapor from convex to concave ice surfaces. Vapor pressure is higher over

Table 3.1 *Summary of major metamorphic processes influencing snowpack density and structure*

Major Metamorphic Process	Physical Principle	Effect	Occurrence
1. Vapor diffusion over crystal surfaces (Equi-temperature Metamorphism)	Vapor pressure higher over convex than concave ice surfaces	Formation of rounded, well-bonded ice grains	Begins soon after snowfall
2. Vapor diffusion among grains (Temperature-gradient Metamorphism)	Vapor pressure higher in warmer than cooler snowpack locations	Formation of faceted, poorly bonded, ice grains and depth hoar	Occurs in subfreezing snowpacks during winter
3. Melt–freeze cycles	Small ice grains have lower melting temperatures than larger grains	As subfreezing snow warms, small grains melt before large grains, refreezing leads to well-bonded, large-grained snowpack	Occurs later in season with surface melting and rain
4. Pressure	Slow compression causes visco-elastic deformation of ice	Pressure of new snow on top of old snow leads to formation of firn and glacial ice	Occurs mainly in perennial older snow

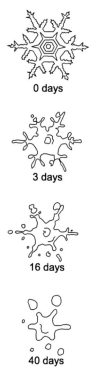

0 days

3 days

16 days

40 days

Figure 3.1 Metamorphism of a steller dry-snow crystal into a rounded grain over time (Doeskin and Judson 1997, copyright 1997 Colorado State University, used with permission).

convex ice surfaces with small radius of curvature, such as points on crystals, than over concave surfaces with large radius of curvature. This causes a vapor pressure gradient over the surface of the ice crystal and the diffusion of vapor from areas of high to low vapor pressure leading to mass being lost from the points and the crystals being rounded (Figures 3.1 and 3.2a). Vapor diffusion also causes growth of necks at concave points of contact among adjacent ice grains which increases the structurally stability of the snowpack. Formation of bonds between adjacent ice grains in dry snow is also known as sintering. Ultimately, this type of vapor diffusion leads to a denser snowpack of well-bonded, rounded ice grains. This type of metamorphism generally occurs soon after ice crystals are deposited during late winter and early spring and occurs more rapidly as snow temperatures approach 0 °C. Vapor diffusion due to radius of curvature effects is known as equitemperature metamorphism because it can occur even when there are no temperature differences within the snowpack. However, temperature gradients within a snowpack are common and can cause even greater vapor diffusion and a totally different ice grain shape and structure.

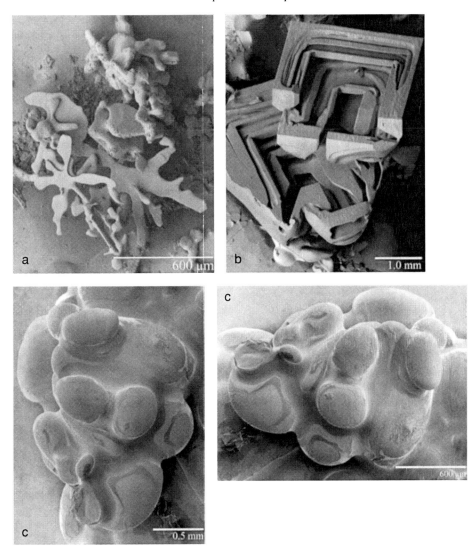

Figure 3.2 Scanning electron microscope pictures of selected metamorphosed snow crystal types occurring in the snowpack: (a) dendritic crystals undergoing equi-temperature metamorphism, (b) faceted crystals formed by temperature-gradient metamorphism, and (c) bonded snow grains after melt–freeze metamorphism. (Courtesy USDA, Agricultural Research Service, Beltsville, MD.)

3.1.2 Formation of faceted grains in dry snow

Vapor diffusion due to temperature gradients within the snowpack can lead to formation of faceted ice grains with multiple plane surfaces. Faceted crystals can take a variety of shapes and patterns, variously described as anisotropic, striated,

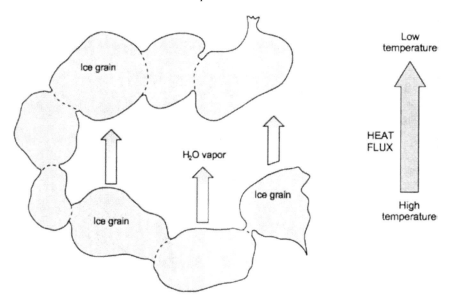

Figure 3.3 Water vapor flowing along a temperature gradient within the snowpack (Perla and Martinelli, 1978, courtesy US Forest Service).

layered, angular, stepped or cup-shaped (Figure 3.2b). Large cup-shaped crystals formed near the base of the snowpack by this process, termed depth hoar, are often associated with structural weakness and snow avalanching.

Since pore space within a snowpack is generally at or near saturation with water vapor and because saturation vapor pressures increases with rising temperature, temperature differences in the snowpack can cause large differences in vapor pressures leading to vapor diffusion. Vapor tends to diffuse through the snowpack pore spaces from warmer locations in the snowpack, typically found near the ground, to colder locations often found near the snowpack surface (Figure 3.3). With this type of vapor diffusion, mass is lost by sublimation from the top of ice grains in the warmer snowpack locales and re-sublimated on the base of cooler ice grains above to form new faceted crystals. Mass and energy transfer by sublimation occurs across individual pore spaces from one ice grain to another in a hand-to-hand process. Since the solid ice has a much higher thermal conductivity and slightly lower temperature than the adjacent air in pore spaces, re-sublimation primarily occurs on the bases of adjacent ice grains rather than at points of contact among crystals. Thus, faceted grains generally are not well bonded to one another, which can lead to structural weakness within the snowpack and avalanche problems. This type of vapor diffusion is also known as temperature-gradient metamorphism.

The rate of vapor diffusion and the size and amount of faceted grains formed by this temperature-gradient process increases with the magnitude of the temperature gradient, porosity of the snowpack and the temperature. Locations with shallow, porous snowpacks and large temperature differences between the ground and the snowpack surface will commonly have well-developed faceted ice grain formation. Continental locations such as the Rocky Mountains in Colorado, USA favor faceted grain growth. Other locations with deep, dense snowpacks and warmer winter air temperatures, commonly found in coastal regions of western North America, do not.

Since saturation vapor pressures are more sensitive to temperature changes than radius of curvature effects in the snowpack, it is common for temperature-gradient vapor diffusion processes to dominate during winter and early spring. Once snowpacks begin to warm and approach $0\,^{\circ}C$, the importance of vapor diffusion due to temperature gradients is lessened. At this point melt/freeze and liquid-water effects begin to dominate snowpack metamorphism. However, brief periods of overnight cooling and localized vapor diffusion at the snowpack surface can occur at any time during the melt season.

3.1.3 Wet-snow metamorphism

Occurrence of melting and freezing cycles and rainfall can also affect ice grain size, shape, and bonding. Metamorphism in wet snow is mainly controlled by the fact that small ice grains have a slightly lower melting temperature than larger ice grains. As a subfreezing snowpack warms, the smaller ice grains with lower melting temperature, melt first. The energy for melting comes from the larger grains, which cool, and refreeze the liberated water. Thus, the larger ice grains cannibalize the smaller grains. Liquid water within snowpack pore space replaces air and greatly increases the rate of energy transfer among ice grains. Large, rounded ice grains grow more rapidly when liquid water is abundant.

Slight reductions in melting temperature can also occur at points of contact between ice grains, especially when pressure is exerted from the weight of the snow above. Preferential melting of bonds among grains can cause weakening of the snowpack structure, but increases the snowpack density as grains move closer together. Refreezing of liquid water held by capillary forces between grains and liquid films over the surfaces of ice grains during melt and rain can greatly increase snowpack strength (Figure 3.2c). Such melt–freeze metamorphism is usually the last major process affecting seasonal snowpacks; however, for perennial snowcover pressure metamorphism may also be significant.

Snowpacks that persist through an entire melt season into another accumulation season can also be affected by the pressure of overlying snow. Densification of snow to firn, the transitional state between snow and ice, can occur due to pressure metamorphism. Snow under pressure behaves as a visco-elastic material. When subjected to slow compression it gradually deforms into firn and then ice. Major differences in the density, thermal conductivity, and shortwave reflectivity exist among snow, firn, and glacial ice.

3.1.4 Other snowpack transformations

In addition to the metamorphic changes described above, several other transformations in snowpack density and structure can occur due to effects of wind, refreezing of liquid water, and sublimation on the snowpack surface. Wind can cause the formation of wind slab, a densified, dry-snow layer at the snowpack surface formed due to compression of fresh, low-density snow or deposition and packing of broken crystals of blowing snow. Wind slab often has enough structure to be lifted off the snowpack surface as an intact layer. Due to its greater density, wind slab has a higher thermal conductivity and altered surface shortwave reflectivity that can affect snowpack heat exchange. Water from surface melt or rainfall can refreeze within the snowpack leading to the formation of ice lenses on or near the surface. Although ice lenses generally are not completely impermeable and continuous, they can affect delivery of liquid water from the snowpack to the ground surface. Finally, sublimation of vapor onto the snowpack surface during cold, clear, calm nights can lead to development of surface hoar crystals or the solid-phase equivalent of dew formation. Surface hoar has an extremely low density and is structural very weak and doubtless has an impact on the reflection of shortwave radiation from the snowpack surface. All of these transformations can have some impacts on snowpack ripening, melt, and water delivery.

3.1.5 Snowpack density and densification

Snowpack density ultimately controls the pore space available for storage and transmission of liquid water and gases. A formal definition of snowpack density that includes the possible effects of liquid water on snowpack mass is:

$$\rho_s = \rho_i(1 - \phi) + \rho_w \phi S_w \tag{3.1}$$

where:

ρ_s = density of snowpack, kg snow per m^3 of snowpack volume
ρ_i = density of ice, 917 kg per m^3 of ice
ρ_w = density of liquid water, 10^3 kg per m^3 of water
ϕ = porosity of snowpack, m^3 of pore space per m^3 of snowpack volume
S_w = water saturation, volume of liquid water per volume of pore space, $m^3 \ m^{-3}$

If the snowpack contains liquid water, then porosity can only be determined from
the density if the water saturation is known. Methods for measuring snowpack
liquid-water content are described later in the next chapter. If the snowpack does
not contain liquid water or liquid-water content is negligible, the second term on
the right of Equation (3.1) is zero and the porosity of the snowpack can be simply
determined from density.

If snowpack liquid-water saturation is zero, then there is a direct relationship between snowpack density and porosity by Equation (3.1). For example,
if the density of a fresh snowpack were 100 kg m^{-3}, then the porosity would
be:

$$100 = 917(1 - \phi)$$
$$\phi = 0.89 \text{ or } 89\%$$

If density of dry snow increased to 300 kg m^{-3} late in the accumulation season, then
the porosity would have been reduced to $\phi = 0.67$ or 67%. Obviously, the porosity
is a very dynamic snowpack parameter that gradually decreases as the snowpack
undergoes metamorphism.

Densification of the snowpack reflects the influence of all the metamorphic processes described above. Initial densities of new fallen snow vary widely with crystal
type and amount of riming. Fresh snowfall densities typically range between 50
and 100 kg m^{-3} at cold continental sites such as the central Rocky Mountains of
Colorado and Wyoming; however, fresh snowfall densities reported in other studies can range between 50 and 350 kg m^{-3} (Judson and Doesken, 2000). Near the
end of the melt season, higher densities in the range of 350–550 kg m^{-3} are often
found in seasonal snowpacks. Figure 3.4 shows seasonal snowpack densification
patterns in seasonal snowpack density at several locations in the USA. Densification is quite rapid during the first month after accumulation. Perennial snow-
cover can transition to even higher densities to create firn and when densities reach
about 800 to 850 kg m^{-3}, it is considered to be glacial ice without interconnected
air spaces. Pure ice at the freezing point and standard pressure has a density of
917 kg m^{-3}.

Vertical density profiles in snowpacks also reflect metamorphic processes (Figure 3.5). A typical density profile in a California snowpack early in the accumulation season on December 23 shows lower density due to new snowfall on the
surface and greater density near the snowpack base due to greater elapsed time
for metamorphism to occur. Later in the spring on April 1 just prior to melt initia-
tion, density had increased substantially at all depths due to further metamorphism
of grains from an average of 311 kg m^{-3} on December 23 to 422 kg m^{-3} on
April 1.

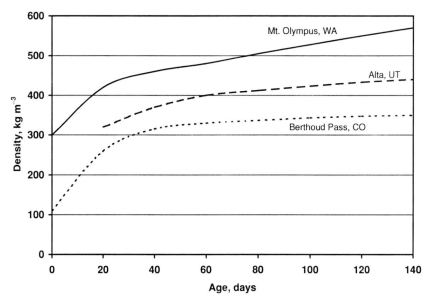

Figure 3.4 Typical seasonal changes in average snowpack density at three locations in the United States (Doeskin and Judson 1997, copyright 1997 Colorado State University, used with permission).

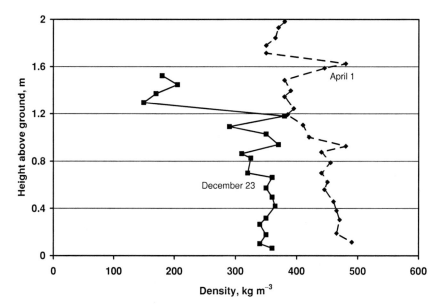

Figure 3.5 Examples of snowpack density profiles measured early and late in the snowpack accumulation season at the Central Sierra Snow Lab, CA, 1953 (data from US Army Corps of Engineers, 1956).

3.1.6 International classification system for seasonal snow cover

An international system has been developed to describe seasonal snow on the ground that has undergone metamorphism (Colbeck *et al.*, 1990; Table 3.2). The system is used to standardize terminology and help describe snowpack conditions in relation to metamorphic processes influencing the snowpack. The classification of precipitation types was previously discussed in Chapter 2. Due to structural differences, the thermal conductivity, liquid-water holding capacity, and other properties important to snow hydrology will vary among snow types.

3.2 Heat conduction in snowpacks

Heat transfer along a temperature gradient in the snowpack occurs by three major processes:

- conduction through the solid portion of the ice matrix
- conduction through air in snowpack pores
- latent heat exchange due to molecular diffusion of water vapor.

Since the thermal conductivity of ice is nearly 100 times greater than that of air (e.g. at $0\,°C\;k(ice) = 2.24$ W m^{-1} K^{-1} while $k(air) = 0.024$ W m^{-1} K^{-1}), pure heat conduction in snow is strongly controlled by conduction through the ice matrix. Bonding of individual ice grains and overall snowpack structure exert major controls on snowpack thermal conductivity. Conduction through air in the snowpack pore space plays an insignificant role, but vapor transfer of latent heat of condensation/ sublimation in the pore space can increase heat transferred along a temperature gradient by up to 50% (Sturm *et al.*, 1997). The thermal conductivity of snow generally is referred to as an effective thermal conductivity due to the combined effects of conduction and latent heat transfer.

Latent heat transfer occurs along a temperature gradient in snowpacks when vapor is sublimated or evaporated from the surface of a warmer ice grain and the vapor either diffuses out of the snowpack through pore spaces or is condensed or re-sublimated onto the surface of a slightly cooler ice grain. Latent heat liberated by condensation/sublimation is conducted through the slightly cooler ice grain and produces further evaporation/sublimation and mass losses from grain surfaces along the temperature gradient. Repeated losses and gains of mass from ice grains along a temperature gradient by this process can transfer significant amounts of latent heat. Because latent heat transfer increases with snow temperature, the effective thermal conductivity of snow generally increases with temperature when vapor transfer is important, even though the thermal conductivity of ice decreases with increasing

Table 3.2 *Grain shape classification for seasonal snow on the ground (Intern. Comm. On Snow and Ice, Colbeck et al., 1990)*

Class of Grains	Symbol	Subclasses	Subclass Description	Process Notes
1. Precipitation particles	+	1a. columns	Short, prismatic	Growth at high supersaturation, $-3\,°C$ to $-8\,°C$ and $<-22\,°C$
		1b. needles	Needle-like, approx. cylindrical	Growth at high supersaturation, $-3\,°C$ to $-5\,°C$
		1c. plates	Mostly hexagonal	Growth at high supersaturation, $0\,°C$ to $-3\,°C$ and $-8\,°C$ to $-25\,°C$
		1d. stellers, dendrites	Six-fold planar or spacial	Growth at high supersaturation, $-12\,°C$ to $-16\,°C$
		1e. irregular	Clusters of very small crystals	Polycrystals growing under varying conditions
		1f. graupel	Heavily rimed particles	Riming by accretion of supercooled water
		1g. hail	Laminar internal structure	Growth by accretion of supercooled water
		1h. ice pellets	Transparent, spheroids	Frozen rain
2. Decomposing and fragmented precipitation particles	/	2a. partly decomposed	Partly rounded, basic shapes recognizable	Recently deposited
		2b. highly broken	Packed shards or rounded fragments	Saltation layer
3. Rounded grains	•	3a. small rounded	<0.5 mm, dry snow	Individual grain growth
		3b. large rounded	>0.5 mm, dry snow	Some grain to grain growth
		3c. mixed forms	Rounded with few facets	Transitional form as temperature gradient increases

Class	Symbol	Subclass	Description	Process
4. Faceted crystals	□	4a. solid faceted	Dry snow, usually hexagonal	Strong grain to grain growth
		4b. small surface faceted	Dry snow, on surface	Near surface temperature gradients
		4c. mixed forms	Transitional form, with some rounding	Temperature gradient decreasing
5. Cup-shaped crystals and depth hoar	<	5a. cup crystals	Cup-shaped, striated, usually hollow	Very rapid growth at large temperature gradients
		5b. depth hoar columns	Large, cup-shaped, arranged in columns	Lateral grain bonds largely gone, almost completely recrystalized
		5c. columnar crystals	Very large columnar crystals, 10–20 mm	Final growth stage, long time needed for growth
6. Wet grains	○	6a. clustered rounded	Wet snow, grain clusters held by ice to ice bonds,	Bonding without melt–freeze cycles, pendular water regime
		6b. rounded polycrystals	Crystals frozen together	Melt–freeze cycles cause grain growth
		6c. slush	Separate, rounded, immersed in water	Poorly bonded, high liquid content
7. Feathery crystals	>	7a. surface hoar	Striated, feathery, usually flat	Rapid transfer of vapor to snow surface
		7b. cavity hoar	Like above, but either flat or feathery	Grown in snowpack cavities, random orientation
8. Ice masses	■	8a. ice layer	Buried horizontal ice layer	Refreezing of draining meltwater or rain
		8b. ice column	Buried vertical ice body	Refreezing of draining water within flow fingers
		8c. basal ice	Ice layer at base of snow cover	Refreezing of meltwater ponded on snowpack substrate
9. Surface deposits and crusts	▽	9a. rime	Soft or hard rime deposits on surface crystals, thin crusts	Snowpack rimed by exposure to fog
		9b. rain crust	Thin, transparent glaze layer	Frozen rain on surface
		9c. sun crust	Thin, transparent glaze layer	Refrozen meltwater on surface, may grow from below

temperature. Mass loss and gain from the surfaces of ice grains can also create
faceted and poorly bonded ice grains (see discussion in Section 3.1.2).

Two other processes can be responsible for energy transfer in snowpacks: radiation transmission and convection. Radiation transmission in snow is discussed in
Chapter 6 in relation to snowmelt. Convection due to air currents in the snowpack
caused by temperature-induced buoyancy effects can occur in snow, but conditions
under which convection occurs are not well understood (Powers *et al.*, 1985). Convection due to wind pressure forces on the snowpack surface, the so-called pressure
pumping of snowpack air, can also occur that transfers gases within the snowpack
(Massman *et al.*, 1995; Colbeck, 1997).

Heat conduction in snow can be described by the equation:

$$q = -k_{eff}dT/dz \qquad (3.2)$$

where:

$q =$ flux density of heat conducted by snow, W m^{-2}
$k_{eff} =$ effective thermal conductivity of snow, W m^{-1} K^{-1}
$dT/dz =$ temperature gradient, K m^{-1}

The effective thermal conductivity of snow has often been related to density,
although the bonding and structure of ice grains and temperature are the controlling
factors. In an extensive review of past studies and analysis of their own data, Sturm
et al. (1997) give the following general relationships for k_{eff}:

$$k_{eff} = 0.138 - 1.01\rho + 3.233\rho^2 \quad [0.156 \le \rho \le 0.6]$$
$$k_{eff} = 0.023 + 0.234\rho \quad \{\rho < 0.156\} \qquad (3.3)$$

where ρ is snow density in g cm^{-3}. The experimental data set used to generate this
relationship ($n = 488$) represented an average snowpack temperature of $-14.7\,^\circ$C
and an average density of 317 kg m^{-3}. As can be seen in Figure 3.6, the effective
thermal conductivity of snow for densities up to 0.6 g cm^{-3} is still well below that
for solid ice (2.24 W m^{-1} K^{-1}), while for very low snow densities it approaches
the thermal conductivity for air (0.024 W m^{-1} K^{-1}). Sturm *et al.* (1997) show
that the density dependence of k_{eff} is greatest for well-bonded and wind-compacted
snow and least, or non-existent, for poorly bonded depth hoar. Sturm and Johnson (1992) found that the k_{eff} for depth hoar was from one-half to one-fourth of
the values normally reported for other snow types of similar density due to thin
and infrequent ice bonds. Sturm *et al.* (1997) recommend using simple average
k_{eff} values for poorly bonded snow and give average values of k_{eff} for various
ICSSG (see Table 3.2) snow types for that purpose. Heat conduction rates under

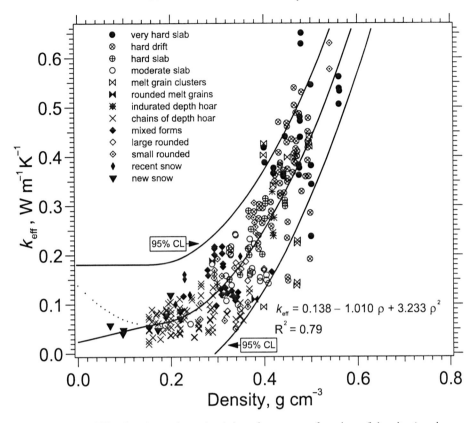

Figure 3.6 Effective thermal conductivity of snow as a function of density (modified from Sturm *et al.* 1997 from *Journal of Glaciology* with permission of the International Glaciological Society).

steady-state conditions in uniform snow can be computed using Equations (3.2)–(3.3). Given a structurally uniform snowpack that is 0.22 m deep with constant surface and base temperatures of $-7\,°C$ (266 K) and $0\,°C$ (273 K), respectively, and a density of 0.35 g cm^{-3}, the rate of heat flow can be computed as:

$$k_{\text{eff}} = 0.138 - 1.01(0.35) + 3.233(0.35)^2 = 0.18 \text{ W m}^{-1} \text{ K}^{-1}$$
$$q = -(0.18 \text{ W m}^{-1} \text{ K}^{-1})[(266 \text{ K} - 273 \text{ K})/(0.22 \text{ m} - 0 \text{ m})] = 5.73 \text{ W m}^{-2}$$

Unfortunately, heat conduction in a subfreezing snowpack is seldom at steady state and the snowpack is seldom structurally uniform, which cause considerable difficulty in modelling snowpack temperatures. Several models exist that can be used to predict the heat conduction, temperature and density distribution within the snowpack over time (Jordan, 1991).

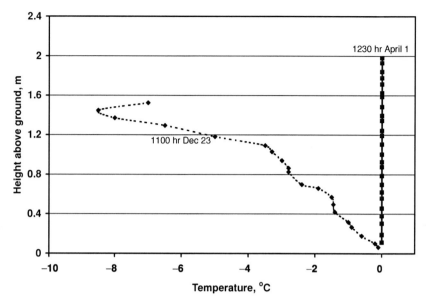

Figure 3.7 Temperature profiles in a California snowpack in the early and late accumulation season at the Central Sierra Snow Lab, CA (US Army Corps of Engineers, 1956).

3.3 Snowpack temperatures and cold content

Snowpack temperatures can vary seasonally and daily as a result of the complex processes of energy transfer within the snowpack. Temperatures profiles in a California snowpack on December 23 and April 1 from the early Corps of Engineers Cooperative Snow Investigations are shown in Figure 3.7 to illustrate typical seasonal differences. In late morning on December 23, profiles show subfreezing temperatures near the snowpack surface due to contact with cold air and temperatures near or at 0 °C at the snowpack base due to soil heat conduction from below. By April 1 just prior to snowpack melting, the snowpack has warmed to an isothermal 0 °C due to heat conduction from above and below. Note that the temperature profiles in Figure 3.7 correspond to the same snowpack represented by density profiles shown in Figure 3.5. Warming of snowpacks from above is greatly enhanced by the infiltration of liquid water from surface melting and rainfall, which refreezes in the colder snow below and releases the latent heat of fusion within the snowpack. Refreezing of liquid water inputs on successive melt days early in the spring season can quickly warm even deep snowpacks. Once the snowpack is warmed to 0 °C and liquid-water holding capacities are satisfied, the snowpack is referred to as a ripe snowpack. See later sections in this chapter for typical liquid-water holding capacities for snow.

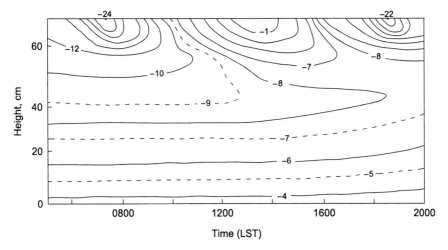

Figure 3.8 Typical diurnal snowpack temperature (°C) variations in a winter Colorado snowpack at local standard time (LST) (Bergen 1968, courtesy US Forest Service) showing periodic cooling and warming near the surface and a relatively stable temperature regime near the ground.

Daily temperature fluctuations with depth in subfreezing snowpacks can be considerable. Due to the low thermal conductivity of snow, heat can not be added quickly by conduction into the snowpack during the daytime or lost quickly by conduction from within the snowpack during nocturnal cooling. Consequently, snowpacks exhibit very large temperature gradients near the surface and large time lags for propagation of daily temperature waves and rapid damping of diurnal temperature fluctuations in deeper snowpack layers. Secondary temperature maxima can occur just below the surface in snow during clear daytime periods due to the combined influence of heat conduction and transmission of shortwave radiation (Koh and Jordan, 1995). Figure 3.8 shows typical diurnal variations in the temperature of a subfreezing winter snowpack in Colorado (Bergen, 1968). Daily amplitudes in temperature are large at the snowpack surface but rapidly diminish with increasing depth. Peak daily temperatures also occur later with increasing depth in the snow. Temperatures near ground level remain relatively stable and often approach 0 °C.

Hydrologists often use snowpack temperature and density profile data to compute snowpack cold content. Cold content represents the amount of liquid water from melt or rainfall that must be frozen in a subfreezing snowpack to warm the snowpack from its current temperature to an isothermal 0 °C. Before liquid water can be released from the snowpack an isothermal condition must generally exist, although pockets of subfreezing snow can exist adjacent to fingers of ripe snow conducting water. Cold content can be calculated as:

$$CC = [\rho_s c_i d(273.16 - T_s)]/(\rho_w L_f) \tag{3.4}$$

where:

CC = snowpack cold content, m of liquid water
ρ_s = average snowpack density, kg m^{-3}
c_i = specific heat of ice, J kg^{-1} K^{-1}
d = snowpack depth, m
T_s = average snowpack temperature, K
ρ_w = density of liquid water = 10^3 kg m^{-3}
L_f = latent heat of fusion, J kg^{-1}

Cold content can also be computed using Equation 3.4 for individual layers in a snowpack and summed using the respective densities and temperatures. Computed cold contents can be compared to the daily predicted melt or forecasted rainfall to help determine when liquid water will be released from snowpacks. Cold contents for ripe snowpacks that are isothermal at 0 °C are zero. Nocturnal cooling can result in small cold contents even during the melt season that must be satisfied before meltwater can be released the next day. Computations of runoff time lags due to cold content are discussed in Chapter 9, Section 9.3.1. A discussion of the snowpack energy balance with consideration of heat storage and use of degree-days to model cold content are given in Chapter 6.

An example computation of cold content using Equation (3.4) can be given for the conditions represented by snowpack conditions for December 23 in Figures 3.5 and 3.7. The depth of snow on that date was 1.524 m, average density was 311 kg m^{-3}, and average temperature was −3.1 °C or 270.1 K. Given the specific heat of ice of 2100 J kg^{-1} K^{-1} and latent heat of fusion of 0.334 × 10^6 J kg^{-1} at the freezing point (see Appendix A1), the cold content would be:

$$CC = [(311)(2100)(1.524)(273.16 - 270.1)]/(10^3)(0.334 \times 10^6)$$
$$= 9.208 \times 10^{-3} \text{ m}$$
$$= 9.2 \text{ mm of liquid water}$$

A cold content of this magnitude could easily be satisfied by surface melt in one day since daily melt frequently exceeds 10 mm.

3.4 Liquid water in snow

Liquid water from melting or rain will first satisfy any snowpack cold content and then the liquid-water holding capacity of the snow. The remaining liquid water will percolate vertically downward through the snowpack to the soil surface unless diverted by less permeable layers and crusts. Once water reaches the base of the snowpack, it can flow downslope through basal layers of the snowpack if the ground surface is impermeable or infiltrate the soil. In preceding sections, the importance of liquid water in snowpack metamorphism and ripening was described. In this section, the storage and movement of liquid water in snow are described. Measurement of snowpack liquid-water content is discussed in Chapter 4.

3.4.1 Liquid-water content of snow

The liquid-water content of a snowpack represents all liquid water in the snow including water in transit from rain or melt and water that is stored against gravitational forces. The latter liquid water is called the irreducible liquid saturation or liquid-water holding capacity of snow. Liquid content of snow is expressed either on a volumetric basis (θ_v) as the volume of liquid water per unit volume of snow or a mass basis (θ_m) as the mass of liquid water per unit mass of snowpack. The two measures are related as:

$$\theta_v = \theta_m(\rho_s/\rho_w) \qquad (3.5)$$

where:

θ_v = volumetric liquid-water content, m^3 of liquid water per m^3 of snow
θ_m = liquid-water content on mass basis, kg of liquid water per kg snow
ρ_s = snowpack density, kg m^{-3}
ρ_w = liquid-water density, 10^3 kg m^{-3}

It should be pointed out that the liquid content of snow is computed differently than the liquid content of soils. In soil, the moisture content by weight is based upon the dry mass of soil, while the liquid content of snow on a mass basis is based upon the total mass of the snow including the mass of liquid water.

Liquid-water contents vary over time due to surface melting and rain. In Figure 3.9, liquid-water contents of snow on a mass basis are shown to vary between about 0–2% at night during drainage, up to about 30% during active melt on the third day. If the snowpack density were 400 kg m^{-3} during active melt, then by Equation (3.5) a 30% liquid content on a mass basis would be equal to a liquid content of 12% on a volumetric basis.

Liquid-water contents affect the amount of water released during melt per unit of energy input. A unit of energy input at times of high liquid-water content will release more water than at times of low liquid content, due to release of water already held in the snowpack as well as water created by melt. To account for the effects of liquid water and cold content when computing melt, hydrologists commonly use a dimensionless parameter called thermal quality (B). Thermal quality (B) is defined as the energy required to melt a unit mass of snow divided by the energy required to melt a unit mass of ice at 0 °C. The energy required to melt a unit mass of snow

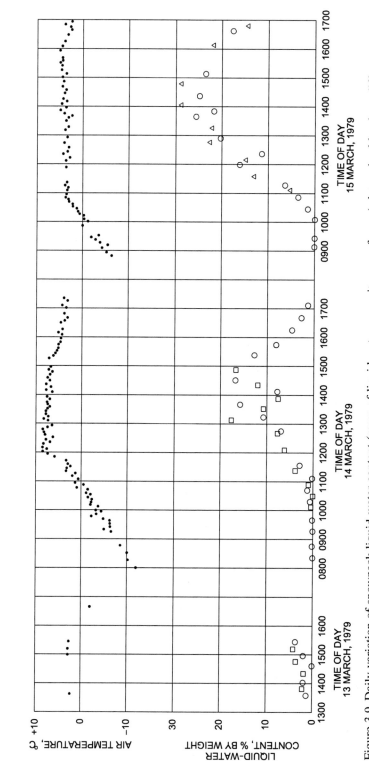

Figure 3.9 Daily variation of snowpack liquid-water content (mass of liquid water per unit mass of snow) determined by three different operators using the freezing calorimetry method at Fraser, Colorado (Jones *et al.*, 1983, copyright 1983 IWA Publishing with permission).

is the sum of: (1) energy needed to raise the temperature of the snow to 0°C when the snowpack is subfreezing and (2) the energy needed to melt the solid portion of the snowpack. Thermal quality is computed as:

$$B = [(1 - \theta_m)L_f + c_i T]/L_f \qquad (3.6)$$

where:

B = thermal quality, dimensionless
θ_m = liquid-water content, mass liquid per unit mass snow
L_f = latent heat of fusion, 0.334×10^6 J kg^{-1} at $0\,°$C
c_i = specific heat of ice, 2.1×10^3 J kg$^{-1}\,°$C^{-1}
T = snowpack temperature depression below $0\,°$C

When snow is ripe and contains some liquid water, the numerator is less than L_f and B is <1. When snow is subfreezing, the numerator is greater than L_f and B is >1. For ripe snow with liquid water present, the thermal quality is simply the fraction of snowpack mass that is ice. The computation ignores the minor contribution of air to snowpack mass. Thermal quality ranges from about 0.8 for ripe snow to about 1.1 for subfreezing snow. The use of thermal quality in melt computations is discussed in Chapter 6.

An example computation can be used to illustrate the two snowpack thermal quality regimes.

For unripe snow with zero liquid water and an average snowpack temperature of $-4.5\,°$C, the thermal quality would be:

$$B = [(1 - 0)(0.334 \times 10^6 \text{ J kg}^{-1})$$
$$+(2.1 \times 10^3 \text{ J kg}^{-1}\,°\text{C}^{-1})(4.5\,°\text{C})]/(0.334 \times 10^6 \text{ J kg}^{-1})$$
$$= 1.028$$

For ripe snow at $0\,°$C and liquid-water content on a mass basis of 0.15, the thermal quality would be:

$$B = [(1 - 0.15)(0.334 \times 10^6 \text{ J kg}^{-1})]/(0.334 \times 10^6 \text{ J kg}^{-1})$$
$$= 1 - 0.15 = 0.85$$

3.4.2 Theory of water movement in snow

Liquid-water movement in snow can be generally described with classical theory of flow in porous media using Darcy's law (Wankiewicz, 1978). Due to the relatively uniform and coarse-grained nature of snow, effects of capillarity on water movement in snow are generally ignored and gravitational forces are considered dominant. Figure 3.10 shows relationships between capillary pressure and liquid

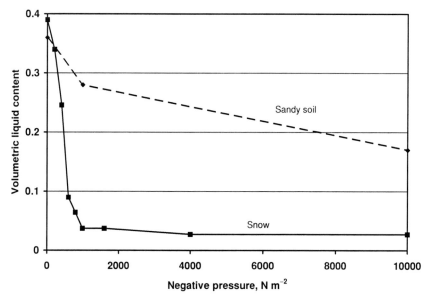

Figure 3.10 Comparison of liquid content vs. capillary pressure curves for snow and sandy soils (data from Colbeck, 1974 and Oke, 1987) showing some snow samples at least have even less capillary attraction for liquid water than sands for a given negative pressure in Newtons (N) m^{-2}.

content for snow in comparison to sandy soil. Over 80% of the snowpack pore space is drained with a pressure of only -800 N m^{-2} or a suction force equivalent to only 8 cm of water, while typical sandy soils would retain more moisture at that same capillary pressure. Wankiewicz (1978) showed similar data for a snow sample taken in February, but a response more similar to sandy soil with a snow sample collected in July. Due to the small effects of capillarity, capillary rise in snow is limited generally to a few centimeters (Coleou *et al.* 1999).

With capillary effects ignored, Darcy's law describing liquid-water movement in snow can be written as:

$$u_w = \rho_w k_w g / \mu_w = \alpha k_w \tag{3.8}$$

u_w = volume flux of liquid water, m^3 m^{-2} s^{-1} or m s^{-1}
ρ_w = density of water, 10^3 kg m^{-3}
k_w = intrinsic permeability of water in snow, m^2
g = acceleration due gravity, 9.80 m s^{-2}
μ_w = dynamic viscosity of water, 1.79×10^{-2} kg m^{-1} s^{-1} at 0 °C
$\alpha = \rho_w g / \mu_w = 5.47 \times 10^5$ m^{-1} s^{-1}

As in any porous media, the permeability of snow during two-phase or unsaturated flow is related to the level of liquid saturation described by the effective saturation (S^*) where:

$$S^* = (S_w - S_{wi})/(1 - S_{wi}) \qquad (3.9)$$

S^* = effective saturation, dimensionless
S_w = liquid saturation, volume of liquid water per volume of pore space, $m^3\ m^{-3}$
S_{wi} = irreducible liquid saturation, irreducible volume of liquid water per unit volume of pore space, $m^3\ m^{-3}$

Effective saturation ranges from zero to one, that is, liquid-water content varies from 0% to 100% of the pore space available to water flow. Typically, S^* ranges between 0.1 and 0.2 during normal melting.

The irreducible volumetric water content is defined by the condition when water films within the snowpack become discontinuous or flow due to gravity drainage is negligible. Colbeck (1974) found a value of 0.07 and Jordan (1991) successfully used a value of 0.04. The commonly used liquid-water holding capacity of snow (θ_{vi}) expressed as the irreducible volume of liquid water per unit volume of snowpack is related to S_{wi} as:

$$\theta_{vi} = S_{wi}\phi \qquad (3.10)$$

where ϕ = porosity of the snowpack (volume of pore space per unit volume of snow). Thus a liquid-water holding capacity of $\theta_i = 0.03$ in a snowpack with porosity $\phi = 0.6$ is equivalent to $S_{wi} = 0.03/0.6 = 0.05$.

Liquid permeability in snow is related to effective saturation (S^*) by:

$$k_w = kS^{*^n} \qquad (3.11)$$

where k equals the intrinsic permeability (m^2) or the permeability with single phase flow at liquid saturation. The exponent n is an empirical constant ranging from two to four, with a typical value being three. According to the form of Equation (3.11), the permeability of snow will obviously be strongly controlled by the level of liquid saturation and hence the melt or rainfall rate. The intrinsic permeability of snow (k) appearing in Equation (3.11) has been related to snowpack grain size (d) and porosity (ϕ) by Shimizu (1970) as:

$$k = 0.077\ d^2 \exp[-7.8(1 - \phi)\rho_i/\rho_w] \qquad (3.12)$$

where:

k = intrinsic permeability, m^2
d = mean grain diameter, m

ϕ = porosity, m^3 m^{-3}

ρ_i = density of ice = 917 kg m^{-3}

ρ_w = density of water = 1×10^3 kg m^{-3}

Equation (3.12) applies to ripe snow where grain size and porosity are reasonably stable. The theory would not apply to fresh snow that would rapidly metamorphose in the presence of liquid water.

Equations (3.8)–(3.12) can be used to compute flow rates of liquid water in ripe snow under gravitational forces. If a melting snowpack has a representative grain size of 0.0015 m (1.5 mm) and porosity of 0.55, then the intrinsic saturated permeability would be:

$$k = 0.077(0.0015)^2 \exp[-7.8(1 - 0.55)(917/1000)] = 6.91 \times 10^{-9} \text{ m}^2$$

If the volumetric liquid-water content of the melting snow layer was 0.15 and the irreducible liquid-water content of the snow was 0.04, then the effective saturation (S^*) would be:

$$S^* = (0.15 - 0.04)/(1 - 0.04) = 0.114$$

The snowpack permeability (k_w) for this level of effective saturation and a value of $n = 3$ in Equation (3.11) would be:

$$k_w = (6.91 \times 10^{-9})(0.114)^3 = 1.02 \times 10^{-11} \text{ m}^2$$

If the volumetric water content of snow doubled to 0.228, then the permeability of snow would increase by nearly a factor of ten to 8.91×10^{-11} m^2, which illustrates the strong control of permeability by liquid-water content. Finally, the rate of liquid-water movement in snow can be computed as:

$$u_w = (10^3)(1.02 \times 10^{-11})(9.8)/(1.79 \times 10^{-2})$$
$$= 5.58 \times 10^{-6} \text{ m s}^{-1} \text{ or 2 cm hr}^{-1}$$

Under these conditions, water would drain slowly from the snowpack. Given a higher water content and higher equivalent permeability (k_w) of 8.91×10^{-11} m^2 from above, the rate of water movement would increase to 17.6 cm hr^{-1}.

Application of the above theory in modelling liquid-water movement in snowpacks has generally agreed with experimental data. Theory confirms the observed rapid gravity drainage of liquid water from natural snow cores (Denoth *et al.*, 1979). In Figure 3.11, over 85% of the accumulated volume was collected within the first two hours of drainage; however, a very prolonged drainage of water occurs thereafter. This behavior is consistent with observed drainage of liquid water from melting snow. Rapid drainage occurs during daytime when melting and liquid-water

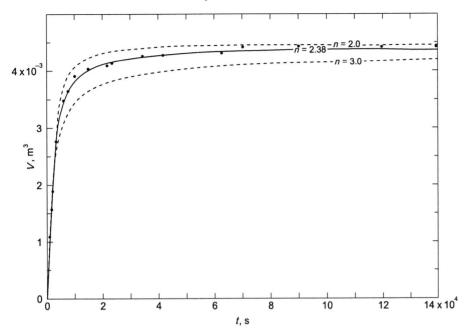

Figure 3.11 Accumulated drainage from melting snow core over time modelled with Darcy's law with the assumption of negligible capillary effects (Denoth *et al.*, 1979, courtesy US Army, Cold Regions Res. Engin. Lab.) showing rapid initial and very slow later drainage of meltwater. Lines show fitted model with varying exponents "*n*" for the snowpack permeability function in Equation (3.11).

contents are highest and a long gradual drainage occurs overnight when liquid inputs and snowpack water contents are low.

Further extension of this theory in combination with the continuity equation allows prediction of the rate of water movement in a snowpack for a given value of water flux as a function of time at various depths in the snow as (Colbeck, 1972, 1977):

$$\mathrm{d}z/\mathrm{d}t|_{u} = 3\alpha^{1/3}k^{1/3}[\phi(1 - S_{wi})]^{-1} u_{w}^{2/3} \tag{3.13}$$

Here the speed of the wetting front in a snowpack, $\mathrm{d}z/\mathrm{d}t|_{u}$, is directly related to the flux of water $u_{w}^{2/3}$ due to melting or rainfall. In Figure 3.12 a plot of predicted volume flux at various depths in snow for an assumed surface water flux distribution is shown. The surface volume flux is translated with only slight modification in shape to a depth of 0.6 m within two hours. At greater depths modification of the melt wave is seen with a more abrupt leading edge and elongated tail, due to the fact that faster moving meltwater (from times when surface melt rates are greater) can overtake slower moving meltwater from other times to create a shock wave type behavior. Simulations indicate that for shallow seasonal snow covers, runoff

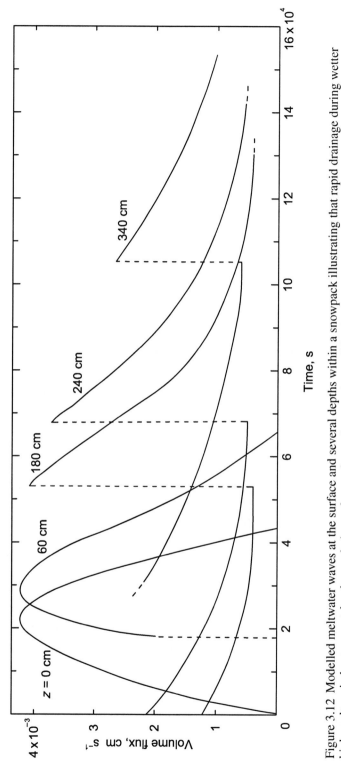

Figure 3.12 Modelled meltwater waves at the surface and several depths within a snowpack illustrating that rapid drainage during wetter high-melt periods can overtake slower drainage from prior melt to produce a "shock wave" type behavior of the wetting front within the snowpack (Colbeck, 1972, with permission of the International Glaciological Society).

Figure 3.13 Flow fingers for preferential liquid-water movement that formed during initial day's melting of a homogeneous 28-cm depth snowpack resulting from a single snowfall in Pennsylvania (photo D. R. DeWalle). Water mixed with food coloring was uniformly sprayed over the snowpack surface to trace water movement. Also note evidence of meltwater movement downslope to the right in the surface layer and at the soil–snow interface. See also color plate.

delays due to liquid-water movement in ripe snow will be minimal. In deeper snow or snow with significant cold content, the delays are much greater. Lag times in snowmelt runoff due to liquid-water movement in snow are discussed further in Chapter 9.

Application of theory of flow in porous media to snow is made difficult by spatial variation in snowpack properties that develop over time and cause erratic liquid-water behavior. Liquid water in snow often drains along preferred flow fingers (Figure 3.13). How such flow fingers develop is not understood, but surface depressions in melting snow on flat terrain are thought to be physical manifestations of such flow channels. Ice columns in snowpacks have also been observed where liquid water in these flow fingers has refrozen. Preferred flow channels may speed release of liquid water from a snowpack. Ice layers and layers of snow with varying density can also delay initial water release from snowpacks and cause horizontal shunting of water. On slopes, liquid water may also travel through the base of the snowpack in pipes. All of these variations in flow paths make modelling of liquid-water release from the snowpack more difficult. A solution to this problem is to define a reference volume of snow that behaves according to theory. One suggestion of such a reference volume is to model a snowpack surface area equal to Z^2, where Z is the snowpack depth (Male and Gray, 1981). If such a snowpack reference area can be defined, then models can be used to represent average conditions for these reference volumes, rather than snow at a smaller scale. Marsh (1999) discussed spatial variations of flow paths within snowpacks and the challenges this presents to modelling meltwater delivery.

3.5 References

Bergen, J. D. (1968). *Some Observations on Temperature Profiles of a Mountain Snow Cover*, Res. Note RM-110. Rocky Mtn. For. and Range Exp. Sta.: US Forest Service.

Colbeck, S. C. (1972). A theory of water percolation in snow. *J. Glaciol.*, **11**(63), 369–85.

Colbeck, S. C. (1974). Water flow through snow overlying an impermeable boundary. *Water Resour. Res.*, **10**(1), 119–23.

Colbeck, S. C. (1977). Short-term forecasting of water run-off from snow and ice. *J. Glaciol.*, **19**(8), 571–88.

Colbeck, S. C. (1997). Model of wind pumping for layered snow. *J. Glaciol.*, **43**(143), 60–5.

Colbeck, S., Akitaya, E., Armstrong, R., Gruber H., Lafeuille, J., Lied, K., McClung, D., and Morris, E. (1990). *The International Classification for Seasonal Snow on the Ground*. Internat. Comm. Snow and Ice, IASH.

Coleou, C., Xu, K., Lesaffre, B., and Brzoska, J.-B. (1999). Capillary rise in snow. *Hydrol. Processes*, **13**, 1721–32.

Denoth, A., Seidenbusch, W., Blumthaler, M., Kirchlechner, P., Ambach, W., and Colbeck, S. C. (1979). *Study of Water Drainage from Columns of Snow*, CRREL Report 79–1. Cold Regions Res. and Engin. Lab: US Army, Corps of Engineers.

Doesken, N. J. and Judson, A. (1997). *The Snow Booklet: A Guide to the Science, Climatology, and Measurement of Snow in the United States*, 2nd edn. Ft. Collins, CO: Colorado Climate Center, Dept. Atmospheric Sci., Colorado State University.

Jones, E. B., Rango, A., and Howell, S. M. (1983). Snowpack liquid water determinations using freezing calorimetry. *Nordic Hydrol.*, **14**, 113–126.

Jordan, R. (1991). A one-dimensional temperature model for a snow cover. In *Technical Documentation for SNTHERM.89*, Special Rpt. 91–16. Cold Regions Res. Engin. Lab: US Army Corps of Engineers.

Judson, A. and Doesken, N. (2000). Density of freshly fallen snow in the central Rocky Mountains. *Bull. Amer. Meteorol. Soc.*, **81**(7), 1577–87.

Koh, G. and Jordan, R. (1995). Sub-surface melting in a seasonal snow cover. *J. Glaciol.*, **41**(139), 474–82.

Male, D. H. and Gray, D. M. (1981). Snowcover ablation and runoff. In *Handbook of Snow: Principles, Processes, Management and Use*, ed. D. M. Gray and D. H. Male. Toronto: Pergamon, pp. 360–436.

Marsh, P. (1999). Snowcover formation and melt: recent advances and future prospects. *Hydrol. Processes*, **13**, 2117–34.

Massman, W. R., Sommerfeld, R., Zeller, K., Hehn, T., Hudnell, L., and Rochelle, S. (1995). CO_2 flux through a Wyoming seasonal snowpack: diffusional and pressure pumping effects. In *Biogeochemistry of Seasonally Snow-covered Catchments*, Publ. No. 228, ed. K. A. Tonnessen, M. W. Williams and M. Tranter. IAHS, pp. 71–80.

McClung, D. and Schaerer, P. (1993). *The Avalanche Handbook*. Seattle, WA: The Mountaineers Books.

Oke, T. R. (1987). *Boundary Layer Climates*, 2nd edn., New York: Methuen and Co.

Perla, R. I. and Martinelli, Jr., M. (1978). Avalanche Handbook. In *Agriculture Handbook 489*. Washington, DC: US Printing Office.

Powers, D., Colbeck, S. C., and O'Neill, K. (1985). Experiments on thermal convection in snow. *Annals Glaciol.*, **6**, 43–7.

Shimizu, H. (1970). Air permeability of deposited snow. *Low Temp. Sci.*, **22**(Series A), 1–32.

Sturm, M. and Johnson, J. B. (1992). Thermal conductivity measurements of depth hoar. *J. Geophys. Res.*, **97**(B2), 2129–39.

Sturm, M., Holmgren, J., Konig, M., and Morris, K. (1997). Thermal conductivity of seasonal snow. *J. Glaciol.*, **43**(143), 26–41.

US Army Corps of Engineers (1956). *Snow Hydrology – Summary Report of the Snow Investigations*. Portland, OR: US Army, Corps of Engineers, N. Pacific Div.

Wankiewicz, A. (1978). A review of water movement in snow. In *Proceedings, Modelling of Snow Cover Runoff*, ed. S. C. Colbeck and M. Ray. Cold Regions Res. Engin. Lab: U.S. Army, Corps of Engineers, pp. 222–68.

4

Ground-based snowfall and snowpack measurements

4.1 Measurement of snow at time of fall

4.1.1 Precipitation gauges

Several complications arise when using precipitation gauges to measure snowfall that are not as much of a problem when measuring rainfall. Gauge catch losses are more severe for snow than rain, so that precipitation gauge shields are much more of a necessity. Additionally, subfreezing temperatures during winter, and usually during a snowfall event, dictate that an antifreeze solution be used to prevent collected water from freezing (an evaporation suppressant might be necessary to minimize evaporative losses); and, to insure that bridging of snow over the gauge orifice does not happen, the gauge is sometimes heated. The positioning of the snow-precipitation gauge on the landscape must also be considered in order to get a more representative catch of snowfall.

Nonrecording gauges

There are two classes of nonrecording gauges, those read on a daily or more frequent basis and storage gauges that are read only after a period of time has elapsed, like once a month or once a season. A subtype of the daily nonrecording gauge is the gauge that is only read after a snowfall event has been observed. If a gauge is to be read on a 6- or 24-hour interval, antifreeze, orifice heating, and evaporation suppressants are not as critical as for a storage gauge. For those nonrecording gauges, the funnel and collection tube, used for rainfall events, are removed from the gauge and snow falls directly into the outside container. When the measurement is made, the collected snow must be melted and poured into the collection tube and the depth of precipitated water measured with a calibrated measuring stick.

The storage gauge, which is left unattended for periods up to several months, needs to have extra precautions such as antifreeze to melt collected snow and an evaporation suppressant. McCaughey and Farnes (1996) report that storage gauges can be charged with an antifreeze solution of 50% propylene glycol and 50% ethanol and topped with mineral or transformer oil to retard evaporation. In general, the snow precipitation gauge orifice is most commonly placed 2 m above the ground surface (Goodison *et al.*, 1981). However, in deep snowfall areas where storage gauges are likely to be located, the storage gauge must be elevated so the height of the orifice is 1 m above the maximum expected snowpack to prevent the precipitation gauge from being inundated by snow accumulation (Goodison *et al.*, 1981).

Recording gauges

Weighing-type recording precipitation gauges measure snow using a spring balance. The capacity of the weighing gauge and the amount of snowfall combine to determine how frequently the gauge site must be visited to remove accumulated water from the gauge. The weight of the snowfall can be recorded at the measurement site, or a signal can be telemetered to a central collection facility. Some recording weighing gauges can operate for up to 1 year unattended; however, it is recommended that they be serviced once every 3 months to insure reliable, continuous operation (Goodison *et al.*, 1981). The advantage of a telemetered gauge is that the need for maintenance is immediately evident through regular examination of the telemetered data. Some additional modifications of the weighing gauge have been suggested; vertical extension of the gauge orifice to prevent snow bridging over the orifice and installation of rubber seals on the housing of the recording mechanism to prevent the entry of water (Tabler *et al.*, 1990; Winter and Sturges, 1989; Morris and Hanson, 1983).

There has always been considerable interest in using the vast number of tipping-bucket rain gauges as snow gauges in the winter season. The snow collected needs to be first melted after it enters the orifice of the gauge, and the tipping-bucket mechanism must also be heated to prevent freezing of water in the bucket. Unfortunately, heating the gauge causes some convection of the warmer air, which can prevent some snow crystals from entering the gauge and possibly increase evaporative loss (hence, the reason that weighing and nonrecording gauges are not generally heated). In a study by Hanson *et al.* (1983), a comparison of heated, tipping-bucket gauges against weighing-type gauges found that the tipping buckets recorded 22% less snow water than did the weighing gauges. In an innovative approach to using the tipping-bucket gauges without producing thermal currents near the orifice, McCaughey and Farnes (1996) experimented with a modified storage gauge, charged with antifreeze and an evaporation suppressant, as the collector device and attached a tipping-bucket recorder used as the measurement device. These modified tipping-bucket gauges collected 2% less snow water than recorded by weighing gauges at a research site in Montana. Additional work with this system in North Dakota indicated that it was a viable way to get reliable data using a tipping-bucket approach (Carcoana and Enz, 2000).

4.1.2 Measurement errors

Although the measurement of precipitation with gauges seems straightforward, errors can arise that need to be accounted for, especially with snowfall as opposed to rainfall. The height of the gauge orifice in relation to the maximum snowpack height must be considered so the gauge is not covered in high accumulation years. A snowcap or bridge over the gauge orifice can form, especially in areas prone to rime ice formation (Tabler *et al.*, 1990). Wind can deflect precipitation particles on trajectories that cause an undercatch of actual precipitation, which is especially true for snow particles compared to rain particles and more so as wind speed increases. Wetting the interior walls of the container or collector and subsequent evaporation can cause a loss of a greater magnitude than errors arising from splash of raindrops out of gauge or evaporation of water accumulated in the container (Sevruk, 1986). Associated with this is the inability of gauges to accurately measure small snowfall amounts. Also to be considered are errors arising from the effects of surrounding topography and structures and the collection of snow blowing from the ground or off nearby vegetation (Tabler *et al.*, 1990).

Minimizing or correcting for errors

Considering that wind can reduce the catch of total snowfall by 50 to 80% (Goodison *et al.*, 1981; Larson and Peck, 1974; Sevruk, 1986; Goodison, 1978; Tabler *et al.*, 1990), this error would be very important to reduce or at least correct for using the results of gauge-catch studies. The first solution is to locate the gauge to minimize the impact of wind, such as in small clearings in the forest snow zone or in brush. Unfortunately, the snow zone does not often have such barriers to the wind. Also, where possible, the gauge orifice should not be elevated more than necessary as the wind speed increases with height and, thus, increases the catch deficit. Because most gauges have to be exposed to the wind, other means to reduce wind speed across the gauge orifice must be used.

Studies show that utilization of various types of gauges with wind shields increases the amount of snowfall caught by the gauge. Many different types of shields have been used around the world (see Figure 4.1). Some of the more common shields employed around the world today are shown in Figure 4.2. Tabler *et al.* (1990) provides details of the Nipher, Alter, Double-Fence, Wyoming, and Dual-Gauge shielding systems. There should be very strong consideration given to shielding all precipitation gauges used to measure winter snowfall due to the extreme catch deficit that can develop as wind speed increases.

Figure 4.1 Shielded snow gauge in the northwest United States used to register snow fall (courtesy National Oceanic and Atmospheric Administration (NOAA), NOAA Central Library, NOAA Photo Library, wea 00901, Historic NWS Collection, original photo, US Weather Bureau, 1917 in "Boy with Weather Men," p. 224)

If a wind shield is not used on a precipitation gauge in snowfall areas, it is puzzling why coefficients to correct for the catch deficit are not widely utilized. In fact, very few snowfall records are corrected and published, while there has been much published on methods to correct such data (Sevruk, 1986; 1989; Goodison *et al.*, 2002). Studies have been done in a variety of countries to come up with wind correction factors that vary according to wind speed and can be used to reconstruct precipitation amounts. For example, Bryazgin and Radionov (2002) provide wind correction factors for liquid, mixed, and solid precipitation events. They also provide correction factors for recalculating precipitation amounts affected by wetting the interior of the precipitation gauge. Such corrections need to be made in order to generate correct data for climate studies and runoff forecasting if precipitation gauges are to be used in such analyses. Unfortunately, several other problems arise of which the user of the data needs to be aware. First, the same correction methods are not used in Canada and the northern United States. Second, about half of the US synoptic stations adjust the observed amount collected using a correction factor, but this correction is not noted in the published records. As a result,

Figure 4.2 Precipitation gauges at the Reynolds Creek Experimental Watershed (RCEW) in southwest Idaho, USA (See Figure 4.3) (courtesy of the Northwest Watershed Research Center, Agricultural Research Service, US Department of Agriculture).

(a) Belfort Universal Recording Gauge with Wyoming Shield;
(b) DFIR Gauge;
(c) TRET Gauge with Shield;
(d) Canadian Nipher-Shielded Snow Gauge;
(e) in foreground is US National Weather Service 8-in. Nonrecording Gauge, and in background are two Belfort Universal Recording Gauges with orifices at 3.05 m, one with constrained-baffle alter-type shield;
(f) Belfort Universal Recording Gauge with unconstrained-baffle alter shield.

historical precipitation data do not have a consistent relation to actual snowfall (Peck, 1997).

Placement of precipitation gauges in natural openings in a forested area is desirable so the wind effect on catch is minimized, but care must be taken not to place the gauge too close to the forest or the wind might blow snow off the trees and into the gauge. The recommendation of the SCS (Soil Conservation Service, 1972) is to place a gauge no closer in horizontal distance than four times the average height of the trees. Although different agencies make different recommendations (see Tabler *et al.*, 1990), a similar "rule of thumb" should be employed when locating gauges. When locating gauges near man-made structures, the same "rule of thumb" should also be employed.

Case study intercomparison in Idaho

A site was established by the USDA/ARS on the Reynolds Creek Experimental Watershed in southwestern Idaho (see Figure 4.3) in fall 1987 and operated through spring 1994 to compare precipitation catch between nine precipitation measuring systems (some with shields, some unshielded) (Hanson *et al.*, 1999). This site was established as part of the World Meteorological Organization's (WMO) program to compare current national methods of measuring snowfall with precipitation gauges. A schematic of the location of the precipitation gauges and other instruments on the WMO comparison site is shown in Figure 4.4.

The following nine gauging systems were used in this study (see Figure 4.2): (1) The Belfort universal recording gauge (orifice at 2.2 m) with a Wyoming shield (WYO) (Rechard, 1975); (2) the Russian double-fence intercomparison reference (DFIR) gauge, consisting of a shielded Tretyakov (TRET) gauge (orifice at 3.0 m) with two concentric wooden outer shields; (3) the Russian TRET gauge (orifice at 2.0 m) with shield; (4) the Canadian Nipher shielded snow gauge (CAN) (orifice at 1.6 m) (Canadian Department of Environment, 1985); (5) the US National Weather Service 8-inch nonrecording gauge (orifice at 0.94 m) without a shield (NATUNSHLD); (6) the Belfort universal recording gauge (orifice at 1.4 m) with an Alter shield with unconstrained baffles (NATSHLD) (Alter, 1937); (7) the Belfort universal recording gauge (orifice at 3.05 m) without a shield (BELUNSHLD); (8) the Belfort universal recording gauge (orifice at 3.05 m) with an Alter-type shield with the shield's baffles individually constrained at an angle of 30° from the vertical (BELSHLD); and, (9) the dual-gauge (DUAL-GAUGE) (Hamon, 1973; Hanson, 1989) configuration that is discussed in the next section.

Table 4.1 is a summary of gauge catch for the study period of 1987–94. The WYO catch amounts were used as the base values in this study because WYO had the greatest total catch during this study. These results agree with a previous study

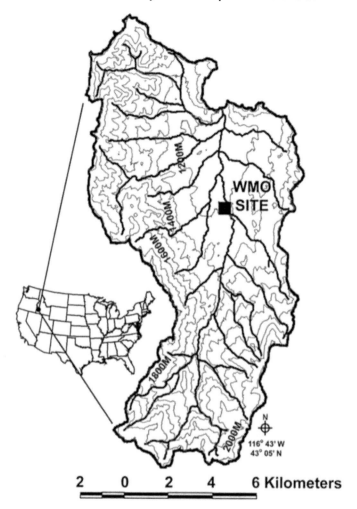

Figure 4.3 World Meteorological Organization (WMO) precipitation gauge catch
site on at the RCEW in southwest Idaho (adapted from Hanson *et al.* [1999]).

at this same site by Hanson (1989), which found that the WYO had a slightly greater
total precipitation catch than the DUAL-GAUGE system.

Snow

As shown in Table 4.1, the computed snow catch from the DUAL-GAUGE system
and the CAN catch were greater than the WYO catch by 4 and 1%, respectively.
The greater catch of snow by the DUAL-GAUGE system supports previous findings
by Hanson (1989). The DFIR and NATSHLD catches were only 2 and 4% less,
respectively, than the WYO catch. The TRET and BELSHLD caught approximately

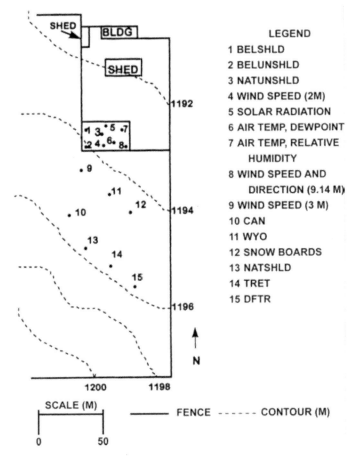

Figure 4.4 Location of precipitation gauges and other meteorological instrumentation on the RCEW test site (adapted from Hanson *et al.* [1999]).

1% less snow than the WYO. The NATUNSHLD and the BELUNSHLD gauges caught 12 and 24% less, respectively.

Total precipitation

For all precipitation types and events combined (total in Table 4.1), the WYO caught more precipitation than any of the other gauges; however, the undercatch was only 3% or less for the DFIR, CAN, NATSHLD, and DUAL-GAUGE. The catch by the WYO, CAN, and DUAL-GAUGE were not significantly different. These results indicate that these five gauging systems may be used with little or no correction, but there may be a small undercatch relative to the WYO. The TRET, NATUN-SHLD, and BELSHLD caught between 8 and 9% less total precipitation than the WYO. It should be noted here that the BELSHLD and the NATSHLD gauges are the

Table 4.1 *Total gauge catch (mm) for 1987–1994 winter periods, Reynolds Creek Experimental Watershed, Idaho*
(Hanson et al., 1999)

					Gauge				
Precipitation type (1)	WYO (2)	DFIR (3)	TRET (4)	CAN (5)	NATUNSHLD (6)	NATSHLD (7)	BELUNSHLD (8)	BELSHLD (9)	DUAL-GAUGE (10)
Snow (N = 49)	211.8b	207.2c	191.6d	213.3b	187.3d	202.3c	160.4e	192.1d	220.9a
Ratio		0.98	0.90	1.01	0.88	0.96	0.76	0.91	1.04
Mixed (N = 23)	118.7a	115.40a	110.50b	115.10a	107.90b	115.20a	97.60c	107.50b	115.50a
Ratio		0.97	0.93	0.97	0.91	0.97	0.82	0.91	0.97
Rain (N = 37)	178.7a	170.0b	164.3cd	167.5bc	167.0bc	174.5b	160.6d	165.3c	168.7bc
Ratio		0.95	0.92	0.94	0.93	0.98	0.90	0.93	0.94
Total (N = 109)	509.2a	492.60b	466.40c	495.90ab	462.20c	492.00b	418.60d	464.90c	505.10ab
Ratio		0.97	0.92	0.97	0.91	0.97	0.82	0.91	0.99

Note. The mean values of rows with the same letter were not significantly different at the 0.05 probability level. Ratio = gauge catch values/WYO catch.

same type of gauge and both were shielded; however, the BELSHLD orifice was at 3.05 m, which most likely accounted for most of the difference in catch between these two gauges. The BELUNSHLD, with its orifice at 3.05 m, caught 18% less precipitation than the WYO gauge, which was significantly less than any of the other gauging systems.

In general, the results of this case study emphasize that precipitation gauge shielding and the location of the orifice above the ground are critical to reliable precipitation measurements. The results also support the concept of correcting precipitation catch when different gauging systems are used.

4.1.3 Manual snow-depth determinations

Ruler measurements and snowboards

A ruler measurement of snow depth provides a physical depth observation that can be converted to snow water equivalent if a representative density can be determined for the area in question. In Canada, a standard ruler measurement is the average of several snow-depth readings over a representative area (Goodison *et al.*, 1988). The major factor that must be considered is selection of the representative area so that the effects of drifting snow are avoided. The determination of the amount of snowfall deposited in a specific storm can be difficult as the entire snowpack, and especially the new snow, is continually settling. The use of snowboards is one way to measure an incremental depth of snow input to the snowpack and avoid measuring the old snow. A snowboard is a square piece (about 40 cm × 40 cm; Goodison *et al.*, 1981) of plywood or lightweight metal, painted white in order to minimize any melting of the snow. After each depth observation (with a ruler), the snowboard is swept clear of snow and set on top of the snow surface in preparation for the next measurement. Most accurate measurements are acquired if the snowboard reading is made after snowfall ceases or several times during a long lasting snowstorm. To convert a recent snowfall depth to a snow water equivalent value, multiply the snow depth by 0.1, which is the commonly used average density of new snow. However, this value can vary widely from region to region and storm to storm, and errors in calculating the snow water equivalent in this way can be large. The actual snow water equivalent can be obtained by taking a sample core off the snowboard, melting the sample, and measuring the melted depth.

Snow stakes and aerial markers

Where snow depth measurements of the entire snowpack are required from some distance, snow stakes can be used, painted with a bright color and contrasting numbers for ease of observation and avoidance by snowmobiles (Trabant and Clagett, 1990) or skiers. Placing the site on level ground helps prevent horizontal pressure

Figure 4.5 Oblique aerial photography of simulated aerial snow marker in Buck pasture, Black Forks Basin, Uinta Mountains, Utah at SNOTEL site (courtesy Natural Resources Conservation Service, US Department of Agriculture).

from downslope motion on the stake, and the use of round, as opposed to rectangular, cross section stakes helps minimize drifting or melt out around the stake. By being able to make a reading from the outside of the snow stake plot, destruction or disturbance of the accumulated snow cover is avoided. If aerial surveys of snow depth are required, aerial snow markers that contrast with the snow and are easy to read from a rapidly moving airplane are necessary. These are the best approaches to use in remote areas with difficult winter access. The aerial markers are commonly made of wood and steel with an established horizontal bar interval. Long horizontal bars are placed every even foot (30.5 cm) (and shorter bars are sometimes placed at the odd foot [30.5 cm]) with taller markers used in deep snowpack areas, being reinforced by diagonal bracing crossing at one-half the measurement interval selected (Trabant and Clagett, 1990).

When visually observing snow depth from a low-flying aircraft, the observation of the number of bars visible from the top down is recorded, with subtraction from the total bars included on a particular marker to be done after the flight. To assure a record of the aerial snow marker, oblique photography of the marker can be obtained during the flight for later analysis. Visual readings of the snow-depth marker are usually accurate to ± 50 mm (Trabant and Clagett, 1990). Photographing the marker with high speed film in a 35 mm camera with a 500 mm telephoto lens has been used by Schumann (1975) to determine the snow depth to ±7.5 mm. Figure 4.5 shows an aerial snow-marker photograph from an aerial snow-survey flight in Utah. Any such snow-depth data must be multiplied by representative snow-density values to obtain the snow water equivalent. The density values, if chosen incorrectly, can cause large errors in estimating the water volume.

4.1.4 Automatic weather stations

The places in a snow basin at which we would like to acquire measurements are often remote and/or experience severe weather conditions. Many times, the only way to acquire the necessary weather data is to make measurements automatically. Strangeways (1984) has devoted significant time to designing an automatic weather station (AWS) that can operate unattended for long time periods. He has come up with an AWS that can operate in cold environments where high winds and icing of instruments are problems. It needs only an aqualung to de-ice the sensors by shock induction and flexing (Strangeways, 1984), and the AWS can measure incoming and reflected solar radiation, temperature, humidity, wind speed and direction, and precipitation. Still, the measurement of the solid precipitation component remains a significant problem, and the most recent advances of accurate solid precipitation component measurement must be incorporated into the AWS concept. More will be said about remote, automated measurements when snowpack measurements are discussed later in this chapter.

4.1.5 Other methods for measuring falling snow

Goodison *et al.* (1988) report on an acoustic ranging instrument developed in Canada to measure the incremental snow depth over any selected period of time. The most recent version measures the time elapsed for an acoustic pulse to travel from a transducer to the target (snow surface) and back again. Air temperature must be measured to account for the fact that speed of sound in air is temperature dependent. The potential resolution of the sensor for measuring incremental snow depth is 1.0 mm. Because the snowpack is dynamic, it is not realistic to expect this measurement accuracy. Over the course of the winter season, the automated acoustic snow sensor measures to ±2.5 cm of the total new snow-depth accumulation (Goodison *et al.*, 1988; Tanner and Gaza, 1990).

   The light attenuation snow-rate sensor is a device that measures the change in light transmission over a path length as snow falls through the light beam. This instrument gives a detailed time history of snow rate and will operate for days unattended. In contrast to a conventional collecting gauge, it has no problems with snow bridging across the orifice. The major disadvantage is that it gives no measurement of snow totals directly, but rather a history of instantaneous rates. An instrument like this is used in the National Weather Service Automated Surface Observing System (ASOS) (Doesken and Judson, 1997).

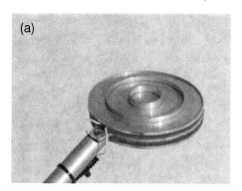 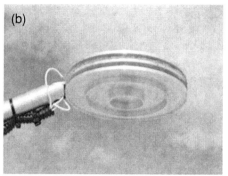

Figure 4.6 Hotplate snow gauge operation where both upper (a) and lower (b) plates are heated and maintained at 75 °C. The power difference between the top and bottom plates is proportional to the precipitation rate (courtesy R. Rasmussen, National Center for Atmospheric Research).

A "hotplate snow gauge" has been developed to do away with oil, glycol, or an expensive wind shield, now typically required for current weighing precipitation gauges. The system has an upper and lower plate heated to constant, identical temperatures (Rasmussen *et al.*, 2002). The two plates are maintained at constant temperatures during wind and precipitation conditions by increasing or decreasing the current to the plate heaters (see Figure 4.6). During precipitation conditions, the top plate has an additional cooling effect due to melting and evaporation of precipitation. The difference between the power required to heat the top plate compared to the bottom plate is proportional to the precipitation rate (Rasmussen *et al.*, 2002).

As noted by Singh and Singh (2001), investigators in Norway (Bakkehoi *et al.*, 1985) have developed a vibrating wire-strain gauge that can measure falling precipitation with a resolution of about 0.1 mm. The weight of solid precipitation on the wire causes a change in frequency vibration, which corresponds to a certain snowfall amount. This gauge can be continuously recording or preset to read at certain specific time intervals (Singh and Singh, 2001).

Ground-based weather radar can be used to estimate areal precipitation in snowfall events in a way similar to rainfall events. Because of several factors, the estimation of areal snowfall is more difficult than estimating areal rainfall amounts. Most weather radars use S-band, but Giguere and Austin (1991) indicate the use of X-band would improve the measurement of areal snowfall due to an increased range of observation and more sensitivity to lower intensity snowfalls. In snowfall situations, it is common to have mixed precipitation, which further confuses the radar determination due to part of the reflected wave being from snow and part from rain. In general, ground-based radar determinations of snowfall can be as much as

50% in error. However, when combined with conventional gauge measurements (and sometimes satellite determinations), the value of the radar lies in its areal coverage and nearly continuous operation.

4.1.6 Snowfall records and data

In the United States, the greatest snowfall depth in a 24-hour period was 192.5 cm at Silver Lake, CO, on April 14–15, 1921. The greatest snowfall in one storm occurred at the Mt. Shasta Ski Bowl in California and totaled 480 cm from February 13–19, 1959. The largest total snowfall to occur in one snow season was 2,475 cm at Thompson Pass in Alaska during 1952–53. The greatest snow depth ever measured on the ground was 1145.5 cm at Tamarack, CA on March 11, 1911. Measuring snow on the ground, however, presents an entirely different challenge, as we shall cover in the next section.

4.2 Measurement of snow on the ground

Measurement of falling snow amounts and recently accumulated snow depth is complicated enough, but measuring the water equivalent of snow on the ground may be even more difficult because of the need to find a representative location and not to disturb the snowpack, making return visit measurements possible. The most common way to assess the snow water equivalent of snow on the ground in the western United States, where snow makes up over 50% of the water supply, is to measure snow water equivalent with a calibrated snow sampler at a number of points along an established transect called a snow course.

4.2.1 Snow-course surveys

Originally, snow courses were established to assess the snow water equivalent of a snowfield to come up with an estimate of the amount of resulting runoff when the snowpack melts in the spring and summer. However, accessibility of representative snowpack areas is often difficult, especially in winter, when the snow courses not only extend over a mere fraction of the snow fields but are often not located in certain key snow accumulation and ablation areas. As a result, snow courses have usually been used to obtain a consistent index to the quantity of runoff that the snow will generate during melt. In more singular cases, the snow course can also be located to represent the accumulation of snow over fairly large, adjacent areas.

The layout of a snow course is usually in one or more lines of about 10 points (ranging from 5–15 measurement points), similar to that shown by the Natural Resources Conservation Service (2003d) in Figure 4.7. Most snow courses established by the Natural Resources Conservation Service in the United States are about

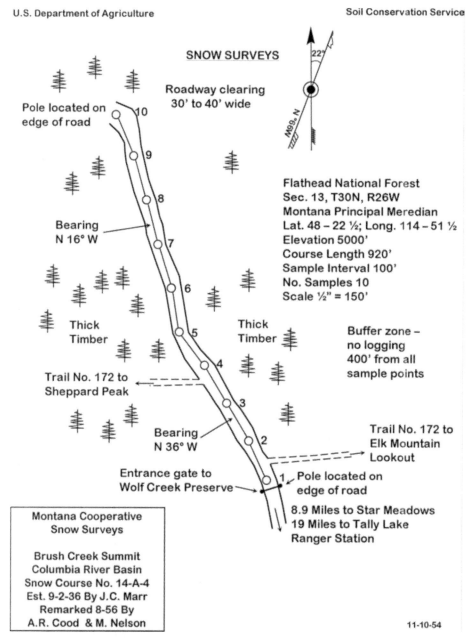

Figure 4.7 Map and instructions for conducting snow surveys at the Brush Creek, Montana snow course (Natural Resources Conservation Service, 2003d).

1000 ft (305 m) long and, if possible, protected from the wind by locating them in small meadows (Julander, 2003). The snow course can be laid out between two stakes, marked by signs on trees, or described by written directions and compass headings (Goodison *et al.*, 1981). Notes can be kept so that measurements are not taken in the same location more than once during the season. In the United States, measurements are taken on the first of every month starting on January 1 and continuing until May 1 or June 1, depending on the snow conditions. Mid-month surveys are made in Canada, as well as at the beginning of the month, and, in special situations, in the United States to help evaluate changing conditions.

The snow sampler equipment is simple in concept. A graduated tube with a cutter on the lower end, allowing easy insertion into the snowpack, is used to obtain a snow core. The tube and core are then weighed on a spring balance. The balance is scaled to read directly in snow-water-equivalent units. Most snow tubes are made of aluminum, but some are made out of plastic or fiberglass. In the western part of North America, the aluminum snow tube sections are made so they can be screwed together, thus, making determination of snow water equivalent in deep snowpacks possible. In the eastern part of North America where snowpacks are relatively shallow, the snow tubes used are typically only one section, like the Adirondack or Canadian MSC snow samplers. Goodison *et al.* (1981) present interesting comparison information on the snow samplers used in North America (Table 4.2).

Before making measurements, the surveyor makes sure the tube is completely clear of any snow, ice, or soil. The manual snow surveys are conducted with a team of two or more surveyors for both safety and technical reasons. One surveyor takes the measurements and a second surveyor serves as the recorder. At the designated measurement point on the course, the snow sampling tube(s) is inserted into the snow all the way to the ground surface, the snow depth is recorded from the scale on the outside of the tube, and a core is extracted with soil plug at the end. The soil plug verifies the tube reached to the ground surface (Julander, 2003). The soil is cleared from the tube and the length of the core is verified to roughly correspond to the outside of the tube depth measurement. The weight of the tube and snow core measured on the spring balance and the tare weight of the empty tube are subtracted to give the snow water equivalent of the snowpack. All the snow water equivalent values are averaged to get one overall value for the snow course.

The average snow-course value is most often used in an empirical regression, along with other independent variables such as precipitation, antecedent streamflow and other snow-course average values, to produce a seasonal snowmelt volume forecast as close as possible to the first of the month, corresponding to when the snow course measurement was made. All streamflow forecasts generated are for

Table 4.2 Snow sampler properties (Goodison et al., 1981)

Material	Standard[a] Federal	Federal[b]	Bowman[c] L-S	McCall[d]	Canadian[e] MSC	Adirondack[f]
	Aluminum	Aluminum	Plastic or Aluminum	Heavy Gauge Aluminum	Aluminum	Glass Fiber
Length of tube[g] (cm)	76.2	76.2	76.2	76.2	109.2	153.7
Theoretical ID of cutter (cm)	3.772	3.772	3.772	3.772	7.051	6.744
Number of teeth	16	8	16	16	16	None[h]
Depth of snow that can be sampled (m)	>5	>5	>3.5	>5	1.0	1.5
Retains snow cores readily	Yes	Yes	Yes	Yes	No	No

[a] Standard sampler used in the western United States and Canada.
[b] Identical to "Standard Federal" but has an 8-tooth cutter.
[c] Cutter has alternate cutter and raker teeth and may be mounted on plastic or standard aluminum tubing. It is more an experimental rather than operational sampler.
[d] Used in dense snow or ice. It is a heavy gauge aluminum tube with 5-cm cutter with straight flukes. It may be driven into the pack with a small slide drop hammer producing an ice-pick effect.
[e] Atmospheric Environment Service large diameter sampler used in shallow snowcover.
[f] Large diameter fiberglass sampler commonly used in eastern United States.
[g] Most snow samplers in North America use inches and tenths as their basic units of measurement. Values in this table are corresponding metric equivalents.
[h] Stainless steel circular cutter edge or small teeth.

streamflow volumes that would occur naturally without any upstream influences. A set of forecasts is made with the most probable forecast being one that has a 50% chance streamflow volume will exceed the forecast value and a 50% chance streamflow volume will be less than the forecast value. The most probable forecast will rarely be correct because of errors resulting from future weather conditions and from errors in the forecast equation itself (Natural Resources Conservation Service, 2003a).

4.2.2 SNOwpack TELemetry (SNOTEL) system

The SNOTEL system is a logical extension of the manual snow course network. NRCS installs, operates, and maintains an extensive, automated SNOTEL system used to collect snowpack and climate data (including related fire-hazard data), especially in remote regions and in response to an economic need to reduce the number of costly, manual snow courses. SNOTEL data are collected and relayed using meteor burst communications technology. Radio signals are reflected off an ever-present band of ionized meteorite trails existing 50–75 miles above the Earth's surface. SNOTEL sites are generally located in remote, high elevation basins with difficult access. These sites are designed to operate unattended for a year. Solar power is used to recharge batteries. The approximately 730 SNOTEL sites are polled daily by three master stations in the western United States, and data are transmitted to the NRCS National Weather and Climate Center in Portland, OR, where the data are stored (Natural Resource Conservation Service, 2003b).

Basic SNOTEL sites have a pressure-sensing snow pillow, a sonic snow depth sensor, storage precipitation gauge, and an air-temperature sensor. The four snow pillows, shown in Figure 4.8, are envelopes of stainless steel or synthetic rubber, containing an antifreeze solution, about 4 ft (1.22 m) on a side and arranged to form a square approximately 8 ft (2.44 m) on a side. As snow accumulates on the pillows, a pressure is exerted on the antifreeze solution. The snow load on the pillow is sensed either by a pressure transducer or by the fluid level in a standpipe. These devices convert the weight of the snow to an electrical reading of the snow water equivalent. Larger surface area pillows are used in areas where bridging of the snowpack over the pillows is a possibility. The precipitation gauge measures all precipitation in any form that falls during the year, while the temperature sensor records maximum, minimum, and average daily readings (Natural Resources Conservation Service, 2003c). The SNOTEL system has a great capability to incorporate additional sensors at particular sites that can be used for other environmental problems. Additional variables that have been measured include wind speed and direction, soil temperature, soil moisture, snow depth, barometric pressure, humidity, and radiation.

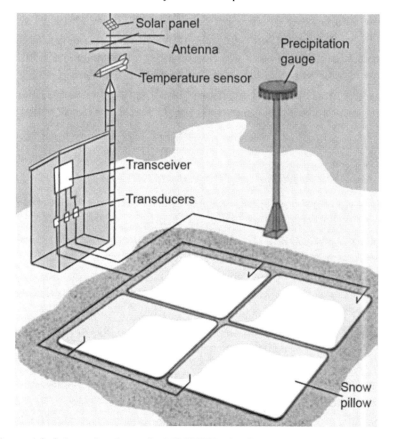

Figure 4.8 Schematic of a typical SNOTEL site. Pressure pillows are used for measuring snow water equivalent, a storage gauge provides seasonal precipitation at the site, and a temperature sensor measures the existing air temperature (adapted from Natural Resources Conservation Service, 2003c).

Both snow course and SNOTEL data are available on the Natural Resources Conservation Service, National Water and Climate Center website, www.wcc.nrcs.usda.gov/. Figure 4.9 shows the location of the SNOTEL sites in Utah, and by putting the cursor on a point, the name of the SNOTEL site is displayed. By clicking on the SNOTEL site, various options for data retrieval are presented. Table 4.3 shows the data for Wolf Creek Summit SNOTEL in Colorado for several days.

4.2.3 Other methods of measuring snow water equivalent (SWE)

Cosmic radiation, particularly the gamma portion, passes through the atmosphere and penetrates the winter snowpack. The greater the snow water equivalent, the

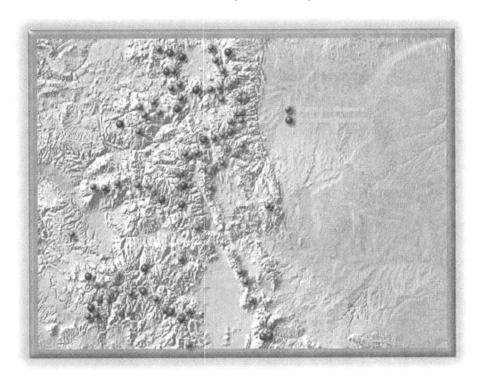

Figure 4.9 Colorado SNOTEL sites (courtesy Natural Resources Conservation Service, US Department of Agriculture). See also color plate.

greater the attenuation of the cosmic gamma radiation. One such cosmic radiation instrument has been tested in California and was found to be an accurate, reliable, noninvasive method for measuring SWE (Osterhuber *et al.*, 1998). The relationship of the gamma sensor SWE to that obtained from manual snow coring for three winters closely coincided, with an $R^2 = 0.97$.

The isotopic-profiling gauge also employs gamma radiation to make measurements, using a radioactive source and a detector to make density profiles through the snowpack (Kattelmann *et al.*, 1983). The snow water equivalent of the snowpack is determined by multiplying the average (integrated) density by the total depth (Wheeler and Huffman, 1984). Although an excellent research tool, this type of gauge has not found any significant operational application, perhaps because of problems with using the radioactive source (e.g., Cesium^{137}) in unattended locations.

Snowmelt lysimeters have been designed primarily to measure the liquid outflow from the bottom of the snowpack as a result of snowmelt or rainfall falling on the snowpack (Kattelmann, 1984). A few of the lysimeters were also designed to weigh

Table 4.3 *Wolf Creek^a summit SNOTEL data report – daily readings (Natural Resources Conservation Service, 2004)^b*

Date	Time (PST)	Snow water equivalent (inches)	Snow depth (inches)	Year-to-date precipitation (inches)	Current temp (°F)	Previous day's temp (°F)		
						Max	Min	Avg
5/26/2004	0	25	39.6	35.2	35.2	48.9	32.5	39.7
5/27/2004	0	24.5	37.5	35.2	36.2	52.1	29.5	40.2
5/28/2004	0	23.5	35.7	35.2	37.7	58.9	30.9	43.2
5/29/2004	0	22.4	33	35.2	41.4	59.5	36.5	46.8
5/30/2004	0	22.1	32.2	35.2	26.1	45.7	25.9	35.6
5/31/2004	0	21.7	31.1	35.2	29	47.4	21.1	33.3
6/1/2004	0	21	28.9	35.2	36.7	56.5	28.5	41.7

[a] Basin: Upper San Juan (HUC 14080101) Elevation: 11 000
[b] Provisional data (As of: Tue Jun 01 13:00:26 PDT 2004), subject to revision
[c] Precip(YTD) = Precipitation from October 1 to current date.

the snowpack in place (e.g. Cox and Hamon, 1968). For determining the snow water equivalent, however, devices such as snow pillows have received more emphasis in the last 20 years than lysimeters.

Some attempts have been made to develop an alternative to the snow pillow. Moffitt (1995) reported on the development of a $1 m^2$, circular, perforated aluminum plate mounted on three load cells. The weight on the snow plate is converted to snow water equivalent. The instrument is accurate and portable. However, it still has snow bridging problems similar to snow pillows because of its relative small size. If coupled with an ultrasonic snow-depth sensor, the combination of sensors can be used to indirectly measure snow density.

4.2.4 Liquid-water measurements

Liquid water in a snowpack, although recognized for some time as a quantifiable parameter of hydrologic significance, is not a regularly reported snowpack characteristic due to difficulties with its measurement in the field. The amount of liquid water representing the liquid phase of snowpack is an important parameter in forecasting runoff, predicting wet slab avalanche release, and interpreting snowpack remote sensing data using microwave techniques. Variation in snowpack liquid water produces changes in the dielectric properties of the snowpack for most of the microwave region, which in turn will have an effect on any microwave techniques being used to estimate water equivalent or snow depth.

The liquid-water content of a snowpack, sometimes called free water, snow water content, or liquid-phase water, includes gravitational water moving downward through the snowpack and capillary water held by surface tension between the individual snow grains.

Various techniques have been used or proposed to measure the liquid-water content of a snowpack. Most of these techniques can be broadly categorized as centrifugal, dielectric, dilution, and calorimetric. Some investigators have tested other methods, including Shoda (1952) who used a measurement of volume expansion upon freezing and Bader (1948) who used the concept of dilution of a solution by the liquid water in the snowpack. Morris (1981) has reported improvement of the solution method for field use. These and other techniques are summarized in Table 4.4.

Kuroda and Hurukawa (1954) and Carroll (1976) described centrifugal separation of the liquid water from a snow sample. Langham (1974, 1978) subsequently improved the method by isolating the amount of melt that occurs during the centrifuging process. The National Bureau of Standards (Jones, 1979) compared the centrifuge technique with the calorimetric technique. It was concluded that the centrifuge and freezing calorimeter methods do not measure the same phenomena and that water may be present in the snowpack in a form which is detectable by

Table 4.4 *Methods for determination of the liquid-water content of snow*

Method	Operating principle	Reference
Melting calorimetry	Energy required to melt mass of ice is measured when warm water is added to snow sample	Radok *et al.*, 1961; Fisk, 1982, 1983; Ohmura, 1980; Kawashima *et al.*, 1998
Freezing calorimetry	Energy liberated when the liquid water in snow is frozen by a freezing agent is measured	Leaf, 1966; Fisk, 1982, 1983; Boyne and Fisk, 1987, 1990; Jones *et al.*, 1983
Alcohol calorimetry	Temperature depression measured as snow sample melts in methanol at 0 °C	Fisk, 1983, 1986
Dilution	Reduction in electrical conductivity of a solution due to dilution by snowpack liquid water is measured	Davis *et al.*, 1985; Boyne and Fisk, 1987, 1990
Dielectric constant	Capacitance or TDR measurement in snow varies due to large difference in dielectric constants of liquid water and ice	Boyne and Fisk, 1987, 1990; Denoth, 1994; Ambach and Denoth, 1974; Schneebeli *et al.*, 1998
Freezing-point depression	Freezing-point depression in a salt solution added to a snow sample due to liquid water is measured	Bader, 1950; Morris, 1981
Centrifuge	Centrifugal force is used to separate liquid water from ice in a snow sample	Kuroda and Hurukawa, 1954; Jones, 1979

the freezing calorimeter method but not by the centrifuge method. Colbeck (1978) analyzes this in more detail.

The dielectric method for measuring liquid-water content relies on the contrast between the dielectric properties of dry and wet snow. Ambach and Denoth (1974, 1980) have developed capacitance techniques and used them extensively in the field. These dielectric sensors have generally been calibrated with freezing calorimetry. Linlor *et al.* (1974, 1975) and Linlor and Smith (1974) have used various dielectric techniques, including capacitance and microwave beam attenuation, in the laboratory with some field testing. The capacitance method is discussed for determining free water in glacial snowfields (Østrem and Brugman, 1991) (Figure 4.10).

The dilution method, as described by Davis *et al.* (1985), mixes snow with free water with an aqueous solution (like 0.1N hydrochloric acid) and measures the change in concentration of the aqueous solution. One advantage of the approach

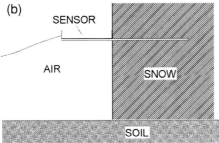

Figure 4.10 Capacitance probe/dielectric device being used on the Devon Ice Cap, Nunavut, Canada (a) to measure snow wetness. Bottom diagram (b) shows the so-called "Denoth probe" inserted into a snow pit wall (photograph courtesy J. Sekerka, and M. Demuth, Geological Survey of Canada; drawing adapted from Østrem and Brugman, 1991).

is that as many as 10–15 snow samples per hour can be processed (Trabant and Clegett, 1990).

Calorimetric methods employ the latent heat of fusion to measure the liquid water in a snow sample from the measured heat balance in a calorimeter. Melting calorimetry using warm water (Yoshida, 1940, 1960, 1967) or electrical energy (de Quervain, 1946; Hansen and Jellinek, 1957) as the heat source has been tested. These techniques are difficult to utilize in the field and require considerable time for complete operation. Freezing or cold calorimetry using liquid freezing agents has been described by Radok *et al.* (1961) and used in the field by Leaf (1966). Howell *et al.* (1976) and Bergman (1978) used the freezing calorimeter with toluene as the freezing agent. Although toluene is satisfactory for freezing the snow liquid water, its toxic properties and relatively high flash point make its use somewhat hazardous. Silicone oil has been used successfully as the freezing agent (Jones *et al.*, 1983), and it has significantly improved safety features over toluene.

Freezing calorimetry has been used both in operational-type fieldwork and snow hydrology research. Because freezing calorimetry has gained acceptance as a standard method for field measurements of liquid-water content and as a method for calibrating new measurement methods, it is presented in more detail here to illustrate an easily duplicated field measurement method.

Calorimetric analysis

The freezing or cold calorimeter technique has been selected by investigators primarily because it is relatively inexpensive, simple to use, and based on known and documented physical phenomena. In addition, it appears that, because a small amount of liquid water is frozen in the freezing calorimetric approach as opposed to a large amount of ice melted in the melting calorimetric method, the freezing calorimetric method is more sensitive to variations in liquid-water content.

A calorimetric analysis is based on the concept that heat gained by the system must equal the heat lost. The calorimeter and the freezing agent gain heat; the snow sample and the liquid water lose heat. If the heat gained terms are set equal to the heat lost terms, the following equation results:

$$(m_3 - m_2)(T_3 - T_2)c_i + m_w L_f = [(m_2 - m_1) + E](T_2 - T_1)c_f \qquad (4.1)$$

where:

m_w = mass of the liquid water, kg
L_f = latent heat of fusion, J kg^{-1}
c_f = specific heat of freezing agent at $(T_1 + T_2)/2$, J kg^{-1} °C^{-1}
c_i = specific heat of ice at $(T_2 + T_3)/2$, J kg^{-1} °C^{-1}
E = calorimeter constant, kg
m_1 = mass of calorimeter, kg
m_2 = mass of calorimeter and freezing agent, kg
m_3 = mass of calorimeter, freezing agent and snow, kg
T_1 = initial temperature of the freezing agent, °C
T_2 = initial temperature of the snow at sample point, °C
T_3 = Final temperature of snow and freezing agent, °C

If θ_m is the fraction of liquid water in the sample, then:

$$m_w = \theta_m(m_3 - m_2) \qquad (4.2)$$

Substituting in Equation (4.1), the expression for the percent of liquid water in the snow sample is

$$\theta_m = \frac{[(m_2 - m_1) + E](T_2 - T_1)c_f}{L_f(m_3 - m_2)} - \frac{c_i(T_3 - T_2)}{L_f} \tag{4.3}$$

Snow thermal quality, B as defined by Bruce and Clark (1966) is the fraction (by weight) of water in the solid state or $(1 - \theta_m)$ and can, therefore, be determined by calorimetric methods (see Section 3.4.1).

The type of container generally used for mixing the freezing agent and the snow sample in cold calorimetry is a vacuum-insulated bottle with a temperature probe and a tightly fitted rubber stopper. Experience indicates that a commercial, 1-qt (946 ml), stainless steel, wide-mouth vacuum bottle works well. The temperature probe is used to monitor changes in temperature that occur in the vacuum bottle during the mixing process. The heat-balance equation requires accurate measurements of temperature changes inside the bottle during the mixing process. Because the vacuum-insulated bottle is not a perfect system and the bottle itself gains some heat, the heat-balance equation contains a calorimeter constant, E. Each calorimeter bottle will have its own constant that must be determined independently. For convenience in using the heat-balance equation, the calorimeter constant is expressed in terms of equivalent weight of freezing agent. Various methods may be used to determine this constant (Jones *et al.*, 1983).

Freezing-calorimeter equipment and materials

Certain basic equipment is required for utilizing the freezing-calorimeter method described in this section. First is the calorimeter which, as previously stated, is a wide-mouth vacuum bottle made of stainless steel. Scales capable of weighing up to 2000 gm with accuracy to the nearest tenth of a gm are required for determining the various weights. Finally, 30 cm long, stainless-steel thermocouple probes accurate to 0.2 °C (used with a multi-channel digital readout) are needed for measuring the temperature changes occurring in the calorimeter. In addition to these basic requirements, several ancillary equipment items are recommended. They are a cold chest for storing the freezing agent, thermal containers for transporting the snow samples, timing devices, and miscellaneous tools, including shovels, trowels, and spoons. As a result, the system is not easily portable in the field.

Cold calorimetry requires using a freezing agent to freeze the liquid water in the snow sample. It is recognized that a small portion of the capillary water may not be frozen, but the resulting error is considered to be very small.

Analysis of error associated with freezing calorimetry by various investigators has indicated that the approach is accurate to within ± 1.0–2.0% by weight. To minimize errors, special care must be taken during determination of weights and temperatures.

Melt-calorimetry approach

Basic theory and computations involved with the melt-calorimetry method are reviewed here to illustrate another commonly applied method. In this method, a snow sample is added to water in a vacuum bottle hot enough to melt all the ice and the temperature of the resulting mixture is measured. The hot water is cooled in proportion to the fractional mass of ice relative to liquid water in the snow sample. An energy balance can be established for the vacuum bottle to compute the mass of liquid water where the energy lost from the hot water and vacuum bottle (left side of Equation 4.4) is equal to the energy required to melt all the ice in the sample and warm the entire sample from 0 °C to the final temperature of the mixture (right side of Equation 4.4) as:

$$(m_{hw} + E)(T_{hw} - T_f)c_w = (m_s - m_w)L_f + m_s(T_f - T_s)c_w \tag{4.4}$$

where:

m_{hw} = mass of hot water, kg
m_w = mass of liquid water, kg
m_s = mass of snow, kg
T_{hw} = initial temperature of hot water, °C
T_s = initial temperature of snow, °C
T_f = final temperature of snow–hot water mixture, °C
c_w = specific heat of water, 4.188×10^3 J kg^{-1} °C^{-1}
L_f = latent heat of fusion, 0.334×10^6 J kg^{-1}
E = calorimeter constant, kg

The calorimeter constant represents heat gained by the vacuum bottle from its environment during the measurement and must be experimentally determined. E is expressed in equivalent mass of hot water for simplicity. For reliable results, this method requires temperature measurements to within 0.1 °C and mass within 0.001 kg.

A simple example ignoring the calorimeter constant can be used to illustrate the use of the melting calorimetry method. If 0.2 kg of snow at 0 °C were added to a vacuum bottle filled with 0.3 kg of hot water at 90 °C and the resulting mixture achieved a final temperature of 25 °C, then the mass of liquid water in the snow could be computed from Equation (4.4) as:

$$(0.3)(90 - 25)(4.188 \times 10^3) = (0.2 - m_w)0.334$$
$$\times 10^6 + (0.2)(25 - 0)(4.188 \times 10^3)$$
$$m_w = 0.018 \text{ kg}$$

Thus, the liquid water content on a mass basis would be $0.018/0.2 = 0.091$ or 9.1% liquid water.

Intercomparison of techniques

Boyne and Fisk (1987) compared four snow cover liquid-water measurement techniques – alcohol calorimetry, freezing calorimetry, dilution, and capacitance. Each method has advantages and disadvantages and should be chosen based on the specific project and resources available. However, Boyne and Fisk (1987) conclude that all methods should be capable of giving comparable accuracies within 1–2% on mass basis.

Liquid-water outflow measurements at the base of the snowpack with snow lysimeters

Several different designs of lysimeters have been used to record the outflow of liquid water at the base of the snowpack. A snowmelt lysimeter consists of a collector, a flow-measuring device and a connector between the two major components (Kattelmann, 1984). The most widely used is the ground-based, unenclosed lysimeter, but this design allows for collection of lateral flows of melt water as well as melt generated in the column of snow directly over the lysimeter. In most cases, it is assumed that lateral flow into the column of snow equals lateral flow out of the column. Where this is not the case, a rim 10–15 cm high around the collector–snow column interface is used to provide a more accurate monitoring of snowmelt in the column of snow (Kattelmann, 1984). Because of problems that can arise by storage of water above the collector and restrictions in proper drainage of the lysimeter, much care in the design, construction, and installation of snowmelt lysimeters is necessary to ensure accurate measurements (Kattelmann, 1984).

4.2.5 Snow pits

Digging a snow pit to determine the properties of a snowpack has important applications for avalanche forecasting, glaciology, remote-sensing data interpretation and snow hydrology. Snow pits can be shallow (if others have been dug on a regular basis) or deep (all the way to the ground surface). In addition, snow-pit data can be compared against many of the measurements previously discussed. Snow-pit observations typically yield information on the snowpack temperature profile, density profile, layers, grain type, grain size, layer hardness, free water content, height, ram number (if a ram penetrometer is used near the snow pit), and snow water equivalent (both for the total vertical snowpack and by horizontal layers in the snowpack). Figure 4.11 is an adaptation of a snow-profile schematic of a snow pit from Perla and Martinelli (1976).

Normally, the snow pit will have a square or rectangular cross section. The southern (shaded) wall of the snow pit should remain untouched, as this is the side that will be used for measurements. There should be a confined area around this shaded wall where no one can step and where no removed snow is thrown. Once

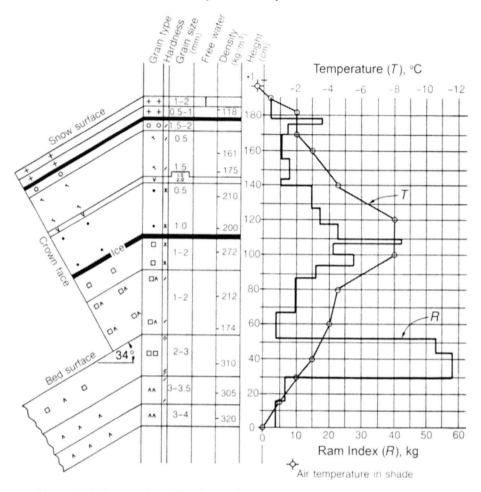

Figure 4.11 Snowpack profile plotted from snow-pit field notes (adapted from Perla and Martinelli, 1976).

the snow pit is excavated, the shaded wall should be freshly cut back several cm and the temperature, wetness, and density measurements made in a timely manner.

The depth of the snowpack should be established first with zero being the ground surface. When using a ram penetrometer to measure the hardness of each layer, it can be left in place when the ground is reached. The snow pit can be dug adjacent to the ram and it can serve as a measure of height. The shaded pit wall should be shaved with a shovel and then brushed horizontally with a soft brush to accentuate the layers (Perla and Martinelli, 1976). The layer thickness should be recorded before additional work is done. Snow temperature can be measured with a 30-cm-long, digital thermometer. This can be measured every 10 cm and in every layer as well. Inserting density tubes horizontally into each layer and weighing them on a spring balance can measure snow density. The hardness of the snow or ice making up layers should be recorded. A hand hardness test should be performed on each layer and recorded as in Figure 4.11. Grain type in each layer can also be recorded along with an estimate of grain size. A more detailed delineation of the crystal type can also be determined depending on the application. Wetness of the snowpack can also be estimated with an empirical test as shown in Figure 4.12. Depending on the purpose for digging a snow pit, all or just some of these measures may be necessary. If time is not a limiting factor, some of the more rigorous and quantitative tests already covered in this chapter may be employed in the snow pit.

4.3 Snow-crystal imaging

Relatively general classifications of snow crystals are made during measurements in snow pits. The same is true of falling snow. When more detailed measurements of snow crystals are necessary, the use of light microscopy (LM) has enabled investigators during the last hundred years to observe, photograph, describe, and classify numerous types of snow crystals and ice grains (Bentley and Humphreys, 1931; Nakaya, 1954). One of the most noteworthy attempts to illustrate natural snow crystals was undertaken by Wilson Bentley, a Vermont dairy farmer and amateur meteorologist. Bentley set up an outdoor laboratory and spent nearly 40 years photographing with the light microscope over 6000 snow crystals mostly consisting of dendrites and plates (Blanchard, 1970). About 20 years later, Nakaya (1954) established a laboratory with a controlled environment to experimentally determine the effects of temperature on the formation and growth of all forms of snow crystals, including dendrites, plates, columns, needles, and irregular crystals. However, the light microscope limited magnification of these forms to about 500×. Poor depth of field prevented resolution of all but the very flat crystals. Another approach was attempted by Cross (1969) who used a scanning electron microscope (SEM) to examine evaporating ice. However, because the sample was imaged in the vacuum of the instrument and not maintained at below freezing temperatures, sublimation and melting limited the observations. These problems were solved by equipping the SEM with a cold stage that maintained the temperature of the ice sample near that of liquid nitrogen.

Recently, studies of snow crystals have been undertaken with SEMs that were equipped with cold stages (Wergin and Erbe, 1994a, b, c; Wolff and Reid, 1994). Because the SEM has resolution and depth of field that exceeds those of the LM, this instrument provides detailed structural features of surfaces and allows three-dimensional imaging of the snow crystals. Furthermore, the cold stage prevents melting and sublimation; therefore, snow crystals and ice grains remain stable for

International symbols and measurements

Grain Type		Hardness		Hand Test*	Ram Number (kg)**	Free Water Content		
Symbol	Description	Symbol	Description			Symbol	Term	Description
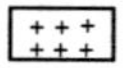 [+ + + / + + +]	Freshly deposited snow. Initial forms can be easily recognized.	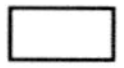 [□]	Very soft	Fist	0–2	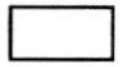 [□]	Dry	Snow usually, but not necessarily, below 0 °C. Grains have little tendency to stick together in a snowball when lightly pressed in gloved hand.
 [ˇ ˇ ˇ / ˇ ˇ ˇ]	Irregular grains, mostly rounded but often branched. Structure often feltlike. Early stages of ET metamorphism.	 [▨]	Soft	Four fingers	3–15	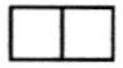 [▢]	Moist	Snow at 0 °C. No water visible even with hand lens. Snow makes good snowball.
 [• • •]	Rounded, often elongated, isometric grains. End stages of ET metamorphism. Grains usually less than 2 mm in diameter.	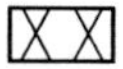 [▨]	Medium hard	One finger	16–50	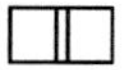 [▤]	Wet	Snow at 0 °C. Water visible as a meniscus between grains but cannot be pressed out by moderate squeezing in the hand.
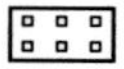 [□ □ □]	Angular grains with flat sides or faces. Early stages of TG metamorphism.	 [▨]	Hard	Pencil	51–100	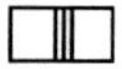 [▤]	Very wet	Snow at 0 °C. Water can be pressed out by moderate squeezing in the hand. There is still an appreciable amount of air confined within the snow.
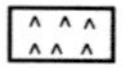 [∧ ∧ ∧ / ∧ ∧ ∧]	Angular grains with stepped faces; at least some hollow cups. Advanced stages of TG metamorphism.	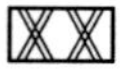 [▨]	Very hard	Knife	Over 100	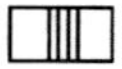 [▤]	Slush	Snow at 0 °C. Snow flooded with water and containing relatively small amounts of air.
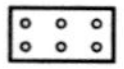 [○ ○ ○]	Rounded grains formed by MF metamorphism. Grains usually larger than 1 mm and often strongly bonded.	 [■]	Ice	--	--			
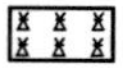 [⊠ ⊠]	Graupel. Occasionally appreciable layers of this form of solid precipitation can be identified in the snow cover.							
 [▨]	Ice layer, lens, or pocket.							

*In the hand test the specified object can be pushed into the snow in the pit wall with a force of about 5 kg. In hard snow, for example, a pencil can be pushed into the snow, but with the same pressure, a finger cannot.

**Ram number is from a cone penetrometer with a 60° apex and 4 cm diameter.

Figure 4.12 International symbols and measurements for snow grain types, snow hardness hand test, and free water content hand test (adapted from Perla and Martinelli, 1976).

indefinite periods of time. Using this technique, Wergin and Erbe (1994a, b, c) captured snow crystals near their laboratory, quickly transferred them to the SEM and were able to observe numerous forms of intact crystals. Since that time, snow crystals and ice grains have been collected from numerous remote locations (Rango *et al.*, 1996a, b, c; 2000) (Wergin *et al.*, 1995; 1996a, b, c; 1998a, b; 2002a, b). The following provides brief descriptions of the equipment and procedures that were fabricated and successfully used to collect, ship, store, and image snow and ice with low temperature SEM (LTSEM).

4.3.1 Materials and methods

The equipment and techniques described below enable collecting, shipping, storing, and observing snow crystals and ice grains that are maintained at near-liquid nitrogen temperatures ($-196°C$). At this temperature, specimens remain stable during observation in the LTSEM where they can be imaged and photographed. Many of the devices that are described are unique and must be fabricated in the laboratory because they are currently unavailable commercially. The devices, which were originally designed for studies with LTSEM, can also be used for observations with light microscopy.

The term "cryo-system" is used generically in this study; it refers to an accessory that is available from several manufacturers and can be retrofitted to most SEMs. A cryo-system typically consists of: (1) a specimen holder (Figure 4.13a); (2) a vacuum transfer assembly mechanism, consisting of a rod that attaches to the specimen holder and is used to move the holder through the various stages or compartments of the cryo-system; (3) a freezing chamber, where specimens are typically plunged into liquid nitrogen (LN_2) or some other cryogen; (4) a pre-chamber, where the specimen can be fractured and/or coated with a heavy metal; and, (5) a cold stage, which is mounted in the microscope and can be maintained near LN_2 temperatures.

Secondary specimen holders (sampling plates)

Close inspection of Figure 4.13, including the caption (Erbe *et al.*, 2003), reveals the details of the various components that are needed for field sampling and laboratory processing of snow crystals for LTSEM photography. A commercial cryo-system normally comes with a single universal specimen holder; replacements or spare holders of this type frequently cost several hundred dollars each. When collecting snow or ice at remote sites, several dozen specimens are frequently desired. Therefore, numerous specimen holders are essential. To solve this problem, a simple, inexpensive secondary specimen holder or sampling plate was devised. The plate is used in the field to collect the sample. In the laboratory, a modified commercial specimen holder that is compatible with the cryo-system accommodates the plate.

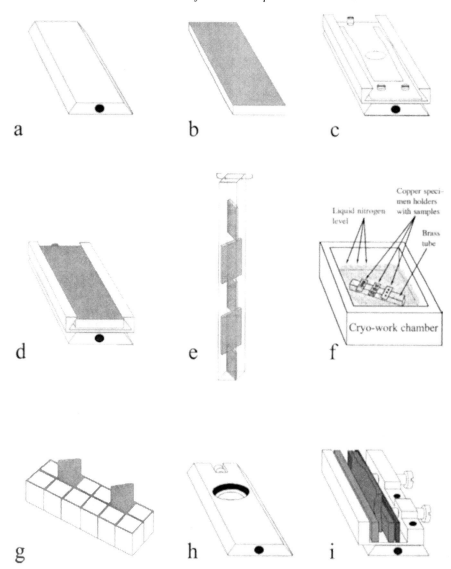

Figure 4.13 Drawings of devices that were fabricated in the laboratory for use with snow-crystal imaging using a Low-Temperature Scanning Electron Microscope (LTSEM) (Erbe, *et al.*, 2003).

(a) Drawing of the standard specimen holder that is supplied with the Oxford CT 1500 HF Cryo-system. The rod of the specimen transfer device screws into the holder (black hole) and enables the specimen to be moved through the pre-chamber to the cold stage in the SEM.

(b) Drawing of a sampling plate, 15 × 29 mm, that is cut from stock sheet of copper, 1.5 mm thick. One surface of the plate is numbered for identification purposes; the opposite surface (shown in the diagram) is roughened to enhance adhesion of the frozen specimen.

Shipping sampling plates

At the collection site, tubes containing the sampling plates are lowered into a lightweight, dry shipping Dewar or Cryopak Shipper (Taylor Wharton, Theodore, AL 36582) that has previously been cooled with LN_2. The Dewar is placed in a backpack or hand carried from the collection site and then either transported by van or sent by priority air express to the laboratory. The shipper, which is designed to maintain LN_2 temperatures for a minimum of 21 days when fully pre-cooled, has been used to transport samples from numerous locations, including remote regions of the states of Washington, North Dakota and Alaska. Upon reaching the laboratory, the samples are transferred to a LN_2 storage Dewar where they remain until being further prepared for observation with LTSEM.

Sampling procedures
Preparing sampling plates

All samples of snow and ice are collected on the sampling plates, which are pre-coated with a thin layer of cryo-adhesive such as Tissue Tek, a methyl cellulose

---

Figure 4.13 (*cont.*)

(c) Drawing of a modified specimen carrier that is fabricated and attached to the standard specimen holder (lower level) depicted in (a).

(d) Drawing of the modified specimen carrier containing the sampling plate depicted in (b). The entire assembly, which attaches to the transfer rod of the cryo-system, can then be moved through the pre-chamber and onto the cold stage in the SEM.

(e) Drawing of the square brass tubing, 13 × 13 mm ID. The bottom end is permanently capped; two holes are drilled in the opposing sides of the upper end. Four plates can be inserted diagonally into the tubes; each plate is alternated at 90°. A paper clip inserted through the holes retains the plates in the tube during shipping and storage.

(f) Drawing of a Styrofoam work chamber, 12 × 20 × 4 cm (depth) where samples collected on the plates are either contact- or plunge-frozen in LN_2 and held until they are transferred to the square brass shipping tubes shown in (e). A Styrofoam cover is used between sample collections to prevent condensation or frost from forming in the LN_2.

(g) Drawing of the staging jig that holds up to 10 plates in the Styrofoam box prior to loading in the square brass tubes. This device is fashioned by soldering together twelve 1-cm segments of the square brass tubing.

(h) Drawing of a standard cylindrical stub holder, which is used to hold a metal tube containing an ice core. The tube is placed in the circular well and held firmly by tightening the setscrew (far end). In the pre-chamber, the core can be fractured to expose an untouched, pristine surface.

(i) Drawing of an indium vise specimen holder. An ice sample can be clamped into the holder where it is held between two sheets of indium metal. Indium, which is thermally conductive and pliable at LN_2 temperatures, helps prevent shattering of the sample. This device has been used to hold glacial ice, icicles, hail, ice lenses, pond ice, and dry ice.

solution. The adhesive and the plates must be cooled to temperatures at or slightly below freezing. However, the adhesive will solidify at temperatures below −3 °C; therefore, it must be protected from freezing in the container, as well as on the sampling plate prior to sampling. To keep the adhesive near its freezing point (−3 °C), the stock bottle can be placed in a pocket of one's parka. Similarly, cold plates, which will also freeze the adhesive prior to sampling, must be maintained at near freezing temperatures. Immediately after a specimen has been collected, the plate containing the adhesive and sample are plunged into a vessel of LN_2. For this purpose, a Styrofoam box, 12 × 20 × 4 cm (depth) containing a 1- to 2-cm layer of LN_2 is used. The box is used to plunge-freeze the plates that contain the samples and to hold them for short-term storage (Figure 4.13f).

In warm or sunny conditions, such as late spring or on glacier surfaces during summer, the temperature of the plates and cryo-adhesive may be well above freezing. When this occurs, the plates and the adhesive can be pre-cooled by placing them in the snowpack and then proceeding as described above. In these cases, excellent preservation of structure has been accomplished if the sampling is done rapidly and the plate is plunged immediately (less than 1 second) into LN_2.

Sampling fresh and falling snow

Falling snow is sampled by allowing it to settle on the surface of the plate containing the cryo-adhesive. Alternatively, a fresh sample can be lightly brushed onto the plate. In either case, the plate containing the sample is then rapidly plunge-frozen in the LN_2 vessel.

Snow-pit samples

To collect samples from snow pits, a pre-cooled (LN_2) scalpel is used to gently dislodge snow crystals from a freshly excavated pit wall. The crystals are allowed to accumulate onto a plate containing the cryo-adhesive and then plunged rapidly into the LN_2. If LN_2 cannot be carried to the snow pit, a brass block, pre-cooled with LN_2 is transported to the pit. After the crystals are sampled, the plate is placed on the brass block and allowed to freeze, then processed as previously described.

Hoar frost

Plates can also be placed in an area where rime, surface hoar or frost are expected to occur and simply allowed to serve as the substrate for the sample that forms. In this case, the cryo-adhesive is not necessary. When a sufficient sample has sublimed on the plate, it is plunge-frozen in LN_2.

Fractured samples

Fracturing of snow clusters or glacial ice is used to reveal the extent of air spaces, details of sintering and biota, such as ice worms, snow algae (*Chlamydomonas nivalis*), fungi and bacteria. This process is accomplished with a pick that is mounted in the pre-chamber of the cryo-system. Prior to coating, the pick is used to randomly fracture or remove a portion of the snow cluster or ice sample. This process exposes a pristine internal surface that is then coated and inserted into the LTSEM for imaging.

Coating samples

All frozen samples are coated with 2 to 10 nm of platinum using a magnetron sputter coating device in a high purity argon environment within the pre-chamber. This process makes the samples electrically conductive and enhances secondary electron emission for imaging in the LTSEM. The argon gas, which is introduced during this process, must be very free of water vapor and other impurities that could contaminate the surface of the specimen. During coating and imaging, any area of the specimen that possesses poor thermal conductivity will etch or sublimate and the continuity of the overlying coating will be compromised. As a result, the sample will be poorly coated and tend to charge.

Recording images with an SEM

All specimens are imaged with a Hitachi S-4100 field emission SEM (Hitachi High-Technology Corp., Tokyo, Japan) equipped with an Oxford CT 1500 HF cryo-system (Oxford Instruments, Enysham, England) (Figure 4.14). The cold stage is maintained at $-130°$ to 185 °C. Accelerating voltages of 500 V to 10 kV are used to observe the samples. However, 2 kV is most commonly used because it minimizes charging and provides adequate resolution. The samples are imaged for as long as 2 hours without observing any changes in the structural features or in the

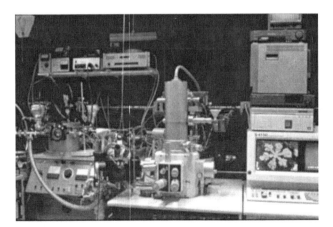

Figure 4.14 LTSEM employed at the Beltsville Agricultural Research Center, Agricultural Research Service, U.S. Department of Agriculture (courtesy USDA Agricultural Research Service, Beltsville, MD).

coating integrity of the snow crystals. Selected images are recorded onto Polaroid Type 55 P/N film (Polaroid, Cambridge, MA, USA).

Stereo pairs are obtained by recording one image, tilting the sample 6°, re-centering the subject and then recording the second image. The two images resulting from this procedure contain the parallax information necessary for three-dimensional observation and study.

For much greater detail on the LTSEM field and laboratory sampling methods, the reader is referred to Erbe *et al.* (2003).

4.4 References

Alter, J. C. (1937). Shielded storage precipitation gauges. *Mon. Weather Rev.*, **65**, 262–5.

Ambach, W. and Denoth, A. (1974). On the dielectric constant of wet snow. In *Proceedings of the Grindelwald Symposium: Snow Mechanics.* IAHS-AISH Publication 114, pp. 136–42.

Ambach, W. and Denoth, A. (1980). The dielectric behaviour of snow: a study versus liquid water content. In *NASA Conference Publication 2153.* Fort Collins, CO, pp. 69–92.

Bader, H. (1948). Theory of non-calorimetric methods for the determination of the liquid water content of wet snow. *Schweiz. Mineral. Petrogr. Mitt.* **28**, 344–61.

Bader, H. (1950). Note on the liquid water content of wet snow. *J. Glaciol.*, **1**, 446–7.

Bakkehoi, S., Oien, K., and Førland, E. J. (1985). An automatic precipitation gauge based on vibrating-wire strain gauges. *Nordic Hydrol.*, **16**, 193–202.

Bentley, W. A. and Humphreys, W. J. (1931). *Snow Crystals.* New York: McGraw-Hill.

Bergman, J. A. (1978). A method for the determination of liquid phase water in a snowpack using toluene freezing calorimetry. Unpublished MS Forestry Thesis, University of Nevada, Reno, NV.

Blanchard, D. C. (1970). Wilson Bentley, the snowflake man. *Weatherwise*, **23**, 260–9.

Boyne, H. S. and Fisk, D. (1987). A comparison of snow cover liquid water measurement techniques. *Water Resour. Res.*, **23**(10), 1833–36.

Boyne, H. S. and Fisk, D. J. (1990). *A Laboratory Comparison of Field Techniques for Measurement of the Liquid Water Fraction of Snow,* Special Report 90–3. US Army, Corps of Engineers, Cold Regions Research and Engineering Laboratory.

Bruce, J. P. and Clark, R. H. (1966). *Introduction to Hydrometeorology.* New York: Pergamon Press.

Bryazgin, N. N. and Radionov, V. F. (2002). Empirical method of correction of monthly precipitation totals for climatic studies. In *Proceedings of the WCRP Workshop on Determination of Solid Precipitation in Cold Climate Regions.* Fairbanks, AK.

Canadian Department of Environment. (1985). *Measuring Snowfall Water Equivalent using the Nipher-Shielded Snow Gauge System,* Report No. IB 04-03-01/1. Ottawa, Canada: Atmospheric Environment Service, Canadian Department of Environment.

Carcoana, R. and Enz, J. W. (2000). An enhanced technique for more reliable daily snow measurements with an antifreeze-based, tipping-bucket gauge. In *Proceedings of the 68th Annual Meeting of the Western Snow Conference,* Port Angeles, WA, pp. 87–91.

Carroll, T. (1976). *Estimation and Prediction of the Liquid Water Distribution in a High Altitude Spring Snowpack.* Avalanche Workshop. University of Calgary, Banff.

Colbeck, S. C. (1978). The difficulties of measuring the water saturation and porosity of snow. *J. Glaciol.*, **20**(28), 189–201.

Cox, L. M. and Hamon, W. R. (1968). A universal surface precipitation gauge. In *Proceedings of the 36th Annual Meeting of the Western Snow Conference*, pp. 6–8.

Cross, J. D. (1969). Scanning electron microscopy of evaporating ice. *Science*, **164**, 174–5.

Davis, R. E., Dozier, J., LaChapelle, E. R., and Perla, R. (1985). Field and laboratory measurements of snow liquid water by dilution. *Water Resour. Res.*, **21**, 1415–20.

de Quervain, M. (1946). Zur bestimmung des wassergehalts von nass-schnee. *Verhl. Schweiz. Naturf. Ges.*, **94**.

Denoth, A. (1994). An electronic device for long-term snow wetness recording. *Annals Glaciol.*, **19**, 104–6.

Doesken, N. J. and Judson, A. (1997). *The Snow Booklet: A Guide to the Science, Climatology and Measurement of Snow in the United States*. Fort Collins, CO: Colorado Climate Center, Department of Atmospheric Science, Colorado State University.

Erbe, E. F., Rango, A., Foster, J., Josberger, E. G., Pooley, C., and Wergin, W. P. (2003). Collecting, shipping, storing, and imaging snow crystals and ice grains with low temperature scanning electron microscopy. *Microsc. Res. Techniq.*, **62**, 19–32.

Fisk, D. (1982). Snow calorimetric measurement at Snow-One. In *Proceedings of Snow Symposium*, I, US Army, Corps of Engineers, Cold Regions Research Engineering Lab, pp. 133–8.

Fisk, D. J. (1983). Progress in methods of measuring the free water content of snow. In *Optical Engineering for Cold Environments*, Society of Photo-Optical Instrumentation Engineers Conference Proceedings. 414, 48–51.

Fisk, D. (1986). Method of measuring liquid water mass fraction of snow by alcohol solution. *J. Glaciol.*, **32**, 538–40.

Giguere, A. and Austin, G. L. (1991). On the significance of radar wavelength in the estimation of snowfall. In *Hydrological Applications of Weather Radar*, ed. I. D. Cluckie and C. G. Collier. London: Ellis Horwood, pp. 117–30.

Goodison, B. E. (1978). Accuracy of Canadian snow gauge measurements. *J. Appl. Meteorol.*, **17**, 1542–8.

Goodison, B. E., Ferguson, H. L., and McKay, G. A. (1981). Measurement and data analysis, Chapter 6. In *Handbook of Snow*, ed. D. M. Gray and D. H. Male. Toronto: Pergamon Press, pp. 191–274.

Goodison, B., Lawford, R., Rudolf, B., and Yang, D. (2002). WCRP Workshop on Determination of Solid Precipitation in Cold-Climate Regions, ECGC-WCRP 2002, Proceedings CD, Fairbanks, AK.

Goodison, B. E., Metcalfe, J. R., Wilson, R. A., and Jones, K. (1988). The Canadian automatic snow depth sensor: A performance update. In *Proceedings of the 56th Annual Meeting of the Western Snow Conference*, Kalispell, MT, pp. 178–81.

Hamon, W. R. (1973). *Computing Actual Precipitation; Distribution of Precipitation in Mountainous Areas*, vol. 1, WMO Report 362. Geneva: World Meteorological Organization.

Hansen, B. L. and Jellinek, H. H. G. (1957). *A Portable Adiabatic Calorimeter*. SIPRE Technical Report No. 49.

Hanson, C. L. (1989). Precipitation catch measured by the Wyoming shield and dual gauge system. *Water Resour. Bull.*, **25**, 159–64.

Hanson, C. L., Johnson, G. L., and Rango, A. (1999). Comparison of precipitation catch between nine measuring systems. *J. of Hydrol. Engineering*, **4**(11), 70–5.

Hanson, C. L., Zuzel, J. F., and Morris, R. P. (1983). Winter precipitation catch by heated tipping-bucket gauges. *T. Am. Soc. Ag. Eng.*, **26**(5), 1479–80.

Howell, S., Jones, E. B., and Leaf, C. F. (1976). *Snowpack Ground Truth (Radar Test Site Steamboat Springs, CO, April 8–16)*. Mission Report (Contract NAS5-22312). Greenbelt, MD: Goddard Space Flight Center.

Jones, E. B., Rango, A., and Howell, S. M. (1983). Snowpack liquid water determinations using freezing calorimetry. *Nordic Hydrol.*, **14**, 113–26.

Jones, R. N. (1979). *A Comparison of Centrifuge and Freezing Calorimeter Methods for Measuring Free Water in Snow*, Report NBSIR 79–1604. Boulder, CO: National Bureau of Standards.

Julander, R. (2003). *Snow Hydrology: Measuring Snow*, www.civil.utah.edu/~cv5450/measuring_snow/snow_measuring.html.

Kattelmann, R. C. (1984). Snowmelt lysimeters: design and use. In *Proceedings of the 52nd Annual Meeting of the Western Snow Conference*, Sun Valley, ID, pp. 68–79.

Kattelmann, R. C., McGurk, B. J., Berg, N. H., Bergman, J. A., Baldwin, J. A., and Hannaford, M. A. (1983). The isotope-profiling snow gauge: twenty years of experience. In *Proceedings of the 51st Annual Meeting of the Western Snow Conference*, Vancouver, WA, pp. 1–8.

Kawashima, K., Endo, T., and Takeuchi, Y. (1998). A portable calorimeter for measuring liquid-water content of wet snow. *Annals Glaciol.*, **26**, 103–6.

Kuroda, M. and Hurukawa, I. (1954). Measurement of water content of snow. *IASH Publication*, General Assembly of Rome. **39** (IV), 38–41.

Langham, E. J. (1974). Problems of measuring meltwater in the snowpack. In *Proceedings of the 1974 Annual Meeting of the Eastern Snow Conference*, Ottawa, Canada, pp. 60–71.

Langham, E. J. (1978). *Measurement of the Liquid Water Content of Snow by Repeated Centrifugal Extraction*, Technical Report. Ottawa, Canada: Glaciology Division, Department of Fisheries and Environment.

Larson, L. W. and Peck, E. L. (1974). Accuracy of precipitation measurements for hydrologic modeling. *Water Resour. Res.*, **10**(4), 857–63.

Leaf, C. F. (1966). Free water content of snowpack in subalpine areas. In *Proceedings of the 34th Western Snow Conference*, Seattle, WA, pp. 17–24.

Linlor, W. I. and Smith, J. L. (1974). Electronic measurements of snow sample wetness. In *Advanced Concepts and Techniques in the Study of Snow and Ice Resources*. Washington, DC: National Academy of Sciences, pp. 720–8.

Linlor, W. I., Clapp, F. D., Meier, M. F., and Smith, J. L. (1975). Snow wetness measurements for melt forecasting. In *NASA Special Publication 391*. Washington, DC: NASA, pp. 375–97.

Linlor, W. I., Meier, M. F. and, Smith, J. L. (1974). Microwave profiling of snowpack free-water content. In *Advanced Concepts and Techniques in the Study of Snow and Ice Resources*. Washington, DC: National Academy of Sciences, pp. 729–39.

McCaughey, W. W. and Farnes, P. E. (1996). Measuring winter precipitation with an antifreeze-based tipping bucket system. In *Proceedings of the 64th Annual Meeting of the Western Snow Conference*, Bend, OR, pp. 130–6.

Moffitt, J. A. (1995). Snow plates: preliminary tests. In *Proceedings of the 63rd Annual Meeting of the Western Snow Conference*, Sparks, NV, pp. 156–9.

Morris, E. M. (1981). Field measurement of the liquid-water content of snow. *J. Glaciol.*, **27**(95), 175–8.

Morris, R. P. and Hanson, C. L. (1983). Weighing and recording precipitation gauge modification. *T. Am. Soc. Civ. Eng.*, **26**(1), 167–78.

Nakaya, U. (1954). *Snow Crystals: Natural and Artificial*. Cambridge, MA: Harvard University Press.

Natural Resources Conservation Service (2003a). *Interpreting Streamflow Forecasts.* www.wcc.nrcs.usda.gov/factpub/interpet.html.

Natural Resources Conservation Service (2003b). SNOTEL Data Collection Network Fact Sheet. www.wcc.nrcs.usda.gov/factpub/sntlfct1.html.

Natural Resources Conservation Service (2003c). *Snow Surveys – SNOTEL.* www.wcc. nrcs.usda.gov/factpub/sect_4b.html.

Natural Resources Conservation Service (2003d). *Snow Survey Sampling Guide: Snow Sampling Procedures – Step 1.* www.wcc.nrcs.usda.gov/factpub/ah169/ah169.htm.

Natural Resources Conservation Service (2004). *Wolf Creek Summit SNOTEL Data Report.* www.wcc.nrcs.usda.gov/nwcc/ sntl-data0000.jsp?site=874&days= 7&state=CO

Ohmura, A. (1980). An economical calorimeter for measuring water content of a snow cover. *Zeitschrift für Gletscherkunde und Glazialgeologie,* **16**(1), 125–30.

Osterhuber, R., Gehrke, F., and Condreva, K. (1998). Snowpack snow water equivalent measurement using the attenuation of cosmic gamma radiation. In *Proceedings of the 66th Annual Meeting of the Western Snow Conference,* Snowbird, UT, pp. 19–25.

Østrem, G. and Brugman, M. (1991). *Glacier Mass-Balance Measurements.* NHRI Science Report No. 4. Saskatoon, Saskatchewan, Canada: National Hydrology Research Institute.

Peck, E. L. (1997). *Improvement in Method to Adjust U.S. Solid Precipitation Records.* Phase 1 Interim Report, GCIP/OGP/NOAA. Hydex Corporation.

Perla, R. I. and Martinelli, M. (1976). Avalanche handbook. In *Agriculture Handbook 489.* Washington, DC: US Printing Office.

Radok, U., Stephens, S. K., and Sutherland, K. L. (1961). On the calorimetric determination of snow quality. In *IAHS Publication 54,* International Association of Hydrological Sciences, 132–5.

Rango, A., Wergin, W. P., and Erbe, E. F. (1996a). Snow crystal imaging using scanning electron microscopy: I. Precipitated snow. *Hydrol. Sciences Journal,* **41**, 219–33.

Rango, A., Wergin, W. P., and Erbe, E. F. (1996b). Snow crystal imaging using scanning electron microscopy: II. Metamorphosed snow. *Hydrol. Sciences Journal,* **41**, 235–50.

Rango, A., Wergin, W. P., and Erbe, E. F. (1996c). 3-D characterization of snow crystals as an aid to remote sensing of snow water equivalent. In *Applications of Remote Sensing in Hydrology,* ed. G. W. Kite, A. Pietroniro and T. J. Pultz. Saskatoon, Saskatchewan, Canada: National Hydrology Research Institute, pp. 295–310.

Rango, A., Wergin, W. P., Erbe, E. F., and Josberger, E. G. (2000). Snow crystal imaging using scanning electron microscopy: III. Glacier ice, snow and biota. *Hydrol. Sciences Journal,* **45**, 357–75.

Rasmussen, R. M., Hallett, J., Purcell, R., Cole, J., and Tryhane, M. (2002). The hotplate snow gauge. In *Proceedings of the WCRP Workshop on Determination of Solid Precipitation in Cold Climate Regions.* Fairbanks, AK.

Rechard, P. A. (1975). Measurement of winter precipitation in wind-swept areas. In *Proceedings of the Symposium on Snow Management on the Great Plains,* Publication 73, Great Plains Agriculture Council. Lincoln, NE, pp. 13–30.

Schneebeli, M., Coleiu, C., Touvier, F. and Lesaffre, B. (1998). Measurement of density and wetness in snow using time-domain reflectometry. *Annals Glaciol.,* **26**, 69–72.

Schumann, H. H. (1975). Operational applications of satellite snowcover observations and Landsat data collection systems operations in central Arizona. In *Proceedings of a Workshop on Operational Applications of Satellite Snowcover Observations,* NASA

SP-391. Washington, DC: National Aeronautics and Space Administration, pp. 13–28.

Sevruk, B. (1986). Correction of precipitation measurements summary report. In *Correction of Precipitation Measurements*, ed. B. Sevruk. ETH/IAH/WMO Workshop on the Correction of Precipitation Measurements, Report ZGS 23. Geographisches Institut, Eidgenössische Technische Hochschule Zürich, pp. 13–23.

Sevruk, B. (ed.). (1989). Precipitation measurement. In *Proceedings of the WMO/IAHS/ETH Workshop on Precipitation Measurement*. Geographisches Institut, Eidgenössische Technische Hochschule Zürich.

Shoda. M. (1952). Methods of measuring snow water content. *Seppyo (Japan)*, **13**, 103–4.

Singh, P. and Singh, V. P. (2001). *Snow and Glacier Hydrology*. Dordrecht: Kluwer Academic Publishers.

Soil Conservation Service (1972). Snow survey and water supply forecasting. In *SCS National Engineering Handbook*, sect 22. Washington, DC.

Strangeways, I. C. (1984). The development of an automated weather station for cold regions. In *Proceedings of the 52nd Annual Meeting of the Western Snow Conference*, Sun Valley, ID, pp. 12–23.

Tabler, R. D., Berg, N. H., Trabant, D. C., Santeford, H. S., and Richard, P. A. (1990). Measurement and evaluation of winter precipitation. In *Cold Regions Hydrology and Hydraulics Monograph*, ed. W. L. Ryan and R. D. Crissman. New York: American Society of Civil Engineers, pp. 9–38.

Tanner, B. D. and Gaza, B. (1990). Automated snow depth and snowpack temperature measurements. In *Proceedings of the 58th Annual Meeting of the Western Snow Conference*, Sacramento, CA, pp. 73–8.

Trabant, D. E. and Clagett, G. P. (1990). Measurement and evaluations of snowpacks. In *Cold Regions Hydrology and Hydraulics Monograph*, ed. W. L. Ryan and R. D. Crissman. New York: American Society of Civil Engineers, pp. 39–93.

Wergin, W. P. and Erbe, E. F. (1994a). Can you image a snowflake with an SEM? Certainly! *Proc. Roy. Microsc. Soc.*, **29**, 138–40.

Wergin, W. P. and Erbe, E. F. (1994b). Snow crystals: capturing snowflakes for observation with the low temperature scanning electron microscope. *Scanning*, **16**, IV88–IV89.

Wergin, W. P. and Erbe, E. F. (1994c). Use of low temperature scanning electron microscopy to examine snow crystals. In *Proceedings of the 13th International Congress of Electron Microscopy*, Publication 3B, pp. 993–4.

Wergin, W. P., Rango, A., and Erbe, E. F. (1995). Observations of snow crystals using low temperature scanning electron microscopy. *Scanning*, **17**, 41–9.

Wergin, W. P., Rango, A., and Erbe, E. F. (1996a). The structure and metamorphism of snow crystals as revealed by low temperature scanning electron microscopy. In *Proceedings of the Eastern Snow Conference*, Publication 53, pp. 195–204.

Wergin, W. P., Rango, A., Erbe, E. F., and Murphy, C. A. (1996b). Low temperature SEM of precipitated and metamorphosed snow crystals collected and transported from remote sites. *J. of the Microscopy Society of America*, **2**, 99–112.

Wergin, W. P., Rango, A., and Erbe, E. F. (1996c). Use of low temperature scanning electron microscopy to observe icicles, ice fabric, rime and frost. In *Proceedings of the Microscopy Society of America Meeting*, Publication 54, pp. 147–8.

Wergin, W. P., Rango, A., and Erbe, E. F. (1998a). Image comparisons of snow and ice crystals photographed by light (video) microscopy and low temperature scanning electron microscopy. *Scanning*, **20**, 285–296.

Wergin, W. P., Rango, A., and Erbe, E. F. (1998b). Use of video (light) microscopy and low temperature scanning electron microscopy to observe identical frozen specimens. *Electron Microsc.*, 1998, 325–6.

Wergin, W. P., Rango, A., Foster, J., Erbe, E. F., and Pooley, C. (2002a). Irregular snow crystals: structural features as revealed by low temperature scanning electron microscopy. *Scanning*, **2**, 247–56.

Wergin, W. P., Rango, A., Foster, J., Erbe, E. F., and Pooley, C. (2002b). Low temperature scanning electron microscopy of "irregular snow crystals". *Microsc. Microanal.*, **8**, 722–3CD.

Wheeler, P. A. and Huffman, D. J. (1984). Evaluation of the isotopic snow measurement gauge. In *Proceedings of the 52nd Annual Meeting of the Western Snow Conference*, Sun Valley, ID, pp. 48–56.

Winter, C. J. and Sturges, D. L. (1989). *Improved Procedures for Installing and Operating Precipitation Gauges and Alter Shields on Windswept Land*, Research Note RM-489. USDA Forest Service.

Wolff, E. W. and Reid, A. P. (1994). Capture and scanning electron microscopy of individual snow crystals. *J. Glaciol.*, **40**, 195–7.

Yoshida, Z. (1940). A method of determining thaw water content in snow layers. *J. of Faculty of Science*, ser. II–III, **4**.

Yoshida, Z. (1960). A calorimeter for measuring the free water content of wet snow. *J. Glaciol.*, **3**, 574–6.

Yoshida, Z. (1967). Free water content of snow. In *Proceedings of the International Conference on Low Temperature Science*, Sapporo, Japan, pp. 773–84.

5

Remote sensing of the snowpack

5.1 The importance of snow

There is such a vast difference in the physical properties of snow and other natural surfaces that the occurrence of snow in a drainage basin can cause significant changes in the energy and water budgets. As an example, the relatively high albedo of snow reflects a much higher percentage of incoming, solar, shortwave radiation than snow-free surfaces (80% or more for relatively new snow as opposed to roughly 15% or less for snow-free vegetation). Snow may cover up to 53% of the land surface in the northern hemisphere (Foster and Rango, 1982) and up to 44% of the world's land areas at any one time. On a drainage basin basis, the snow cover can vary significantly by elevation, time of year, or from year to year. The Rio Grande Basin near Del Norte, Colorado, is 3419 km^2 in area and ranges from 2432 m a.s.l. at the streamgauge up to 4215 m a.s.l. at the highest point in the basin. Figure 5.1 compares the snow-cover depletion curves obtained from Landsat data in 1977 and 1979 in elevation zones A (780 km^2; 2432–2926 m), B (1284 km^2; 2926–3353 m), and C (1355 km^2; 3353–4215 m) of the Rio Grande Basin. In a period of only 2 years from April 10, 1977, to April 10, 1979, a great difference in seasonal snow-cover extent was experienced. Landsat data show that 49.5% or 1693 km^2 were covered by snow on April 10, 1977. Two years later in 1979, 100% or 3419 km^2 were snow covered on April 10.

Snow cover and its subsequent melt supplies at least one-third of the water that is used for irrigation and the growth of crops worldwide (Steppuhn, 1981). In high mountain snowmelt basins of the Rocky Mountains, USA, as much as 75% of the total annual precipitation is in the form of snow (Storr, 1967) and 90% of the annual runoff is from snowmelt (Goodell, 1966). The runoff depth can vary greatly from year to year for mountain basins because of differences in the seasonal snow-cover accumulation. In the Rio Grande Basin near Del Norte, Colorado, the seasonal snow cover in the 2 years cited above, 1977 and 1979, produced vastly different annual

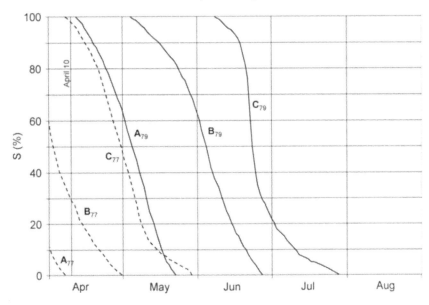

Figure 5.1 Comparison of snow-cover depletion curves for the Rio Grande basin near Del Norte, Colorado derived from Landsat data for elevation zones A, B, and C for 1977 and 1979. On April 10, the snow cover is 100% in all zones for 1979 and 97.5%, 29%, and 0% in zones A, B, and C for 1977.

runoff depths of 7.8 cm and 35.36 cm for 1977 and 1979, respectively (Rango and Martinec, 1997).

Because the physical properties of snow are so much different than other naturally occurring surfaces, it has opened up possibilities of exploiting widely different parts of the electromagnetic spectrum to learn more about the snowpack resource. In the above cited example, the visible portion (0.4–0.7 µm) of the electromagnetic spectrum (shown in Figure 5.2) was used to map the percentages of the basin covered by snow because of the difference in reflectance of snow and snow-free areas. The near infrared portion (0.7–1.1 µm) can also be used for snow mapping and for the detection of near surface liquid water because of the near infrared sensitivity to the liquid water molecule. The thermal infrared portion of the spectrum (8–14 µm) is also a possible area to exploit, but to date little has been done. The capability of detecting snowpack surface temperature variability in space and time can be directly linked to the presence of water in the snowpack and perhaps to snowmelt. Although in widely disparate portions of the spectrum, the gamma radiation (3×10^{-6}–3×10^{-5} µm) and the microwave radiation (1 mm–1 m) bands can be used in similar ways to detect the snowpack water equivalent using the attenuation of the relevant radiation by the snowpack itself. In the gamma radiation portion, the

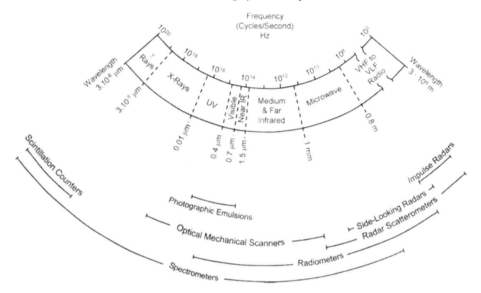

Figure 5.2 Electromagnetic spectrum and instrumentation associated with specific electromagnetic intervals (adapted from Allison *et al.*, 1978; Hall and Martinec, 1985.)

snowpack attenuates the gamma ray signal. In the microwave portion, the snowpack emits and attenuates depending on the size and shape of crystals and the presence of liquid water. The use of active microwaves as opposed to passive microwaves adds the additional complication of surface orientation and reflectance.

5.2 General approach

5.2.1 Gamma radiation

Snow-water-equivalent methodology

The use of gamma radiation to measure snow water equivalent (SWE) is based on the attenuation of natural terrestrial radiation by the mass of water in the overlying snow cover. Gamma radiation is emitted from potassium (40K), uranium (238U), and thorium (208Ti) radioisotopes in the soil with the majority of the emission coming from the top 20 cm. The intensity of gamma radiation is measured using a gamma radiation spectrometer, which is generally flown on an airplane for snow survey measurements. Fritzsche (1982) provides a description of the physics behind the airborne, gamma radiation spectrometer flown by the US National Weather Service (NWS) (see Table 5.1 at end of chapter for acronyms) and methods of calibration.

Airborne gamma measurements of snow water equivalent in a basin are generally acquired along an established network of flight lines. A background (no-snow) flight is conducted in the fall before snow accumulation and then the same lines are flown during the winter when snow covered. When an airborne gamma snow survey is conducted, several flight lines are usually flown within a basin, providing information on the relative spatial variation in SWE over the basin.

Airborne snow-water-equivalent measurements using gamma radiation are made using the relationship given in Equation (5.1) (Carroll, 1990):

$$\text{SWE} = \frac{1}{A}\left(\ln\frac{C_0}{C} - \ln\frac{100 + 1.11M}{100 + 1.11M_0}\right) \tag{5.1}$$

where:

SWE $=$ snow water equivalent, g cm^{-2}
C and C_0 $=$ uncollided, terrestrial gamma counts over snow and bare ground, respectively
M and M_0 $=$ percent soil moisture in snow-covered and bare-soil areas, respectively
A $=$ Radiation attenuation coefficient in water, cm^2 g^{-1}

The use of airborne gamma measurements for snow-water-equivalent retrieval is a reliable technique that has been proven effective for operational hydrological monitoring in many countries. Basin surveys can be completed in a matter of hours, producing information on the spatial distribution of snow water equivalent and a representative estimate of the volume of snow contained in the basin. The method is applicable to all sizes of basins. As demonstrated by the NWS operational airborne gamma snow-survey program (Carroll, 1990), the derived SWE information can be made available to user agencies within several hours of the airborne survey, thus providing timely information for hydrological forecasts. Carroll (1986) presents a cost–benefit analysis of airborne gamma SWE data acquired in support of snowmelt flood forecasting for a major flood event that occurred in the USA in February 1985.

Since gamma radiation is attenuated by water in all phases, the radiation measurements will include the effects of water in the soil as well as the snow-cover mass and, thus, the contribution of soil moisture has to be taken into account. Inaccurate estimates of soil moisture will yield over- or under-estimations of snow water equivalent. By conducting a background airborne gamma flight in the fall before snow accumulation, the soil moisture background is measured, but it will not be representative if subsequent rainfall adds moisture to the soil or if moisture migrates upward from the soil to the snow during winter. Ground-based soil moisture measurements conducted along the flight line during the winter over-snow flight can help contribute to a more accurate snow water equivalent (Carroll *et al.*, 1983). Additionally, atmospheric absorption of gamma radiation limits the area measured to relatively narrow flight line swaths resulting from low-altitude aircraft overpasses.

As discussed earlier, the use of airborne gamma surveys for snow-water-equivalent retrieval in support of hydrological monitoring has been conducted in several northern countries. One of the largest operational programs is maintained by the US National Weather Service. Over 1500 flight lines have been established in the USA and Canada, and each winter from January to April, snow-water-equivalent measurements are collected continuously over many of these lines (Carroll, 1990). After each survey, data are transmitted to the NWS National Operational Hydrologic Remote Sensing Center (NOHRSC) for archiving, processing, and distribution to forecast offices, other government agencies, and the general public. The NOHRSC website, www.nohrsc.nws.gov, provides a description of the NWS airborne gamma survey program, locations of flight lines, and access to data products. Historical data are also available from the US National Snow and Ice Data Center (NSIDC) in Boulder, Colorado. Low-flying aircraft are generally the platforms used to acquire airborne gamma measurements. The NWS airborne gamma survey program uses Aero Commander and Turbo Commander twin-engine aircraft, which fly at 150 m above the ground during the gamma surveys (Carroll, 1990), which prevents airborne gamma measurement in rugged topographic areas. The maximum snow water equivalent that can be obtained using airborne gamma surveys is between 30–50 cm (Vershinina, 1985; Carroll and Vose, 1984).

5.2.2 Visible and near infrared imagery

Snow-extent methodology

Snow cover can be detected and monitored with a variety of remote-sensing devices, but the greatest application has been found in the visible and near infrared region of the electromagnetic spectrum. The red band (0.6–0.7 μm) of the multispectral scanner subsystem (MSS) on the Landsat satellite was used extensively for snow-cover mapping because of its strong contrast with snow-free areas. Originally, snow-extent mapping was performed manually using photointerpretive devices and MSS photographs (Bowley *et al.*, 1981). More recently, digital mapping of the snow cover has been the preferred approach (Baumgartner *et al.*, 1986; 1987; Dozier and Marks, 1987; Baumgartner and Rango, 1995), and the Landsat TM and ETM+ with a red band at 0.63–0.69 μm is used.

The greatest problem hindering Landsat (and SPOT) snow mapping in the past was a poor observational frequency. Depending on the Landsat satellite being used, each study area or drainage basin was only revisited every 16–18 days. In areas with minimal cloud cover during the snowmelt season, like the southwest United States, this was a sufficient frequency of observation. In other mountain snow areas, however, this observational frequency is inadequate because cloud cover will often hide the underlying snow from the satellite sensors.

As a result of the Landsat (and SPOT) frequency of observation problem, many users turned to the NOAA polar orbiting satellite with the AVHRR, which has a resolution of about 1 km in the 0.58–0.68 μm red band. The frequency of coverage is twice every 24 hours (one daytime pass and one nighttime pass). The major problem with the NOAA-AVHRR data is that the spatial resolution of 1100 m may be insufficient for snow mapping on small basins. As the nighttime pass of NOAA-AVHRR cannot be used for snow mapping in the visible spectrum, only one-half of the overflights can be used. However, several NOAA satellites may pass overhead at different times on a given day in specific locations.

Recently, a linear combination of the AVHRR visible (channel 1) and near infrared (channel 2) bands has provided the percentage of snow cover present in each pixel (Gomez-Landesa and Rango, 2002). The method is further improved using the 250 m spatial resolution visible (0.62–0.67 μm) and near infrared (0.841–0.876 μm) bands on the Terra MODIS sensor (Gomez-Landesa *et al.*, 2004). This approach has thus provided subpixel information allowing much smaller basins to be mapped for snow extent. A near infrared, spectral channel (1.55–1.75 μm), currently available on the TM, ETM+, and AVHRR instruments can be used to assist in mapping snow cover when clouds partially cover a drainage basin. In this band, clouds are usually more reflective than snow (Dozier, 1989). As a result, automatic discrimination between snow and clouds is possible. Although useful, this capability does not overcome the problem of a complete cloud cover or an inadequate frequency of observation. A similar but more narrow spectral band (1.628–1.652 μm) is also available on MODIS. Another sensor, the Medium Resolution Imaging Spectrometer (MERIS), was launched in 2002 on ENVISAT and has pertinence to snow-cover mapping (Seidel and Martinec, 2004). It has 15 narrow spectral bands in the visible and near infrared with a spatial resolution of 300 m at nadir.

When a basin is partially snow covered, a method has been developed to estimate the snow cover in the cloud-obscured parts of the basin (Lichtenegger *et al.*, 1981; Baumgartner *et al.*, 1986; Erhler *et al.*, 1997). The method uses digital topographic data and assumes that pixels with equal elevation, aspect and slope have the same relative snow coverage over the entire basin. With this extrapolation method, information from the cloud-free portion of the basin can be used to estimate the snow cover in the cloud-covered parts of the basin.

Despite the various problems mentioned, visible and near infrared aircraft and satellite imagery have been found to be very useful for monitoring both the buildup of snow cover in a drainage basin and, even more importantly, the disappearance of the snow-covered area in the spring. This disappearance or depletion of the snow cover is important to monitor for snowmelt-runoff forecasting purposes. It has been recommended (Rango, 1985) that the optimum frequency of observation of the snow cover during depletion would be once a week. Depending on the remote-sensing data used, it could be very difficult to obtain this frequency. Nevertheless, certain snowmelt-runoff applications have been possible with as few as two to three observations during the entire snowmelt season (Rango, 1985).

According to Seidel and Martinec (2004), depletion curves of snow cover are usually S-shaped. This results from the frequency distribution of snow depths in a basin which is bell-shaped with a peak in the middle elevations of the basin; snowmelt starts in the low parts of the basin, progresses across the middle elevations, and ends in the high elevations. Because the area–elevation curve in a basin is usually steep in the low and high parts and flat in the middle elevations, the resulting depletion curve of snow coverage is steep in the middle portion and fairly flat on both ends (Seidel and Martinec, 2004).

There are several advantages to using visible and near infrared satellite data for snow applications. First, the data are relatively easy to interpret, and it is also relatively easy to distinguish snow from snow-free areas. If necessary, the analysis of visible satellite data for snow mapping can be accomplished on microcomputer-based systems (Baumgartner and Rango, 1995). The visible satellite data is available in a range of resolutions from 0.61 m resolution (QuickBird) to 8 km resolution (NOAA satellites), which allows applications on small headwater subcatchments up to applications on continental-size areas. Because these data have progressed to an operational status in some cases, much snow-extent data are readily available to many users on the World Wide Web free of charge, except for certain private-sector satellites like QuickBird and Ikonos.

There are also several disadvantages to using visible satellite data for snow and ice mapping. Nothing can directly be learned about snow water equivalent from the visible data. Snow-cover extent is valuable in forecasting of snowmelt runoff, but clouds can be a significant problem in restricting observations, especially when the frequency of observation is every 16 days. The recent availability of Terra MODIS with daily coverage and resolution of 250 m fills a significant gap. It would seem that the advantages of Landsat and NOAA-AVHRR are combined in MODIS, and it is the optimum source of snow-cover data.

There is no restriction on acquiring visible or near infrared data for a particular area; however, you must have financial resources available to purchase the more expensive, high-resolution data such as QuickBird, Ikonos, Landsat ETM+, and SPOT. The primary satellites and sensors employed for snow mapping are Landsat-TM, Landsat ETM+, DMSP, NOAA-AVHRR, GOES, and, very recently, ASTER and MODIS on the EOS platforms. Airplanes are infrequently, but effectively, used to provide areal snow-cover data. Table 5.2 compares spatial resolution, swath width, and first year of operation for the sensors that have been primarily used for satellite snow mapping. Dozier and Painter (2004) provide a comprehensive review of visible and near infrared remote sensing of alpine snow properties,

Table 5.2 *Spaceborne imaging systems most applicable to snow cover monitoring using visible and near infrared wavelengths*

SATELLITE	SENSOR	SPATIAL RESOLUTION	SWATH WIDTH	YEAR FIRST LAUNCHED
NOAA	AVHRR	1100 m	2400–3000 km	1972
Landsat	MSS	80 m	185 km	1972
	TM	30 m	185 km	1982
	ETM+	15 m	185 km	1999
NOAA	GOES	1100 m	Geostationary	1975
DMSP	OLS	600 m	3000 km	1982
Terra/Aqua	MODIS	250, 500, 1000 m	2300 km	1999
	ASTER	15, 30 m	60 km	1999

including snow-covered area, albedo, grain size, contaminants, and liquid-water content.

5.2.3 Thermal infrared

Thermal infrared remote sensing currently only has minor potential for snow mapping and snow hydrology. Thermal infrared mapping, like visible mapping, is hindered by cloud cover. In addition, the surface temperature of snow is not always that much different from the surface temperatures of other surfaces in rugged mountain terrain where elevation and aspect differences can cause major changes in temperature. This makes it extremely difficult to distinguish snow cover from other features, especially during the snowmelt period. When clouds permit, the big advantage of thermal snow-cover mapping is that the mapping can be done during the nighttime overpasses.

In hydrological forecasting, monitoring the surface temperature of the snowpack may have some direct application to delineating the areas of a basin where snowmelt may be occurring. If the snow surface temperature is always below 0 °C, no melt is occurring. If the snow surface is at 0 °C during the daytime and below 0 °C at night, the melt–freeze cycle is occurring but it is uncertain whether melt water is being released at the bottom of the snowpack. If the snow surface stays at 0 °C both day and night, then it is highly likely that the snowpack is isothermal and that melt is occurring and being released at the base of the snowpack. Much research needs to be done to determine whether thermal infrared remote sensing can play a useful role in assisting in snowmelt-runoff modelling and forecasting. The primary sensors applicable to this application are AVHRR, MODIS, and ASTER. The best thermal infrared spatial resolution is from ASTER with 90 m.

5.2.4 Passive and active microwave

In general, there are two types of remote-sensing systems: passive and active sensors. A passive system detects or measures radiation of natural origin, usually reflected sunlight or energy emitted by an object. In the case of the microwave spectrum, passive-microwave radiometers are used. Conversely, an active sensor, providing its own source of electromagnetic radiation, transmits a series of signals to the target and detects the reflected signal. Examples of active-microwave sensors are a synthetic aperture radar or a radar altimeter (Kramer, 2002).

Snow water equivalent and associated applications

Snow on the Earth's surface is, in simple terms, an accumulation of ice crystals or grains, resulting in a snowpack, which over an area may cover the ground either completely or partly. The physical characteristics of the snowpack determine its microwave properties; microwave radiation emitted from the underlying ground is scattered in many different directions by the snow grains within the snow layer, resulting in a microwave emission at the top of the snow surface being less than the ground emission. Properties affecting microwave response from a snowpack include: depth and water equivalent, liquid-water content, density, grain size and shape, temperature, and stratification as well as snow state and land cover. The sensitivity of the microwave radiation to a snow layer on the ground makes it possible to monitor snow cover using passive-microwave remote-sensing techniques to derive information on snow extent, snow depth, snow water equivalent and snow state (wet/dry). Because the number of scatterers within a snowpack is proportional to the thickness and density, SWE can be related to the brightness temperature (T_B) (the temperature of a black body radiating the same amount of energy per unit area at the microwave wavelengths under consideration as the observed body) of the observed scene (Hallikainen and Jolma, 1986); deeper snowpacks generally result in lower brightness temperatures.

The general approach used to derive SWE and snow depth from passive-microwave satellite data relates back to those presented by Rango *et al.* (1979) and Kunzi *et al.* (1982) using empirical approaches and Chang *et al.* (1987a) using a theoretical basis from radiative transfer calculations to estimate snow depth from SMMR data. As discussed in Rott (1993), the most generally applied algorithms for deriving depth or snow water equivalent (SWE) are based on the generalized relation given in Equation (5.2):

$$SWE = A + B((T_B(f1) - T_B(f2))/(f2 - f1)) \text{ in mm, for SWE} > 0 \qquad (5.2)$$

where A and B are the offset and slope of the regression of the brightness temperature difference between a high-scattering channel ($f2$, commonly 37GHz) and a low-scattering one ($f1$, commonly 18 or 19 Ghz) of vertical or horizontal polarization. No single, global algorithm will estimate snow depth or water equivalent under all snowpack- and land-cover conditions. The coefficients are generally determined for different climate and land-covered regions and for different snow-cover conditions; algorithms used in regions other than for which they were developed and tested usually provide inaccurate estimates of snow cover. Also, accurate retrieval of information on snow extent, depth, and water equivalent requires dry snow conditions because the presence of liquid water within the snowpack drastically alters the emissivity of the snow, resulting in brightness temperatures significantly higher than if that snowpack were dry. Therefore, an early morning overpass (local time) is preferred for retrieval of snow-cover information to minimize wet snow conditions. Despite the fact that water in the snowpack hinders estimation of snow water equivalent, it is also recognized that knowledge of snowpack state, particularly the presence of melt water, is useful for hydrological applications. Regular monitoring allows detection of the onset of melt or wet snow conditions (Goodison and Walker, 1995).

The accuracy of the retrieval algorithms is a function of the quality of both the satellite data and the snow-cover measurements used in their development. Empirical methods to develop algorithms involve the correlation of observed T_B with coincident conventional depth measurements (as from meteorological stations) or ground SWE data (such as from areally representative snow courses). Goodison *et al.* (1986) used coincident airborne microwave data and airborne gamma data (see Section 5.2.1) collected over the Canadian prairie area, supplemented by special ground surveys, to derive their SWE algorithm, which has now been used for over 10 years in operational hydrological forecast operations. As originally suggested by Rango *et al.* (1979), non-forested, open prairie areas have generally shown the best correlation between areal SWE and brightness temperature (e.g., Kunzi *et al.*, 1982; Goodison *et al.*, 1986; Chang *et al.*, 1987b). Hallikainen and Jolma (1986) and Hallikainen (1989) report on Finnish studies over various landscapes. Rott and Nagler (1993) report on European algorithm development which incorporates information from 19, 37, and 85 GHz channels for snow-cover mapping; the 85 GHz data were of use in mapping very shallow snowcovers (<5 cm depth). They use a decision tree for snow classification, including the separation of snow/no snow and the calculation of depth. On a global scale, Grody and Basist (1996) use a decision tree to produce an objective algorithm to monitor the global distribution of snow cover, which includes steps to separate snow cover from precipitation, cold deserts, and frozen ground. With time, algorithms can be expected to become more sophisticated as they incorporate filters to eliminate or minimize errors or biases.

Several investigators have attempted to use SAR data to map snow-cover area and to infer the snow water equivalent. In the snow-cover aspects, Nagler and Rott (1997) indicate that the European Remote Sensing (ERS) satellite SAR (C-band) cannot distinguish dry snow and snow-free areas. When the snow becomes wet, the backscattering coefficient is significantly reduced and the wet snow area can be detected. In order to derive a snow map for mountain regions, both ascending (northward) and descending (southward) ERS SAR orbit images must be compared with reference images acquired from the same positions. This technique was used to successfully derive snow cover for input to the Snowmelt Runoff Model (SRM) for two drainage basins in the Austrian Alps (Nagler and Rott, 1997). Haefner and Piesberger (1997) also used ERS SAR data to map wet snow cover in the Swiss Alps. Shi and Dozier (2000) are developing multiband-multipolarization methods that have some promise for obtaining snow water equivalent with SAR data; however, the required satellite sensors are not likely to be available in the near future.

Microwave data are used to derive other cryospheric information. Although the resolution of passive-microwave satellite data prevents its use in deriving information on mountain glaciers and river ice, satellite SAR data has been shown to be useful for locating the firn or snowline on glaciers and monitoring ice breakup in rivers (Rott and Nagler, 1993; Leconte and Klassen, 1991; Rott and Matzler, 1987). Glacier snowline mapping is an important input in hydrological models and in the computation of glacier mass balance. Adam *et al.* (1997) used ERS-1 C-band, VV SAR data to map the glacier snowline within 50–75 m of ground-based measurements. The technique could separate wet, melting snow from glacier ice and bedrock but was not applicable when the snow was dry, since dry snow was transparent to C-band SAR. Some research has been done on freeze–thaw applications in permafrost areas (England, 1990; Zuerndorfer *et al.*, 1989) but other sensors with higher resolution (e.g. Landsat) may be more suited to mapping permafrost areas (Leverington and Duguay, 1997; Duguay and Lewowicz, 1995).

Passive-microwave data provides several advantages not offered by other satellite sensors. Studies have shown that passive-microwave data offer the potential to extract meaningful snowcover information, such as SWE, depth, extent, and snow state. SSM/I is a part of an operational satellite system, providing daily coverage of most snow areas with multiple passes at high latitudes, hence allowing the study of diurnal variability. The microwave-based technique generally has all-weather capability (although affected by precipitation at 85GHz) and can provide data during the night. The data are available in near-real time, and hence can be used for hydrological forecasting. SAR has an additional advantage of having resolution of about 25 m conceptually making it very useful for mountain snowpacks.

There are limitations and challenges in using microwave data for deriving snow-cover information for hydrology. The coarse resolution of passive microwave satellite sensors such as SMMR and SSM/I ($\sim$25 km) is more suited to regional and large basin studies, although Rango *et al.* (1989) did find that reasonable SWE estimates could be made for basins of less than $10\,000$ km^2. Heterogeneity of the surface and the snow cover within the microwave footprint results in a mixed signature, which is ultimately represented by a single brightness temperature that is an areally weighted mean of the microwave emission from each surface type within the footprint. Hence, an understanding of the relationship between snow cover, surface terrain, and land cover (e.g., Goodison *et al.*, 1981) is important for developing a better remote-sensing estimate of the snow cover in these mixed-pixel areas.

Another challenge is to incorporate the effect of changing snowpack conditions throughout the winter season. Seasonal aging, or metamorphism, results in a change in the grain size and shape, and this will affect the microwave emission from the snowpack. In very cold regions, depth hoar, characterized by its large crystal structure, enhances the scattering effect on the microwave radiation, resulting in lower surface emission producing an overestimate of SWE or snow depth (Hall, 1987 and Armstrong *et al.*, 1993). The increase in T_B associated with wet snow conditions currently prevents the quantitative determination of depth or water equivalent since algorithms will tend to produce zero values under these conditions. The best way to view the seasonal variability in microwave emission from the snowpack is to compile a time series of satellite data spanning the entire season, which can then be related to changes in the pack over the season (Walker *et al.*, 1995).

SAR data has a major disadvantage of not being able to detect the dry snowpack. Additionally, the SAR data is difficult to process with backscattering from rough surfaces beneath the snow causing interpretation problems. The optimum bands for snow, around 1 cm wavelength, are not represented in any existing or planned SAR instrument.

Passive-microwave data for hydrological and climatological studies are available from sensors operated onboard NASA's Nimbus 7 (SMMR) and the US Defense Meteorological Satellites (DMSP). Kramer (2002) provides comprehensive details of the individual satellites. Data are archived and available in a variety of formats at NSIDC; the EASE-Grid brightness temperature product (Armstrong and Brodzik, 1995), a gridded, 25-km resolution, global dataset (12.5 km at 85.5 GHz) is particularly suited to historical hydrological analyses where the user may wish to run different regional snow-cover algorithms or integrate other geophysical information. SSM/I data are available in near real-time for operational hydrological applications from the NOAA National Environmental Satellite Data and Information Services (NESDIS).

5.2.5 Related applications

Hand-drawn snow maps

It is possible to use a base map, such as a snow-free Landsat image, and map the location of the snow line in a basin during an aircraft over flight. This approach is even possible from the ground in a very small basin (say less than several km^2 in area) by climbing to the highest point and visually transferring the area covered by snow to a base map (Seidel and Martinec, 2004). These techniques are most useful where access to the basin or aircraft flights is easily arranged.

Photointerpretation

Where labor is inexpensive in certain developing countries, it is still possible to take the photos from satellite overpasses and manually map the area covered by snow and save a considerable amount of money. Computer systems have become so reasonable in price, however, that the photointerpretive approach is no longer used very much.

Frequency Modulated-Continuous Wave (FM-CW) radar

In recent years FM-CW radars of various types have found applications in snow and ice. Sturm *et al.* (1996) used an X-band FM-CW radar mounted on a towed sled to make dry snow-depth measurements in Alaska. Much care has to be taken in the data collection with coincident field verification of the snow depths. With frequent field calibration, the radar-determined snow depths were accurate to about 2 cm.

5.3 Current applications

5.3.1 NOHRSC – snow-cover and snow-water-equivalent products

Because areal snow-cover-extent data have been available since the 1960s, various investigators have found many useful applications. A team of scientists from a variety of US government agencies developed plans in the early 1980s for operational snow mapping by the US National Weather Service (NWS) for hydrological purposes. In 1986, NWS adopted these plans and proceeded to develop operational, remote-sensing products, mostly for snow hydrology. The most widely distributed products of the NWS National Operational Hydrologic Remote Sensing Center (NOHRSC) are periodic river basin snow-cover-extent maps from NOAA-AVHRR and the Geostationary Operational Environmental Satellite (GOES). Digital maps for about 4000 basins in North America are produced about once per week and are used by a large group of users, including the NWS River Forecast Centers and individual water authorities. On about 10% of these basins, the mapping is done by elevation zone (Carroll, 1995). Data distribution is possible in real time through a variety of electronic methods such as the Internet and with the assistance of Geographic Information Systems. The 1-km resolution of the product makes it useful on basins or sub-areas greater than 200 km^2 in area (Rango *et al.*, 1985), and various users employ the data to assist in hydrological forecasting using models. NOHRSC products are continually under development, and the latest products and services can be accessed by visiting their website at www.nohrsc.nws.gov.

In addition to producing operational snow-cover-extent data, NOHRSC also produces operational, airborne, gamma radiation, snow-water-equivalent data. The difference between the NOHRSC airborne radiation measurements over bare ground and snow-covered ground is used to calculate a mean areal snow-water-equivalent value with a root-mean-square error of less than 1 cm (Carroll, 1995). Immediately after each airborne snow survey, the snow water equivalent derived for each flight line is used in a GIS to generate a contoured surface of snow water equivalent for the region of the survey. After each survey, users of the data are able to operationally obtain a contour map of snow water equivalent in the region and the mean areal snow water equivalent for each basin in the region by electronic means.

In recent years, NOHRSC has provided daily comprehensive snow information across the USA through its National Snow Analyses (NSA). The NSA are based on modelled snowpack characteristics that are updated every weekday using a variety of data, including operationally available satellite, airborne, and ground-based observations of snow cover, depth, and water equivalent. The model is a multilayer, physically based snowpack model operated at 1 km^2 spatial resolution and up to hourly temporal resolution. Although the products provided are difficult to validate, the integration of all current technologies provides timely information otherwise unavailable (NOHRSC, 2004). Recent versions of this procedure, the SNOw Data Assimilation System (SNODAS) are discussed in Section 5.4.4.

5.3.2 Canadian prairie snow-water-equivalent mapping

In Canada, a federal government program (Climate Research Branch, Atmospheric Environment Service) has been ongoing since the early 1980s to develop, validate, and apply passive-microwave satellite data to determine snow extent, snow water equivalent and snowpack state (wet/dry) in Canadian regions for near real-time and operational use in hydrological and climatological applications. Goodison and Walker (1995) provide a summary of the program, its algorithm research and development, and future thrusts. For the prairie region a snow-water-equivalent algorithm was empirically derived using airborne microwave radiometer data (Goodison *et al.*, 1986) and tested and validated using Nimbus-7 SMMR and DMSP SSM/I satellite data (Goodison, 1989).

With the launch of the first SSM/I on the DMSP F-8 satellite in 1987 came the ability to access passive-microwave data in near real-time and generate snow-cover products for users within several hours after data acquisition. Since 1989, the Climate Research Branch prairie SWE algorithm has been applied to

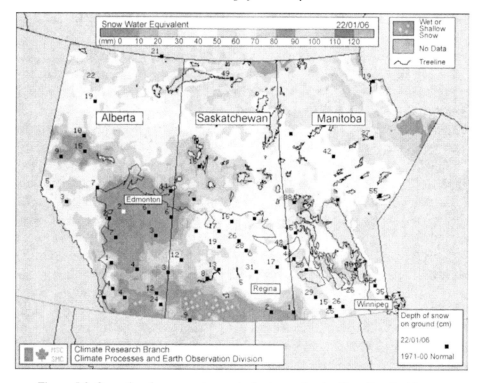

Figure 5.3 Operational snow-water-equivalent map for the Canadian Prairies on January 22, 2006 derived from passive-microwave satellite data. Numbers on the map indicate point measurements of observed snow depth. Maps are distributed by posting on the World Wide Web, www.socc.ca/SWE/snow˙swe.html/. (Reproduced with the permission of the Minister of Public Works and Government Services Canada, 2006.) See also color plate.

near-real-time SSM/I data to generate weekly maps depicting current SWE conditions for the provinces of Alberta, Saskatchewan, and Manitoba in western Canada. Thirkettle *et al.* (1991) describe the procedures for data acquisition, processing and mapping of the SWE information. The prairie maps are disseminated (originally by fax machine, now over the World Wide Web) (see Figure 5.3) to water resource agencies and meteorological offices throughout the prairie region where they are used to monitor snow-cover conditions, plan for field surveys, and make forecasts regarding spring water-supply conditions including potential flooding or drought. In the winter and spring of 1994, the maps were particularly useful for monitoring the high SWE conditions in southern Manitoba and North Dakota preceding the devastating Red River flood (Warkentin, 1997).

After 15 winter seasons in operation, the Canadian prairie SWE mapping program has successfully demonstrated a useful application of SSM/I-derived

snow-cover information for operational hydrological analyses. It is also a cooperative program in that user feedback has served to enhance the validation and
the refinement of the SSM/I SWE algorithm (Goodison and Walker, 1995). One
enhancement has been the development of a wet-snow indicator (Walker and
Goodison, 1993), which overcomes a major limitation of the passive-microwave
technique by providing the capability to discriminate wet-snow areas from snow-
free areas and, hence, a more accurate retrieval of snow extent during melting conditions. A more recent development, starting with December 1, 1999, employs a new
SWE algorithm taking into account forest-cover effects on the passive-microwave
satellite data. This new algorithm is employed in the boreal forest regions of Alberta,
Saskatchewan, and Manitoba, and the original algorithm is used in the true prairie
regions (Goita *et al.*, 2003).

The Climate Research Branch prairie SWE algorithm has been applied to
Nimbus-7 SMMR and DMSP SSM/I data to create a 25-year time series of maps
depicting winter SWE conditions over the Canadian prairie region for the purpose
of investigating seasonal and interannual variability in support of the Branch's
climate-research activities in assessing climate variability and change. Walker
et al. (1995) present the SMMR time series in the form of an atlas.

5.3.3 Snowmelt-runoff-forecast operations

Very few hydrological models have been developed to be compatible with remote-
sensing data. One of the few models that was developed requiring direct remote
sensing input is the Snowmelt-Runoff Model (SRM) (Martinec *et al.*, 1998).
SRM requires remote-sensing measurements of the snow-covered area in a basin.
Although aircraft observations can be used, satellite-derived snow-cover extent is
the most common.

Two versions of SRM are now available. The most commonly used version
employs the degree-day approach to melting the snow cover in a basin (Martinec
et al., 1998). A second version of the model has recently been developed (Brubaker
et al., 1996) that adds a net radiation index to the degree-day index to melt snow
from a basin's hydrologic response units (based on elevation and aspect). This
version (SRM-Rad) is more physically based and, thus, requires more data than the
original version of SRM. If actual net radiation is not measured, SRM-Rad needs
cloud-cover observations to calculate actual net radiation.

Norway power applications

Norway has been using satellite snow-cover data for planning of hydroelectric
power generation since 1980. This approach is a simple digital ratioing, which
is converted to a percentage of snow cover by pixel in the study basin (Andersen,

1991; 1995). NOAA-AVHRR has again been used as the data source for this system because of its daily coverage. Snow-cover maps are produced for the various basins, and snow cover by elevation zone in a basin is also a product. The data are now input to snowmelt-runoff models for the prediction of streamflow (Andersen, 1995).

Spain power applications

Spain is also using NOAA-AVHRR snow-cover data for the forecasting of snowmelt-runoff volume during the spring and summer months in the Pyrenees. Development of subpixel analysis techniques (Gomez-Landesa, 1997) has allowed snow-cover mapping on basins as small as 10 km^2 using the AVHRR data. A pixel-reflectance value greater than or equal to a certain snow threshold corresponds to a fully covered snow area. Conversely, pixel-reflectance values less than or equal to a certain ground threshold corresponds to a completely snow-free area. The pixel reflectances between the thresholds correspond to different percentages of snow cover in each NOAA-AVHRR pixel (Gomez-Landesa and Rango, 2002). The snow-cover data for each basin or zone are input into the Snowmelt-Runoff Model (SRM) for use in forecasting the seasonal snowmelt-runoff volume in the Pyrenees to assist in planning hydropower production. More recently MODIS data have been used to derive the snow cover data for the forecasts (Rango *et al.*, 2003). Figure 5.4 shows a comparison of NOAA-AVHRR and MODIS snow-cover maps for the Noguera Ribagorzana basin (572.9 km^2) in the Central Pyrenees of Spain. The higher resolution of the MODIS data is very evident and allows mapping on basins as small as 2–3 km^2.

India runoff forecasts

Some good examples of operational application of snow-extent data are found in India. Initially, Ramamoorthi (1983, 1987) started to use NOAA-AVHRR data in a simple regression approach for empirical forecasts of seasonal snowmelt runoff in the Sutlej River Basin (43 230 km^2) and the forecasts were extended to other basins. Ramamoorthi (1987) also decided to use satellite data as input to a snowmelt-runoff model for shorter-term forecasts. This idea was developed by Kumar *et al.* (1991), and satellite data were input to SRM (Martinec *et al.*, 1998) for use in operational forecasts of daily and weekly snowmelt runoff on the Beas (5144 km^2) and Parbati (1154 km^2) Rivers in India. Figure 5.5 shows the SRM results of Kumar *et al.* (1991) who produced very accurate simulations on the Beas River at Thalot, India, with very little historical data. Kumar *et al.* (1991) conclude that short-term runoff models, such as SRM, can be effectively put to use in hydroelectric schemes already in operation or can be developed for projects being planned or under construction. Apparently, because of the isolation of these remote basins, the only way forecasts can be made in this region is with the use of remote-sensing data.

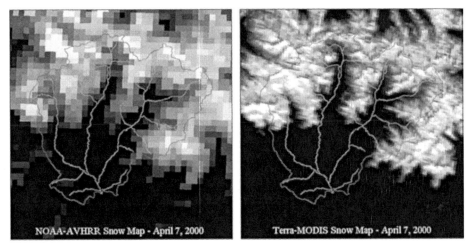

Figure 5.4 Comparison of NOAA-AVHRR and MODIS derived snow cover for the Noguera Ribagorzana Basin (572.9 km^2) in the Central Pyrenees of Spain on April 7, 2000. The different gray levels correspond to different percents of snow cover in each pixel. Snow cover in the Basin totals 181 km^2 from AVHRR and 184 km^2 from MODIS as reported by Rango *et al.* (2003). See also color plate.

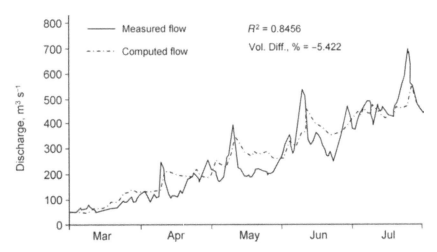

Figure 5.5 SRM computed versus measured snowmelt runoff on the Beas River at Thalot, India for 1987 (Kumar *et al.*, 1991).

United States runoff-forecast applications

NOHRSC has also developed a subpixel snow-mapping algorithm that can be used in the United States and southern Canada. This is a refinement of the river-basin snow-cover mapping that has gone on since the early 1990s. Both river-basin snow-cover maps and fractional snow-cover maps are used by NOAA/NWS River Forecast Centers and other agencies in their forecasting procedures.

5.4 Future directions

5.4.1 Improved resolution in the passive microwave

The most significant problem standing in the way of effective snow-water equivalent-mapping from space is a poor resolution at present in the passive microwave. This changed when the Advanced Microwave Scanning Radiometer (AMSR) on EOS-Aqua satellite was launched in 2002 that improved the resolution at about 0.8 cm wavelength from 25 km to 8 km. It gives investigators previously experienced with the 25 km capabilities the chance to determine the improved resolution advantages over study areas previously used.

5.4.2 Improved algorithms in the passive microwave

It is generally agreed that no single algorithm will produce representative values on a global basis due to spatial variations in land cover, terrain, and snow-cover characteristics. Hence, a regional approach to snow-cover algorithm development has been adopted by several research groups (e.g., Goodison and Walker, 1995; Solberg *et al.*, 1998).

In 1992, a strategy for future snow-cover algorithm development was outlined by the passive-microwave research community at an international workshop – "Passive Microwave Remote Sensing of Land-Atmosphere Interactions." This strategy is outlined in Choudhury *et al.* (1995). The priorities identified at this meeting included: (i) the need for signature research using ground-based microwave radiometry to understand the influence of vegetation cover and physical variations in snowpack structure; (ii) the development of theoretical models to characterize microwave interactions with snowpack properties; (iii) the incorporation of land-cover information (e.g., type and density) into snow-cover retrieval algorithms; (iv) the identification of target areas with good in-situ snow-cover measurement for algorithm development and validation; and (v) investigation of "mixed pixels," characterized by high spatial variations in snow distributions and/or land use, and the associated variations in microwave emission that contribute to the brightness temperature measured for the pixel. Since that meeting, research has focused on these priority areas with a common goal of developing algorithms that will produce consistently representative information on snow-cover parameters over as much of the globe as possible. Examples of recent passive-microwave snow-cover research developments in relation to the above priorities include: Chang *et al.* (1987b), Matzler (1994), Woo *et al.* (1995), Foster *et al.* (1996), De Sève *et al.* (1997), Sun *et al.* (1997), and Kurvonen and Hallikainen, (1997).

5.4.3 Outlook for radar applications

The application of current satellite active-microwave sensors for snow-cover information retrieval is generally limited to wet-snow detection and mapping (Rott, 1993), mainly due to the single frequency (typically C-band) and single transmit/receive polarization characteristics of these sensors. Thus, the outlook for improved snow-cover information retrieval from active-microwave sensors has focused on multifrequency and multipolarization systems (Rott, 1993).

In 1994, a multifrequency and multipolarization SAR system (SIR-C/X-SAR) was flown on two NASA Space Shuttle missions, providing research datasets for investigating the potential of an advanced SAR system for retrieval of snow-cover information. The SIR-C (Spaceborne Imaging Radar-C) operated at L- and C-bands and the X-SAR (X-Band Synthetic Aperture Radar) operated at X-band, thus providing simultaneous data for three radar frequencies (1.25, 5.3 and 9.6 GHz). The SIR-C also operated at four different transmit/receive polarizations to provide polarimetric data. The results of snow-cover investigations using SIR-C/X-SAR data demonstrated a potential for using multifrequency polarimetric SAR data to retrieve information, such as the discrimination of wet-snow areas (Li and Shi, 1996), snow-cover wetness or liquid-water content (Shi and Dozier, 1995; Matzler *et al.*, 1997), and the water equivalent of dry snow (Shi and Dozier, 1996).

Although the availability of routine SAR data from satellites is currently limited to single frequency and single polarization, two polarimetric SAR sensors are planned for launch in the near future. The Advanced Synthetic Aperture Radar (ASAR), launched on the European Space Agency's Envisat-1 platform in 2002, provides C-band images in a number of alternating polarization modes. Radarsat-2, the Canadian Space Agency's C-band SAR follow-on to the current Radarsat-1, was launched in late 2007 and has new advanced capabilities, including dual or quad polarization options on selected beam modes. Radarsat-2 will offer SAR imagery at horizontal, vertical, and cross polarizations over a range of resolutions from 3–100 m with swath widths ranging from 20–500 km. The availability of polarimetric SAR data from these satellite platforms should enhance the use of active-microwave sensors for routine snow-cover monitoring in support of hydrological applications, especially in small alpine basins. The appropriate frequency and polarization combinations for operational monitoring by a single space platform, however, will be unlikely in the future.

5.4.4 Integration of various data types

Combinations of two different remote sensors can at certain times increase the information available about the snowpack. As an example, a method has been developed in Finland that combines several different approaches to measure or estimate snow water equivalent (Kuittinen, 1989). Ground-based point measurements of water equivalent were made as usual about twice a month. One airborne gamma ray flight was made at the beginning of the snowmelt season to give line transect values of snow water equivalent. All available NOAA-AVHRR satellite images in the spring are used to provide areal snow-water-equivalent estimates based on a relationship between the percentage of bare spots in the snow cover and snow water equivalent (Kuittinen, 1989). The point, line, and areal snow-water-equivalent values were used with a method of correlation functions and weighting factors, as suggested by Peck *et al.* (1985), to determine an areal value based on all data. Evaluations in the Finland study have shown that the error of the estimate of areal snow water equivalent is less than 3.5 cm (Kuittinen, 1989). Carroll (1995) also developed a system that combines ground-based snow observations and airborne gamma ray snow-water-equivalent observations to yield gridded snow-water-equivalent values. Satellite snow-cover-extent data are used to constrain the snow-water-equivalent interpolations to regions where snow cover is observed to be present.

Later Carroll *et al.* (2001) developed SNODAS to integrate remote sensing, ground based and modelling into a product available on the National Operational Hydrologic Remote Sensing Center website, www.nohrsc.nws.gov/. The purpose of SNODAS is to provide a physically consistent framework for displaying and integrating the wide variety of snow data available at different times. SNODAS includes (1) data ingest and downscaling procedures, (2) a spatially distributed energy and mass balance snow model that is run once each day, for the previous 24-hour period and for a 12-hour forecast period, at high spatial (1 km) and temporal (1 h) resolutions, and (3) data assimilation and updating procedures (Carroll *et al.*, 2001).

With technological advances in data processing and transmission, data and derived snow and ice products from many of the current sensors are available to the hydrological community in near real-time (e.g., within 6–24 hours of satellite overpass). The development of the Internet and World Wide Web has facilitated the availability of many remote-sensing-derived snow and ice products to users right from their computers. As the technology related to data access continues to advance, the hydrological community can expect to see an expanded variety of satellite-data

Table 5.3 *Relative sensor band responses to various snowpack properties*

		SENSOR BAND		
SNOW PROPERTY	Gamma rays	Visible/near infra-red	Thermal infra-red	Microwaves
Snow-covered area	Low	High	Medium	High
Depth	Medium	Low	Low	Medium
Water equivalent	High	Low	Low	High
Stratigraphy	No	No	No	High
Albedo	No	High	No	No
Liquid water content	No	Low	Medium	High
Temperature	No	No	High	Medium
Snowmelt	No	Low	Medium	Medium
Snow-soil interface	Low	No	No	High
Additional factors				
All weather capability	No	No	No	Yes
Current best spatial resolution from space platform	Not possible	1 m	90 m	8 km passive 10 m active

products available to them for use in hydrological monitoring and modelling. With both the Terra and Aqua EOS platforms in orbit since 2002, NASA has made a variety of snow products available to users from MODIS (daily snow cover and 8-day composite, maximum snow cover at 500 m resolution; daily climate modelling grid (CMG) snow cover, and 8-day composite, CMG maximum snow cover at $\frac{1}{4}° \times \frac{1}{4}°$ resolution) and AMSR (daily global snow-storage index map; pentab (5-day) composite snow-storage index map). The cost associated with satellite-derived snow and ice products varies depending on the data policy associated with the satellite sensor. For commercial satellite platforms, such as Radarsat-1 and 2 SAR or QuickBird, derived products may come with a price tag in the thousands of dollar range, whereas products from federally funded satellite platforms (e.g., NOAA-AVHRR, DMSP, SSM/I, MODIS) are generally available at no cost via the Internet. The various sensors covering different spectral bands have varying applications for measuring snowpack properties. The relative sensor band responses to different snowpack properties are shown in Table 5.3 as updated from Rango (1993).

5.5 References

Adam, S., Pietroniro, A., and Brugman, M. M. (1997). Glacier snowline mapping using ERS-1 SAR imagery. *Remote Sens. Environ.*, **61**, 46–54.

Allison, L. J., Wexler, R. Laughlin, C. R., and Bandeen, W. R. (1978). Remote sensing of the atmosphere from environmental satellites. In *American Society for Testing and Materials: Special Publications 653m*. Philadelphia, PA, pp. 58–155.

Andersen, T. (1991). AVHRR data for snow mapping in Norway. In *Proceedings of the 5th AVHRR Data Users Meeting*. Tromsoe, Norway.

Andersen, T. (1995). SNOWSAT-Operational snow mapping in Norway. In *Proceedings of the First Moderate Resolution Imaging Spectroradiometer (MODIS) Snow and Ice Workshop*, NASA Conf. Publ. CP-3318, Greenbelt, MD: NASA/Goddard Space Flight Center, pp. 101–2.

Armstrong, R. L. and Brodzik, M. J. (1995). An earth-gridded SSM/I dataset for cryospheric studies and global change monitoring. *Adv. Space Res.*, **16**(10), 155–63.

Armstrong, R. L., Chang, A., Rango, A., and Josberger, E. (1993). Snow depths and grain-size relationships with relevance for passive microwave studies. *Annals Glaciol.*, **17**, 171–6.

Baumgartner, M. F. and Rango, A. (1995). A microcomputer-based alpine snowcover and analysis system (ASCAS). *Photogramm. Eng. Rem Sens.*, **61**(12), 1475–86.

Baumgartner, M. F., Seidel, K., Haefner, H., Itten, K. I., and Martinec, J. (1986). Snow cover mapping for runoff simulations based on Landsat-MSS data in an alpine basin. In *Hydrological Applications of Space Technology Proceedings*, IAHS Publ. No. 160, Cocoa Beach Workshop, pp. 191–9.

Baumgartner, M. F., Seidel, K. and Martinec, J. (1987). Toward snowmelt runoff forecast based on multisensor remote-sensing information. *IEEE Trans. Geosci. Remote Sens.*, **25**, 746–50.

Bowley, C. J., Barnes, J. C., and Rango, A. (1981). *Satellite Snow Mapping and Runoff Prediction Handbook*, NASA Technical Paper 1829. Washington, DC: National Aeronautics and Space Administration.

Brubaker, K., Rango, A., and Kustas, W. (1996). Incorporating radiation inputs into the Snowmelt Runoff Model. *Hydrol. Processes*, **10**, 1329–43.

Carroll, T. R. (1986). Cost-benefit analysis of airborne gamma radiation snow water equivalent data used in snowmelt flood forecasting. In *Proceedings of the 54th Annual Meeting of the Western Snow Conference*, Phoenix, AZ, pp. 1–11.

Carroll, T. R. (1990). Airborne and satellite data used to map snow cover operationally in the U.S. and Canada. In *Proceedings of the International Symposium on Remote Sensing and Water Resources*. Enschede, The Netherlands, pp. 147–55.

Carroll, T. R. (1995). Remote sensing of snow in the cold regions. In *Proceedings of the First Moderate Resolution Imaging Spectroradiometer (MODIS) Snow and Ice Workshop*. NASA Conf. Publ. CP-3318, Greenbelt, MD: NASA/Goddard Space Flight Center, pp. 3–14.

Carroll, T. R. and Vose, G. D. (1984). Airborne snow water equivalent measurements over a forested environment using terrestrial gamma radiation. In *Proceedings of the 41st Annual Eastern Snow Conference*. New Carrollton, MD.

Carroll, T. R., Glynn, J. E., and Goodison, B. E. (1983). A comparison of U.S. and Canadian airborne gamma radiation snow water equivalent measurements. In: *Proceedings of the 51st Annual Western Snow Conference*, Vancouver, BC, Canada, pp. 27–37.

Carroll, T., Kline, D., Fall, G., Nilsson, A., Li, L., and Rost, A. (2001). NOHRSC operations and the simulation of snow cover properties for the coterminous U.S., In *Proceedings of the 69th Annual Western Snow Conference*, Sun Valley, ID, pp. 1–10.

Chang, A. T. C., Foster, J. L., Hall, D. K., Goodison, B. E., Walker, A. E., Metcalfe, J. R., and Harby, A. (1987a). Snow parameters derived from microwave measurements during the BOREAS winter field campaign. *J. Geophys. Res.*, **102**(D24), 29 663–71.

Chang, A. T. C., Foster, J. L., and Hall, D. K. (1987b). Nimbus-7 SMMR-derived global snow cover parameters. *Annals Glaciol.*, **9**, 39–44.

Choudhury, B. J., Kerr, Y. H., Njoku, E. G., and Pampaloni, P. (eds.). (1995). Working group A1: Snow. In *Passive Microwave Remote Sensing of Land-Atmosphere Interactions*, Utrecht: VSP, pp. 651–6.

De Sève, D., Bernier, M., Fortin, J. P., and Walker, A. (1997). Preliminary analysis of snow microwave radiometry using the SSM/I passive-microwave data: The case of La Grande River watershed (Quebec). *Annals Glaciol.*, **25**, 353–61.

Dozier, J. (1989). Spectral signature of alpine snow cover from the Landsat Thematic Mapper. *Remote Sens. Environ.*, **28**, 9–22

Dozier, J. and Marks, D. (1987). Snow mapping and classification from Landsat Thematic Mapper data. *Annals Glaciol.*, **9**, 1–7.

Dozier, J. and Painter, T. H. (2004). Multispectral and hyperspectral remote sensing of alpine snow properties. *Annu. Rev. Earth Pl. Sci.*, **32**, 465–94.

Duguay, C. R. and Lewkowicz, A. G. (1995). Assessment of SPOT panchromatic imagery in the detection and identification of permafrost features, Fosheim Peninsula, Ellesmere Island, N.W.T. In *Proceedings of the 17th Canadian Symposium on Remote Sensing*, Saskatoon, Saskatchewan, pp. 8–14.

England, A. W. (1990). Radiobrightness of diurnally heated, freezing soil. *IEEE Trans. Geosci. Remote Sens.*, **28**(4), 464–76.

Erhler, C., Seidel, K., and Martinec, J. (1997). Advanced analysis of snow cover based on satellite remote sensing for the assessment of water resources. In *Remote Sensing and Geographic Information Systems for Design and Operation of Water Resources Systems*, IAHS Publication 242, pp. 93–101.

Foster, J. L. and Rango, A. (1982). Snow cover conditions in the northern hemisphere during the winter of 1981. *J. Clim.*, **20**, 171–83.

Foster, J., Chang, A., and Hall, D. (1996). Improved passive microwave algorithms for North America and Eurasia. In *Proceedings of the Third International Workshop on Applications of Remote Sensing in Hydrology*, Greenbelt, MD, pp. 63–70.

Fritzsche, A. E. (1982). *The National Weather Service Gamma Snow System Physics and Calibration*, Publication NWS-8201. Las Vegas, NV: EG&G, Inc.

Goita, K., Walker, A., and Goodison, B. (2003). Algorithm development for the estimation of snow water equivalent in the boreal forest using passive microwave data. *Int. J. Remote Sens.*, **24**(5), 1097–102.

Gomez-Landesa, E. (1997). Evaluacion de Recursos de Agua en Forma de Nieve mediante Teledeteccion usando satelites de la sine NOAA (Evaluation of water resources in the form of snow by remote sensing using NOAA satellites). Unpublished PhD thesis,Universidad Politenica de Madrid, Madrid, Spain.

Gomez-Landesa, E. and Rango, A. (2002). Operational snowmelt runoff forecasting in the Spanish Pyrenees using the Snowmelt Runoff Model. *Hydrol. Processes*, **16**, 1583–1591.

Gomez-Landesa, E., Rango, A., and Bleiweiss, M. (2004). An algorithm to address the MODIS bowtie effect, *Can. J. Remote Sens.*, **30**(4), 644–50

Goodell, B. C. (1966). Snowpack management for optimum water benefits. In *ASCE Water Resources Engineering Conference Preprint 379*. Denver, CO.

Goodison, B. E. (1989). Determination of areal snow water equivalent on the Canadian prairies using passive microwave satellite data. In *Proceedings of the International Geoscience and Remote Sensing Symposium*, IGARSS. Vancouver, Canada, pp. 1243–6.

Goodison, B. E. and Walker, A. E. (1995). Canadian development and use of snow cover information from passive microwave satellite data. In *Passive Microwave Remote*

Sensing of Land-Atmosphere Interactions, ed. B. J. Choudhury, Y. H. Kerr, E. G. Njoku, and P. Pampaloni. Utrecht: VSP, pp. 245–62.

Goodison, B. E., Ferguson, H. L., and McKay, G. A. (1981). Measurement and Data Analysis. In *Handbook of Snow*, ed. D. M. Gray and D. H. Male. Toranto: Pergamon Press, pp. 191–274.

Goodison, B. E., Rubinstein, I., Thirkettle, F. W., and Langham, E. J. (1986). Determination of snow water equivalent on the Canadian prairies using microwave radoimetry. In *Modelling Snowmelt Induced Processes*, IAHS Publication 155, pp. 163–73.

Grody, N. C. and Basist, A. N. (1996). Global identification of snowcover using SSM/I measurements. *IEEE Trans. Geosci. Remote Sens.*, **34**(1), 237–49.

Haefner, H. and Piesberger, J. (1997). High alpine snow cover monitoring using ERS-1 SAR and Landsat TM data. In *Remote Sensing and Geographic Information Systems for Design and Operation of Water Resources Systems*, Proc. Rabat Symp, IAHS Publication 242, pp. 113–18.

Hall, D. K. (1987). Influence of depth hoar on microwave emission from snow in northern Alaska. *Cold Reg. Sci. Technol.*, **13**, 225–231.

Hall, D. K. and Martinec, J. (1985). *Remote Sensing of Ice and Snow*. New York: Chapman and Hall.

Hallikainen, M. (1989). Microwave radiometry of snow. *Adv. Space Res.*, **9**(1), 267–75.

Hallikainen, M. and Jolma, P. (1986). Development of algorithms to retrieve the water equivalent of snow cover from satellite microwave radiometer data. In *Proceedings of the International Geoscience and Remote Sensing Symposium*, IGARSS. Zurich, Switzerland, pp. 611–16.

Kramer, H. J. (2002). *Observation of Earth and Its Environment: Survey of Missions and Sensors*. Berlin: Springer-Verlag.

Kuittinen, R. (1989). Determination of snow water equivalents by using NOAA-satellite images, gamma ray spectrometry and field measurements. In *Remote Sensing and Large-Scale Global Processes*, IAHS Publication No. 186, pp. 151–9.

Kumar, V. S., Haefner, H., and Seidel, K. (1991). Satellite snow cover mapping and snowmelt-runoff modelling in Beas Basin. In *Snow Hydrology and Forests in High Alpine Areas*, Proc. Vienna Symp, IAHS Publ. 205, pp. 101–9.

Kunzi, K. F., Patil, S., and Rott, H. (1982). Snow cover parameters retrieved from Nimbus-7 Scanning Mutlichannel Microwave Radiometers (SMMR) data. *IEEE Trans. Geosci. Remote Sens.*, GE-**20**(4), 452–67.

Kurvonen, L. and Hallikainen, M. (1997). Influence of land-cover category on brightness temperature of snow. *IEEE Trans. Geosci. Remote Sens.*, **35**(2), 367–77.

Leconte, R. and Klassen, P. D. (1991). Lake and river ice investigations in northern Manitoba using airborne SAR imagery. *Arctic*, **44**(Supp.1), 153–63.

Leverington, D. W. and Duguay, C. R. (1997). A neural network method to determine the presence or absence of permafrost near May, Yukon Territory, Canada. *Permafrost Periglac.* **8**, 205–15.

Li, Z. and Shi, J. (1996). Snow mapping with SIR-C multipolarization SAR in Tienshen Mountain. In *Proceedings of the International Geoscience and Remote Sensing Symposium*, IGARSS. Lincoln, NE, pp. 136–8.

Lichtenegger, J., Seidel, K., Keller, M., and Haefner H. (1981). Snow surface measurements from digital Landsat MSS data. *Nordic Hydrol.*, **12**, 275–88.

Martinec, J., Rango, A., and Roberts, R. (1998). *Snowmelt Runoff Model (SRM) User's Manual*. Geographica Bernensia P35, Department of Geography, University of Berne.

Matzler, C. (1994). Passive microwave signatures of landscapes in winter. *Meteorol. Atmos. Phys.*, **54**, 241–60.

Matzler, C., Strozzi, T., Weise, T., Floricioiu, D. M., and Rott, H. (1997). Microwave snowpack studies made in the Austrian Alps during the SIR-C/X-SAR experiment. *Int. J. Remote Sens.*, **18**(12), 2505–30.

Nagler, T. and Rott, H. (1997). The application of ERS-1 SAR for snowmelt runoff modelling. In *Remote Sensing and Geographic Information Systems for Design and Operation of Water Resources Systems,* Proc. Rabat Symp., IAHS Publication 242, pp. 119–26.

NOHRSC (2004). *Overview of the Center's Website and Products.* Minneapolis, MN: National Operational Hydrologic Remote Sensing Center, National Weather Service, www.nohrsc.nws.gov.

Peck, E. L., Johnson, E. R., Keefer, T. N., and Rango, A. (1985). Combining measurements of hydrological variables of various sampling geometries and measurement accuracies. In *Hydrological Applications of Remote Sensing and Remote Data Transmission*, Proc. Hamburg Symp., IAHS Publ. 145, pp. 591–9.

Ramamoorthi, A. S. (1983). Snow-melt run-off studies using remote sensing data. *Proc. Indian Acad. Sci.*, **6**(3), 279–86.

Ramamoorthi, A. S. (1987). Snow cover area (SCA) is the main factor in forecasting snowmelt runoff from major basins. In *Large-Scale Effects of Seasonal Snow Cover*, Proc. Vancouver Symp., IAHS Publ. 166, pp. 279–86.

Rango, A. (1985). The snowmelt-runoff model. In *Proceedings of the ARS Natural Resources Modeling Symposium*, USDA-ARS-30, Pingree Park, CO, pp. 321–25.

Rango, A. (1993). Snow hydrology processes and remote sensing. *Hydrol. Processes*, **7**, 121–38.

Rango, A. and Martinec, J. (1997). Water storage in mountain basins from satellite snow cover monitoring. In *Remote Sensing and Geographic Information Systems for Design and Operation of Water Resources Systems*, Proc. Rabat Symp, IAHS Publication 242, pp. 83–91.

Rango, A., Chang, A. T. C., and Foster, J. L. (1979). The utilization of spaceborne microwave radiometers for monitoring snowpack properties. *Nordic Hydrol.*, **10**, 25–40.

Rango, A., Gomez-Landesa, E., Bleiweiss, M., Havstad, K., and Tanksley, K. (2003). Improved satellite snow mapping, snowmelt runoff forecasting, and climate change simulations in the upper Rio Grande Basin. *World Resource Review*, **15**(1), 25–41.

Rango, A., Martinec, J., Chang, A. T. C., Foster, J., and van Katwijk, V. (1989). Average areal water equivalent of snow in a mountain basin using microwave and visible satellite data. *IEEE Trans. Geosci. Remote Sens.*, GE-**27**(6), 740–5.

Rango, A., Martinec, J., Foster, J., and Marks, D. (1985). Resolution in operational remote sensing of snow cover. In *Hydrological Applications of Remote Sensing and Remote Data Transmission*, Proc. Hamburg Symp., IAHS Publ. 145, pp. 371–82.

Rott, H. (1993). Capabilities of microwave sensors for monitoring areal extent and physical properties of the snowpack. In *Proceedings of the NATO Advanced Research Workshop on Global Environmental Change and Land Surface Processes in Hydrology*. Tucson, AZ.

Rott, H. and Matzler, C. (1987). Possibilities and limits of synthetic aperture radar for snow and glacier surveying. *Annals Glaciol.*, **9**, 195–9.

Rott, H. and Nagler, T. (1993). Snow and glacier investigations by ERS-1 SAR – First results. In *Proceedings, First ERS-1 Symposium: Space at the Service of Our Environment*. Cannes, France, pp. 577–82.

Seidel, K. and Martinec, J. (2004). *Remote Sensing in Snow Hydrology: Runoff Modelling, Effect of Climate Change*. Berlin Springer-Praxis.

Shi, J. and Dozier, J. (1996). Estimation of snow water equivalence using SIR-C/X-SAR. In *Proceedings of the International Geoscience and Remote Sensing Symposium, IGARSS*, Lincoln, NE, pp. 2002–4.

Shi, J. and Dozier, J. (1995). Inferring snow wetness using C-band data from SIR-C's polarimetric synthetic aperture radar. *IEEE Trans. Geosci. Remote Sens.*, **33**(4), 905–14.

Shi, J. and Dozier, J. (2000). On estimation of snow water equivalence using SIR-C/X-SAR. In *Proceedings of the Fourth International Workshop on Applications of Remote Sensing in Hydrology*. NHRI Symposium Report, Santa Fe, NM, pp. 197–208.

Solberg, R., Hiltbrunner, D., Koskinen, J., Guneriussen, T., Rautiainen, K., and Hallikainen, M. (1998). SNOWTOOLS: Research and development of methods supporting new snow products. In *Proceedings of the XX Nordic Hydrology Conference*. Helsinki, Finland.

Steppuhn, H. (1981). Snow and Agriculture. In *Handbook of Snow: Principles, Processes, Management and Use*, ed. D. M. Gray and D. H. Male. Toronto: Pergamon Press, pp. 60–125.

Storr, D. (1967). Precipitation variations in a small forested watershed. In *Proceedings of the 35th Annual Western Snow Conference*, pp. 11–16.

Sturm, M., Holmgren, J. A., and Yankielun, N. E. (1996). Using FM-CW radar to make extensive measurements of arctic snow depth: problems, promises, and successes. *EOS Trans., AGU*, **77**(46), F196.

Sun, C., Neale, C. M. U., McDonnell, J. J., and Cheng, H. D. (1997). Monitoring land-surface snow conditions from SSM/I data using an artificial neural network classifier. *IEEE Trans. Geosci. Remote Sens.*, **35**(4), 801–9.

Thirkettle, F., Walker, A., Goodison, B., and Graham, D. (1991). Canadian prairie snow cover maps from near real-time passive microwave data: from satellite data to user information. In *Proceedings of the 14th Canadian Symposium on Remote Sensing*, Calgary, Canada, pp. 172–7.

Vershinina, L. K. (1985). The use of aerial gamma surveys of snowpack for spring snowmelt runoff forecasts. In *Hydrological Applications of Remote Sensing and Remote Data Transmission*, Proc. of the Hamburg Symposium, IAH Publication 145, pp. 411–20.

Walker, A. E. and Goodison, B. E. (1993). Discrimination of a wet snow cover using passive microwave satellite data. *Annals Glaciol.*, **17**, 307–11.

Walker, A., Goodison, B., Davey, M., and Olson, D. (1995). *Atlas of Southern Canadian Prairies Winter Snow Cover from Satellite Passive Microwave Data: November 1978 to March 1986*. Atmospheric Environment Service, Environment Canada.

Warkentin, A. A. (1997). The Red River flood of 1997: an overview of the causes, predictions, characteristics and effects of the flood of the century. *CMOS Bulletin*, **25**(5).

Woo, M. K., Walker, A., Yang, D., and Goodison, B. (1995). Pixel-scale ground snow survey for passive microwave study of the Arctic snow cover. In *Proceedings of the 52nd Annual Meeting of the Eastern Snow Conference*, Toronto, Ontario, Canada, pp. 51–7.

Zuerndorfer, B., England, A. W., and Wakefield, G. H. (1989). The radiobrightness of freezing terrain. In: *Proceedings of the International Geoscience and Remote Sensing Symposium, IGARSS*, Vancouver, Canada, pp. 2748–51.

Table 5.1 *Definitions of satellite, instrument, and program acronyms*

ACRONYM	DEFINITION
AMSR	Advanced Microwave Scanning Radiometer
ASAR	Advanced Synthetic Aperature Radar
ASTER	Advanced Spaceborne and Thermal Emission and Reflection Radiometer
AVHRR	Advanced Very High Resolution Radiometer
DMSP	Defense Meteorological Satellite Program
EASE	Equal Area Scalable Earth
ENVISAT	Environmental Satellite
EOS	Earth Observation System
ERS	European Remote Sensing satellite
ESA	European Space Agency
ETM+	Enhanced Thematic Mapper plus
FM-CW	Frequency Modulated-Continuous Wave
GOES	Geosynchronous Operational Environmental Satellite
MERIS	Medium Resolution Imaging Spectrometer
MODIS	Moderation Resolution Imaging Spectroradiometer
MSS	Multispectral Scanner Subsystem
NASA	National Aeronautics and Space Administration
NESDIS	National Environmental Satellite Data and Information Services
NOAA	National Oceanic and Atmospheric Administration
NOHRSC	National Operational Hydrologic Remote Sensing Center
NSA	National Snow Analyses of NOHRSC
NSIDC	National Snow and Ice Data Center
NWS	National Weather Service
OLS	Operational Linescan System
Radar	Radio Detection and Ranging
SAR	Synthetic Aperature Radar
SIR-C	Spaceborne Imaging Radar-C band
SMMR	Scanning Multichannel Microwave Radiometer
SNODAS	SNOw Data Analysis System of NOHRSC
SPOT	Systeme Pour l'Observation de la Terre
SSM/I	Special Sensor Microwave Imager
TM	Thematic Mapper

6

Snowpack energy exchange: basic theory

6.1 Introduction

The exchange of energy between the snowpack and its environment ultimately determines the rate of snowpack water losses due to melting and evaporation/sublimation. Energy exchange primarily occurs at the snowpack surface through exchange of shortwave and longwave radiation and turbulent or convective transfer of latent heat due to vapor exchange and sensible heat due to differences in temperature between the air and snow. Relatively small amounts of energy can also be added due to warm rainfall on the snowpack surface and soil heat conduction to the snowpack base. Changes in snowpack temperature and meltwater content also constitute a form of internal energy exchange. Melting usually represents the major pathway for dissipation of excess energy when the snowpack ripens and becomes isothermal at $0\,^{\circ}C$ (see Chapter 3).

The energy budget for the snowpack can be written as the algebraic sum of energy gains and losses as:

$$Q_i = Q_{ns} + Q_{nl} + Q_h + Q_e + Q_r + Q_g + Q_m \tag{6.1}$$

where:

Q_{ns} = net shortwave radiant energy exchange (≥ 0)
Q_{nl} = net longwave radiant energy exchange ($\pm$)
Q_h = convective exchange of sensible heat with the atmosphere ($\pm$)
Q_e = convective exchange of latent heat of vaporization and sublimation with the atmosphere ($\pm$)
Q_r = rainfall sensible and latent heat (≥ 0)
Q_g = ground heat conduction ($\pm$)
Q_m = loss of latent heat of fusion due to meltwater leaving the snowpack (≤ 0)
Q_i = change in snowpack internal sensible and latent heat storage ($\pm$)

Each of these energy exchange terms is commonly written as an energy flux density expressed as energy exchange per unit surface area per unit time. Flux density is generally expressed as J s^{-1} m^{-2} or W m^{-2}. In older literature, units of cal cm^{-2} min^{-1} or ly min^{-1} were used where 1 ly (langley) = 1 cal cm^{-2}. All terms except for Q_{ns}, Q_r, and Q_m can represent either energy gains or losses to the snow-pack depending upon the time interval involved, as indicated by the signs given in Equation (6.1).

Several reviews of the snowpack energy budget have been previously given. Early comprehensive works on the snowpack energy budget by the US Army Corps of Engineers (1956) and Kuz'min (1961) were followed by more recent review papers and book chapters by Male and Granger (1981), Male and Gray (1981), Morris (1989), Nakawo *et al.* (1998), and Singh and Singh (2001). Although the processes of snowpack energy exchange are reasonably well understood, computation of snowmelt and/or vapor losses are often made difficult by the lack of on-site snow-pack and meteorological data. The physics behind computations of each energy budget component with example calculations are given in this chapter for a horizontal snowpack in open conditions. Approximate analysis of snowmelt and cold content using degree-day or temperature indices is covered in Chapter 10 which deals with modelling snowmelt runoff. Forest cover and topographic effects are covered in the next chapter, Chapter 7.

6.2 Shortwave radiation exchange

The net flux density of shortwave radiation generally represents the major source of energy for snowmelt. Net shortwave radiation (Q_{ns}) represents the sum of incoming and outgoing flux densities of shortwave radiant energy according to:

$$Q_{ns} = K\downarrow - K\uparrow \qquad (6.2)$$

where:

K = shortwave radiation (wavelengths (λ) = 0.4–2 μm)
$\downarrow\uparrow$ = incoming and outgoing radiation flux densities, respectively.

Outgoing or reflected shortwave radiation can be computed simply from $K\downarrow$ as:

$$K\uparrow = \alpha K\downarrow \qquad (6.3)$$

where α is the albedo or fractional reflectivity of the snowpack averaged across the entire shortwave region. In the next several sections, the theory and computation of incoming and outgoing shortwave flux densities for snowpacks are described.

6.2.1 *Incoming shortwave radiation*

Shortwave radiation from the Sun is transmitted, absorbed, refracted, and reflected by atmospheric constituents. Scattering is a term used to describe multiple reflections and refraction in the atmosphere that diffuses shortwave radiation. Consequently, the flux density reaching the Earth's surface, sometimes referred to as global radiation, occurs as the sum of a direct-beam portion transmitted by the atmosphere at a given zenith angle and a diffuse portion emanating unevenly from the entire hemisphere overhead due to scattering in the atmosphere. The flux density of incoming solar radiation on a horizontal surface can then be shown as:

$$K{\downarrow} = D + I_b \cos Z \qquad (6.4)$$

where:

$K{\downarrow}$ = global radiation or incoming shortwave flux density
D = diffuse shortwave radiation flux density
I_b = beam shortwave radiation flux density passing through the atmosphere
   measured at normal incidence
Z = zenith angle

Zenith angle is the angle between the solar beam and a perpendicular to the horizontal snowpack surface. Zenith angle varies with time of day and day of the year. Solar altitude, or the elevation angle of the Sun above the horizon, is the complement $(90 - Z)$ of the zenith angle.

Cloud cover and zenith angle affect the fraction of incoming shortwave radiation received as beam and diffuse components. On clear days, about 80% of incoming shortwave radiation occurs as direct beam ($I_b \cos Z$) and 20% as diffuse shortwave (D) radiation for solar altitudes >40° above the horizon. At lower solar altitudes and in morning and evening, the diffuse fraction is larger because the solar beam must pass through a greater path length in the atmosphere and more scattering occurs. In contrast, on days with thick, complete cloud cover, total incoming shortwave radiation is reduced and 80% or more occurs as diffuse or scattered radiation. High-latitude stations that have large zenith angles throughout the accumulation and melt season will also experience higher fractions of diffuse shortwave radiation than mid-latitude stations with smaller zenith angles.

Incoming shortwave radiation received on unobstructed, horizontal surfaces is ideally obtained by direct measurement using radiometers at nearby index stations. Pyranometers are radiometers used to measured total incoming shortwave radiation. Shortwave instruments generally respond to wavelengths between 0.4 and about 2 microns, which includes the majority of solar radiant energy received at the Earth's surface. Pyranometers can also be shaded using moving occulting discs or fixed shading bands that eliminate the direct beam radiation to record diffuse or scattered

Figure 6.1 Radiation instruments at the Penn State, Pennsylvania SURFRAD Network site: (a) pyranometer used to measure incoming shortwave radiation on a horizontal surface, (b) a continuously shaded pyranometer (near) and pyrgeometer (far) used to measure incoming diffuse shortwave and incoming longwave radiation, respectively, (c) tubular pyrheliometer (see arrow) mounted on a solar tracker to measure direct-beam solar radiation at normal incidence, and (d) inverted pyranometer and pyrgeometer on tower to measure reflected shortwave and outgoing longwave radiation from the ground surface (photographs by D. DeWalle). See also color plate.

shortwave from the atmosphere. Pyrheliometers are radiometers that are rotated to receive the solar beam at normal incidence on the sensing surface that allows independent measurement of direct-beam shortwave radiation. Inverted pyranometers are generally used to measure the flux density of reflected shortwave radiation. Examples of some shortwave radiation instruments commonly employed are shown in Figure 6.1. Networks for radiation measurements exist in many locations that

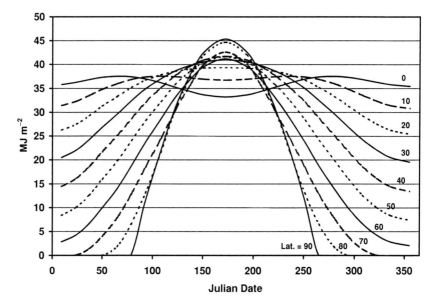

Figure 6.2 Potential solar irradiation of an unobstructed horizontal surface $(MJ\ m^{-2}\ d^{-1})$ by Julian Date (Jan 1 = 1) for latitudes 0–90° North. Data were generated with equations in Appendix B using 1360 W m^{-2} or 4.896 $MJ\ h^{-1}\ m^{-2}$ as the solar constant.

allow for extrapolation of data to the application site. For example the SURFRAD network provides measurements of incoming shortwave total, direct-beam, and diffuse radiation at a number of sites in the United States, www.srrb.noaa.gov/ surfrad.

If measured incoming shortwave radiation data are lacking, computations can be based upon potential solar radiation theory (Lee, 1963; Frank and Lee, 1966). Potential solar radiation theory involves computations to adjust energy received from the Sun for time of day, time of year, slope, aspect, and latitude for the watershed or site of interest, without including any atmospheric effects. Figure 6.2 shows the annual variations of total daily potential solar radiation in MJ m^{-2} received on a horizontal surface for various latitudes computed with equations given in Appendix B. Seasonal variations in daily potential irradiation are a minimum at the equator, where day length is constant. Maximum seasonal variations occur at the poles. Absolute symmetry between summer and winter does not occur between hemispheres due to slight differences in Earth–Sun distance or radius vector with time of year. When computations are needed for a single latitude it is possible to empirically derive polynomial expressions relating potential solar irradiation values to day number for ease of computation (Shutov, 1997). Once daily potential solar radiation is obtained, corrections for atmospheric effects can be incorporated into the estimates for snowmelt computation.

6.2.2 Empirical estimation of incoming shortwave radiation

Empirical methods are available for estimation of incoming shortwave radiation ($K\downarrow$) when direct measurements are unavailable. All of these models involve potential solar irradiation calculations (I_q see Figure 6.2 and Appendix B) to account for astronomical variations, plus use of either the fraction of maximum possible hours of sunshine during a day (n/N), the fractional coverage of the sky by clouds (C), or the daily range of air temperature (ΔT). Using these approaches and some derived values of empirical coefficients (Linacre, 1992; Bristow and Campbell, 1984), the daily global radiation can be computed using:

$$K\downarrow = I_q(0.25 + 0.5\, n/N) \tag{6.5}$$
$$K\downarrow = I_q(0.85 - 0.47\, C) \tag{6.6}$$
$$K\downarrow = I_q\{0.7[1 - \exp(-0.01\Delta T^{2.4})]\} \tag{6.7}$$

Although empirical coefficients are given for these models, parameters are somewhat to very site-specific due to variations in air pollution, clouds, and the range of solar altitudes encountered. For clear skies, $n/N = 1$ and $C = 0$, Equations (6.5) and (6.6), show that $K\downarrow/I_q = 0.75$ to 0.85; thus on clear days anywhere from 75 to 85% of potential radiation is received at the surface. Under conditions of complete cloud cover, e.g. $n/N = 0$ or $C = 1$, these equations show that $K\downarrow/I_q = 0.25$ to 0.38. Using Equation (6.7), a cloudy day with $\Delta T = 5\,°C$ would give $K\downarrow/I_q = 0.26$, while a clear day with $\Delta T = 20\,°C$ would give $K\downarrow/I_q = 0.7$. All these models agree in general; however, errors using any of these relationships to estimate $K\downarrow$ on individual days can be as great as 30% and local measurement of $K\downarrow$ is recommended wherever possible.

Ideally, parameters for these models should be determined for conditions where they are to be applied. According to Linacre (1992), Equation (6.5) is preferred over Equation (6.6), since hours of sunshine can be measured with less error than cloud cover can be visually estimated. However, cloud-cover data can be easily observed in field campaigns and published cloud-cover records are more readily available. Shutov (1997) employed a cloud-cover function in Russia that made use of total and low-level cloud-cover amounts. Equation (6.7) was originally proposed by Bristow and Campbell (1984) and several different methods of deriving coefficients and defining ΔT have been developed (Weiss et al., 1993; Thornton and Running, 1999;

Donatelli *et al.*, 2003; Bellocchi *et al.*, 2002). The air-temperature approach is particularly well suited for snow hydrology applications when only air-temperature
data are available, but coefficients vary for high latitudes and snow-covered
ground.

As a simple example of use of these empirical equations and given potential solar
irradiation (Figure 6.2) for a latitude of 55° N on February 7 or Julian Day $= 38$ of
approximately 9.6 MJ m^{-2} d^{-1} or 111 W m^{-2} averaged over a 24-hour period, cloud
cover can be used to estimate total incoming shortwave radiation using Equation
(6.6). If the fraction of cloud cover was $C = 0.3$, representing relatively clear sky,
the incoming shortwave radiation estimate would be:

$$K\downarrow = (111 \text{ W m}^{-2})[0.85 - (0.47)(0.3)] = 78.7 \text{ W m}^{-2}$$

Equation (6.6) was employed to account for cloud cover here to facilitate later
comparisons with estimated incoming longwave radiation functions that also
involve C.

6.2.3 Transmission of shortwave radiation

Shortwave radiation can be transmitted through snow, especially as snow densi-
fies and approaches solid ice in structure. Reflection of shortwave radiation is not
a simple surface phenomena. Transmission, absorption, and multiple reflections
of shortwave radiation can occur within the snowpack and transmission of some
shortwave radiation to the ground below is possible. Warming of underlying soil
or objects buried in the snow by shortwave radiation transmitted through the snow
can accelerate the melting of shallow snowcover and reduce the effective albedo of
the snowpack. Transmission through snow can be approximated by Beer's Law for
radiation penetration through a homogeneous media:

$$K\downarrow_z = K\downarrow_0[\exp(-\nu z)] \tag{6.8}$$

where $K\downarrow_z$ is the shortwave flux density at depth z in the snowpack, $K\downarrow_0$ is the
shortwave flux density at the surface, and ν is an extinction coefficient averaged across the shortwave wavelength spectrum. Extinction coefficients range
from about $\nu = 0.4$ cm^{-1} for low-density snow to about $\nu = 0.1$ cm^{-1} for
high-density snow (Anderson, 1976). Pure ice is a relatively good transmitter
and has a shortwave extinction coefficient of only about 0.01 cm^{-1}. The transition from low-density fresh snow to older high-density snow is thus accompanied by a large increase in the ability of the snowpack to transmit shortwave
radiation.

To illustrate the depth to which shortwave radiation can penetrate snow, Equation (6.8) can be used to compute the depth of high-density snow ($v = 0.1$ cm^{-1}) needed to reduce the incoming shortwave by 90% or to a level equal to 10% of its surface value as:

$$K{\downarrow}_z/K{\downarrow}_0 = 0.10 = \exp[(-0.1 \text{ cm}^{-1})z]$$
$$\ln(0.10) = -0.1\ z$$
$$z = -2.30/-0.1 = 23 \text{ cm}$$

Thus, the flux density of shortwave radiation transmitted is only 10% of the surface value at a snowpack depth of 23 cm.

Penetration of shortwave radiation to the ground could affect melt rates for the last 20–25 cm depth of dense snowpacks. In comparison, only about 6-cm depth is required to cause a 90% reduction in shortwave radiation penetration for fresh, low-density snow ($v = 0.4$ cm^{-1}). Even though shortwave radiation is not easily transmitted through low-density snow, shortwave radiation absorption helps to speed metamorphism in surface layers of fresh snow (Koh and Jordan, 1995).

6.2.4 Outgoing or reflected shortwave radiation

Outgoing or reflected shortwave radiation as given in Equation (6.3) depends upon the snowpack albedo. Snowpack albedo (α) ranges from above 0.95 for a fresh snowfall down to below 0.40 for shallow, dirty snow. Albedo of a snowpack is affected by many factors (Warren, 1982; Marks and Dozier, 1992; Kustas *et al.*, 1994; Winther, 1993) that make precise prediction of snowpack albedo difficult (see Table 6.1). In the absence of measured reflected shortwave radiation data, some investigators have estimated snowpack albedo based upon theory with grain size and zenith angle of the Sun (Marks and Dozier, 1992; Kustas *et al.*, 1994). Due to lack of data for theoretical approaches, others have stressed development of more empirical approaches to predict snowpack albedo using accumulated air temperatures or time since last snowfall as indices (Pluss and Mazzoni, 1994; Winther, 1993). Shutov (1997) applied a correction to snowpack albedo for particulate pollution. Regardless of the approach employed to estimate snowpack albedo, net solar energy is one of the most important energy sources for snowmelt and the choice of α becomes critical in any melt-prediction scheme.

Early work by the US Army Corps of Engineers (1956) showed snowpack albedo could be related to accumulated maximum daily air temperatures (Figure 6.3). A simple polynomial expression fit to these data is given. Winther (1993) used temperature, solar radiation, and snowpack water equivalent to predict albedo at a site in Norway. Anderson (1976) developed an expression for α as a function of

Table 6.1 *Factors controlling snowpack albedo*

Factor	Theory	Effect
Ice grain size	Albedo decreases as grain size increases	Albedo declines after snowfall with metamorphosis and crystal growth
Zenith angle	Albedo decreases as the zenith angle for direct-beam shortwave radiation decreases (solar altitude increases)	Albedo lower during mid-day and later in the melt season; effect pronounced on clear days with higher direct beam radiation
Visible vs. near-infrared radiation	Albedo lower in near-infrared (0.7–2.8 μm) than visible (0.4–0.7 μm) region	Albedo lower on cloudy days when relatively more near-infrared radiation is received
Snowpack contaminants	Albedo decreased by particulates and larger organic debris deposited or exposed on the snowpack surface	Albedo lower in forests and regions with aeolian dust or polluted air
Snowpack depth	Albedo affected by transmission of shortwave radiation within a snowpack layer whose thickness varies with snow density	Albedo declines as snowpack becomes shallow and dense and transmitted shortwave radiation is absorbed by ground below

snow surface density, that was computed with a finite-difference model, for snow at an open site at the Sleepers River Research Watershed near Danville, VT, USA (see Figure 6.4). In a related empirical model, Kuchment and Gelfan (1996) estimated albedo (α) from snow density (ρ_s, g cm^{-3}) as $\alpha = 1.03 - \rho_s$. All of these schemes suggest that snowpack α will decline over time, but can rapidly increase with fresh snowfall onto the snowpack.

A simple approach to modelling snowpack albedo given by Rohrer and Braun (1994) based upon the US Army Corps of Engineers' (1956) work and tested on two other data sets is:

$$\alpha = \alpha_0 + K \exp(-nr) \tag{6.9}$$

where:

α = snowpack albedo
α_0 = minimum snowpack albedo $\cong 0.4$
K = constant $\cong 0.44$
n = number of days since last major snowfall
r = recession coefficient

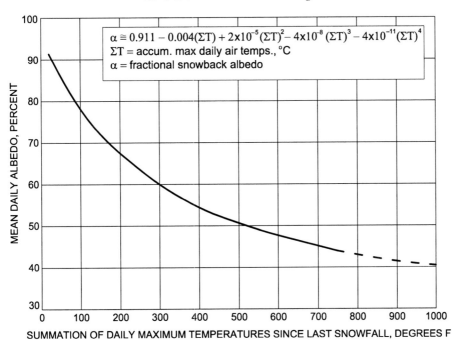

Figure 6.3 Snowpack albedo variations with an accumulated temperature index (modified from US Army Corps of Engineers, 1956).

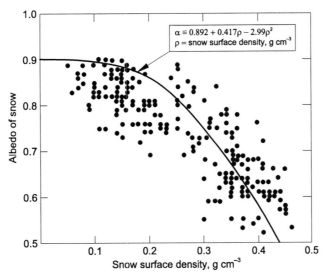

Figure 6.4 Snowpack albedo as a function of snow surface density (modified from Anderson, 1976).

The recession coefficient was found to be 0.05 for air temperatures less than $0\,°C$ and 0.12 for temperatures greater than $0\,°C$. The definition of a major snowfall needed for determining n was a snowfall depth of at least 3 cm over as many days. The maximum albedo when $n = 0$ is 0.84 using values of the constants given above. If n were 3 days and air temperatures had been above freezing, then the computed albedo would be:

$$\alpha = 0.4 + (0.44)(2.72)^{-(3)(0.12)} = 0.71 \text{ or } 71\%$$

The equation worked well as long as snow depths exceeded 10 cm. Pluss and Mazzoni (1994) found the model underpredicted slightly at low temperatures and overestimated at temperatures above freezing.

6.2.5 Net shortwave radiation exchange

Once the albedo of the snowpack has been resolved by measurement or computation, it is a simple matter to compute net shortwave radiation exchange. Given a representative high mid-winter albedo of 0.8 and the estimated incoming shortwave radiation flux density for relatively clear sky of 78.7 W m^{-2}, computed previously, the reflected shortwave flux density would be:

$$K\!\uparrow = \alpha K\!\downarrow = 0.8(78.7) = 63.0 \text{ W m}^{-2}$$

and net shortwave radiation exchange would be:

$$Q_{ns} = K\!\downarrow - K\!\uparrow = 78.7 - 63.0 = 15.7 \text{ W m}^{-2}$$

As illustrated here, under conditions of mid-winter solar irradiation and high snowpack albedo, the absorbed shortwave radiation available for melt can be quite low. However, small changes in albedo when the albedo is high can disproportionately affect net shortwave radiation exchange. If the albedo in the above example were reduced by 25% from 0.8 to 0.6, then net shortwave radiation exchange would double (31.5 W m^{-2}). As incoming shortwave radiation increases during the spring melt season and the snowpack albedo declines due to combined effects of increasing grain size, greater solar altitude, snowpack surface contamination, and possibly transmission of radiation to the ground, net shortwave exchange can become increasingly important as an energy budget component.

6.3 Longwave radiation exchange

All substances above absolute zero in temperature radiate or emit radiant energy and at terrestrial temperatures materials such as atmospheric gases, clouds, forest vegetation, and snowpacks radiate in the longwave region ($\lambda = 2$–100 µm).

Unlike shortwave radiation exchange that is restricted to daytime, longwave radiation exchange occurs both day and night, which increases the importance of the longwave component. Measurements of incoming longwave radiation are much less common than for shortwave radiation, but in recent years radiometers have been developed that can be used for routine monitoring. Regardless, computations of longwave exchange are often possible using other available environmental data as described below.

Net longwave radiation exchange (Q_{nl}) in Equation (6.1) for an unobstructed, horizontal snowpack surface can be defined as the sum of incoming $(L\downarrow)$ and outgoing $(L\uparrow)$ longwave radiation flux densities as:

$$Q_{nl} = L\downarrow - L\uparrow \tag{6.10}$$

Outgoing longwave radiation can be computed using the Stefan–Boltzmann equation derived to describe emission from a perfect or black-body radiator as:

$$L\uparrow = \varepsilon\sigma T_s^4 + (1 - \varepsilon)L\downarrow \tag{6.11}$$

where:

ε = emissivity of snowpack surface
σ = Stefan–Boltzmann constant $= 5.67 \times 10^{-8}$ W m^{-2} K^{-4}
T_s = surface temperature of the snowpack, K, (K $= °C + 273.16$)

Emissivity is the ratio of radiation emitted by a substance to that emitted by a perfect or black-body radiator at the same temperature. The ratio or emissivity equals one for a black body and snowpacks come fairly close to being a perfect radiator with $\varepsilon = 0.97 - 1.00$ (Kondrat'yev, 1965; Anderson, 1976). Since, by Kirchhoff's law, emissivity is equal to absorptivity at a given wavelength, the term $(1 - \varepsilon)$ represents the small fraction of incoming longwave radiation that is reflected from the snow surface. The loss of longwave radiant energy is primarily a function of snow surface temperature, but does depend slightly on the incoming longwave radiation flux density from the atmosphere due to reflection.

Since the surface temperature of a snowpack during melt is $0\,°C$, the flux density of longwave radiation that could be lost from the snowpack for an assumed $\varepsilon = 1$ would be:

$$L\uparrow = (1)(5.67 \times 10^{-8} \text{ W m}^{-2} \text{ K}^{-4})(273.16 \text{ K})^4 = 316 \text{ W m}^{-2}$$

More precise calculation of outgoing longwave depends upon knowing the snowpack emissivity and the concurrent flux of incoming longwave radiation. In addition, estimating the snowpack surface temperature during non-melting conditions, when the snowpack is at less than $0\,°C$, becomes problematical for many operational snowmelt simulations.

Incoming or downward longwave radiation ($L\downarrow$) from the atmosphere can also be measured using a radiometer called a pyrgeometer (see Figure 6.1), although such measurements are less common than measurement of incoming shortwave radiation. Pyrgeometers are equipped with interference filters to prevent shortwave radiant energy from reaching the sensors. Inverted pyrgeometers can be used to measure outgoing longwave radiation.

Incoming longwave radiation from the atmosphere may also be estimated by using the Stefan–Boltzmann equation. Atmospheric gases radiate in defined spectral bands rather than radiating uniformly across all wavelengths and that causes the average emissivity to be quite a bit lower than that for a perfect black body. In addition, the effective radiating temperature of atmospheric gases and cloud bases is not easily determined and ground-level air temperatures measured in a standard weather shelter are often used as an index to upper-level temperatures. Generally the incoming longwave radiation flux density for a clear sky is calculated and then a correction is added for effects of cloud cover, if any (see discussions by Kondrat'yev 1965; Oke 1987). Variations in clear-sky incoming longwave radiation depends primarily on air temperature and the atmospheric water vapor pressure. Many approximately equivalent forms of empirical equations exist and the expression given by Brutsaert (1975) for clear-sky conditions is given here because of its simplicity:

$$L\downarrow \text{ clear sky} = \left(0.575\, e_a^{1/7}\right) \sigma T_a^4 \qquad (6.12)$$

where:

e_a = atmospheric water vapor pressure, mb (1 mb = 100 pascals or Pa)
T_a = air temperature in a standard weather shelter, K

Equation (6.12) can be used to compare incoming longwave radiation from a clear sky with longwave losses from a snowpack calculated above. If air temperature was $2\,°C$ (275.16 K) and vapor pressure was 3.53 mb (50% relative humidity at this temperature), then by Equation (6.12) the incoming longwave radiation would be:

$$L\downarrow \text{ clear sky} = [0.575(3.53)^{1/7}](5.67 \times 10^{-8})(275.16)^4 = 224 \text{ W m}^{-2}$$

Under clear skies the incoming longwave radiation is much less than the longwave radiation losses from a melting snowpack, consequently the net longwave exchange would be:

$$Q_{nl} \text{ clear sky} = (L\downarrow \text{ clear sky}) - L\uparrow = 224 \text{ W m}^{-2} - 316 \text{ W m}^{-2} = -92 \text{ W m}^{-2}$$

Under clear-sky conditions, the net longwave exchange at the surface of a melting snowpack is commonly negative and represents a major loss of energy from the snowpack.

Table 6.2 *Cloud-cover factors used to compute*
incoming longwave radiation with Equation (6.13)
(Kondrat'yev, 1965; Oke, 1987)

Cloud Type	a_c
Cirrus	0.04
Cirrostratus	0.08
Altocumulus	0.17
Altostratus	0.20
Cumulus	0.20
Stratocumulus	0.22
Stratus	0.24
Fog	0.25

Incoming longwave radiation with cloud cover is simply estimated by adjusting clear-sky estimates using an equation such as:

$$L\!\downarrow \text{ cloudy sky} = (L\!\downarrow \text{ clear sky})(1 + a_c C^2) \qquad (6.13)$$

where C is fractional cloud cover and the empirical coefficient a_c varies with cloud type (see Table 6.2).

Given the range of cloud-cover factors in Table 6.2, complete cloud cover or fog ($C^2 = 1$) can increase incoming longwave by about 24–25% for a given air temperature and humidity value. If in our example a cumulus cloud cover of 0.3 occurred ($a_c = 0.2$), then by Equation (6.13) we would be increasing incoming clear longwave radiation computed by a factor of $[1 + (0.20)(0.3)^2] = 1.018$ or about 2% from $L\!\downarrow = 224$ W m^{-2} to $L\!\downarrow = 228$ W m^{-2}. Net longwave exchange over melting snow would therefore be increased from –92 to –88 W m^{-2} or by about 4%. Complete cumulous cloud cover ($C = 1.0$, $a_c = 0.2$) would have increased incoming longwave radiation to 269 W m^{-2} and net longwave radiation exchange at the snowpack surface by 50%. Obviously, clouds can significantly add incoming longwave radiation to the snowpack and significantly reduce net longwave radiation losses.

6.4 Sensible heat convection

Whenever a temperature difference exists between the atmosphere and the snow-pack surface, the potential exists for significant convective transfer of sensible heat to or from the snowpack depending upon the magnitude and sign of the tempera-ture difference, wind speed, surface roughness, and stability of the air. Late-lying

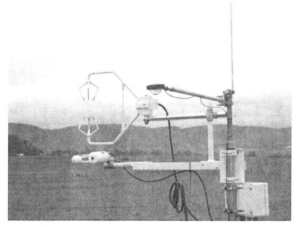

Figure 6.5 Solar-powered climate stations equipped with data loggers (left) are used to provide data (wind speed, air temperature, humidity) to support computations of snowpack convective exchange. The addition to climate stations of sonic anemometers (top right), fine-wire thermocouples, and high-speed open-path gas analyzers for CO_2 and water vapor (bottom right) allows computation of convective exchange using the eddy covariance method (photographs by D. DeWalle). See also color plate.

melting snowpacks in late spring and early summer can exist when air temperatures are over 20 °C greater than the snow, causing large energy gains to the snowpack. In contrast, during winter or at night during the melt season, the snowpack can be warmer than air, causing convective energy losses from the snowpack to air.

In this section an approach to computation of sensible heat flux densities is presented. The bulk aerodynamic approach for computation of sensible heat and water vapor flux in the atmosphere will be stressed here because it often becomes the method of choice due to lack of meteorological data. The bulk aerodynamic method uses measurements of wind speed, air temperature, and humidity in the air at a fixed height above the snow surface (Figure 6.5). Marks and Dozier (1992) applied a bulk aerodynamic method corrected for stability adapted from Brutsaert (1982) that iteratively solves for friction velocity and the sensible and latent heat flux densities over snow. Application of bulk aerodynamic theory to computation of sensible heat exchange involves several simplifying assumptions (Moore, 1983). Despite these theoretical limitations, bulk theory has been used to approximate sensible heat exchange to and from snowpacks within acceptable error limits (see Anderson, 1976; Kustas *et al.*, 1994).

Several other approaches exist to compute atmospheric convective fluxes on an operational level (Oke, 1987; Arya, 1988). The eddy covariance method, which requires measurements of the fluctuations of vertical wind speed, air temperature, and humidity is rapidly becoming feasible for snow hydrology applications with continuing improvements in sensor technology (Figure 6.5); however, greater costs, more complex data manipulation and harsh winter weather conditions still present considerable challenges for snow hydrologists.

According to bulk aerodynamic theory the sensible heat flux density can be computed as (Kustas et al., 1994):

$$Q_h = \rho_a\, c_p\, C_h\, u_a(T_a - T_s) \tag{6.14}$$

$$C_{hn} = k^2[\ln(z_a/z_0)]^{-2} \tag{6.15}$$

where:

ρ_a = density of air, kg m^{-3}

c_p = specific heat of air, J kg^{-1} K^{-1}

C_h = bulk transfer coefficient for sensible heat, dimensionless,

C_{hn} = bulk transfer coefficient for sensible heat for neutral stability atmosphere, dimensionless,

k = von Karmin's constant = 0.4

z_0 = aerodynamic roughness length for the snow surface, m

u_a = wind speed at height z_a, m s^{-1}

T_a = air temperature at height z_a, K

T_s = air temperature at the snowpack surface, K

The aerodynamic roughness length indexes the mean height of snowpack surface obstacles that control the friction between the moving air and snow. Roughness length is evaluated as the height above the ground where the wind speed (u_a) extrapolates to zero when plotted against ln(height) as defined by the logarithmic wind profile equation:

$$u_z = u_*/k\, \ln(z/z_0) \tag{6.16}$$

where u_* is the friction velocity in m s^{-1} and u_*/k defines the slope of the line at neutral stability. The friction velocity is related to the shear stress or flux density of horizontal momentum at the snow surface (τ, kg m^{-1} s^{-2}) where $u_* = (\tau/\rho_a)^{1/2}$. Detailed evaluations of roughness lengths for surfaces are generally done with at least four anemometers that are spaced logarithmically with height under neutral stability atmospheric conditions.

Values for snowpack roughness lengths reported in the literature generally fall between 0.001 and 0.005 m (1 and 5 mm, respectively) for average seasonal snow-pack conditions with a very wide range from about 0.0002 m to 0.02 m reported for very smooth to very rough snowpack and ice conditions (see reviews by Moore, 1983 and Morris, 1989). Given the very wide range of roughness lengths for snow reported in the literature, and the relative importance of this parameter, an objective method of selecting the appropriate z_0 is needed. A relationship originally developed by Paeschke (1937) can help somewhat (see discussion by Brutsaert, 1982), where roughness length (z_0) is related to the average height of roughness obstacles (h_0) as:

$$z_0 = 0.136\, h_0 \qquad\qquad (6.17)$$

Using this relationship for commonly observed snowpack roughness lengths of 1 and 5 mm, the average heights of roughness obstacles would be 7.4 and 37 mm, respectively. These mean heights of roughness obstacles are greater than the typical dimensions of individual ice grains and suggest that measurements of the larger-scale surface features of the snowpack roughness elements are needed to estimate h_0.

Stability conditions in the air above a surface can greatly affect the amount of turbulence and rates of sensible heat exchange. Neutral stability conditions that are assumed by Equations (6.14) and (6.15) refer to an atmosphere with no buoyancy effects on turbulence. Under neutral stability conditions all turbulence is created by forced convection due to horizontal wind movement over a rough surface. Free convection refers to turbulence due to rising or sinking air caused by density differences in the air near the ground. Air in contact with a relatively warm surface will warm, expand, and rise due to reduced density and can greatly enhance turbulent mixing. This type of atmospheric condition is referred to as "unstable" and is not common over a cold snow surface. In contrast, air in contact with a relatively cool surface like snow will cool, become more dense, sink, and retard turbulent mixing. This atmospheric condition, referred to as "stable," commonly occurs over snow (Anderson, 1976; Shutov, 1993). In the atmosphere, turbulence is generally caused by a mixture of forced and free convection processes. Corrections for the effects of atmospheric stability on turbulence are made to the bulk transfer coefficient (C_{HN}) using an index to atmospheric stability.

Atmospheric stability conditions can be indexed using the dimensionless bulk Richardson number (Ri_B) that is calculated as:

$$Ri_B = \left[g\, T_m^{-1}\right] z_a (T_a - T_s) / \left(u_a^2\right) \qquad\qquad (6.18)$$

where g is the acceleration due to gravity, 9.8 m s^{-2}, $T_m = (T_a + T_s)/2$ in K, and all other parameters have been previously defined. The bulk Richardson number relates the consumption of energy by buoyancy forces to the generation of energy by shear stress forces. Positive values of Ri_B indicate stable conditions and negative values indicate unstable conditions. Magnitudes of the computed Ri_B, either positive or negative, indicate the degree of instability or stability in the air. Values of Ri_B from about -0.01 to $+0.01$ indicate neutral stability. Values of Ri_B greater than about 0.2 to 0.4 are believed to represent conditions where turbulence is completely damped and no convection would occur.

Empirical functions to correct C_h for the effects of stability can be based upon Ri_B. The functions predict the ratio of the bulk transfer coefficient for the actual conditions (C_h) to that for neutral stability conditions (C_{hn}) as given by Equation (6.19). This ratio (C_h/C_{hn}) is then multiplied by C_{hn} given by Equation (6.15) to correct for stability effects. Equations derived from field data to correct for stability appearing in the literature vary widely and remain controversial (Morris, 1989; Male and Gray, 1981; Oke, 1987; Anderson, 1976; Hogstrom, 1988; Brutsaert, 1982; Shutov, 1993; Granger and Male, 1978; Kustas *et al.*, 1994). General stability correction equations given by Oke (1987) for unstable conditions and stable conditions are given here:

$$\text{Unstable: } C_h/C_{hn} = (1 - 16Ri_B)^{0.75} \tag{6.19a}$$
$$\text{Stable: } \quad C_h/C_{hn} = (1 - 5Ri_B)^2 \tag{6.19b}$$

Equation (6.19b) gives $C_h/C_{hn} = 0$ when Ri_B equals 0.2 as expected for fully damped turbulence. Anderson (1976) used Equation (6.19b) for stable conditions over snow. Equation (6.19a) for unstable conditions gives similar results to a function derived from data over snow used by Granger and Male (1978) for Ri approaching -0.02. Once the ratio C_h/C_{hn} is obtained with the appropriate function (Equation 6.19a or 6.19b), then C_h is computed as $C_h = (C_h/C_{hn})C_{hn}$. Atmospheric stability conditions over snowcover and effects on convection of sensible heat to and from the snowpack are summarized in Table 6.3.

Example calculations will help illustrate the use of this theory and supporting empirical relationships. Assuming air temperature of $2\,°C$ and wind speed of 3.2 m s^{-1} at 2 m height over melting snow with a small roughness length of 0.0005 m and using $c_p = 1.005 \times 10^3\text{ J kg}^{-1}\text{ K}^{-1}$ and $\rho_a = 1.27\text{ kg m}^{-3}$, the sensible heat exchange for neutral stability can be computed. Refer to Appendix A1 for specific heat and air density data. First the bulk transfer coefficient for sensible heat exchange under neutral stability is computed with Equation (6.15) as:

$$C_{hn} = (0.4)^2\ [\ln(2\text{ m}/0.0005\text{ m})]^{-2} = 0.00232 \text{ dimensionless}$$

If the roughness length were increased in the above computation to $z_0 = 0.005$ m, C_{hn} would increase to 0.00446 an increase of 92%. Obviously, roughness length is a major control on the bulk transfer coefficient.

Table 6.3 *Summary of atmospheric stability and snowpack convection conditions*

Atmospheric Stability	Meteorological Conditions	Occurrence	Bulk Richardson Number	C_h/C_{hn}
Unstable-free convection	$T_s \gg T_a$, u_a relatively low	Rare over snow	< -1	$\gg 1$
Unstable-mixed convection	$T_s > T_a$, u_a relatively high	Occasional; windy, cold winter accumulation season, night time	< -0.01	> 1
Neutral-forced convection	$T_s \sim T_a$, u_a relatively high	Common; windy, cool periods, melt initiation	-0.01 to $+0.01$	~ 1
Stable-damped forced convection	$T_s < T_a$, u_a relatively high	Common; windy, warm melt periods	$> +0.01$	< 1
Stable-fully damped forced convection	$T_s \ll T_a$, u_a relatively low	Common; calm, warm melting periods; nights with fog and cold air drainage	$\rightarrow$critical value $\sim +0.2$ to $+0.4$	$\rightarrow 0$

Next, the bulk Richardson number is needed to compute the correction for stability using Equation (6.18):

$$Ri_B = [(9.8 \text{ m s}^{-2})/((275.16 + 273.16)/2 \text{ K})]$$
$$\times (2 \text{ m})(275.16 - 273.16 \text{ K})/(3.2 \text{ m s}^{-1})^2$$
$$= 0.014$$

This small, positive value of Ri_B indicates neutrally stable to slightly stable conditions. The ratio C_h/C_{hn} computed from Ri_B using Equation (6.19b) for stable conditions becomes:

$$C_h/C_{hn} = [1 - (5)(0.014)]^2 = 0.865$$

Turbulence is damped by the slightly stable atmospheric conditions. This correction factor is then multiplied by C_{hn}, previously derived, to obtain the corrected bulk transfer coefficient as $C_h = (C_h/C_{hn})\, C_{hn} = (0.865)\,(0.00232) = 0.00201$.

Finally, the sensible heat flux density is computed using Equation (6.14) as:

$$Q_h = (1.27 \text{ kg m}^{-3})(1.005 \times 10^3 \text{ J kg}^{-1} \text{ K}^{-1})(0.00201)(3.2 \text{ m s}^{-1})(2 - 0\,^\circ\text{C})$$
$$= 16.4 \text{ J s}^{-1} \text{ m}^{-2} = 16.4 \text{ W m}^{-2}$$

Since air temperature is greater than snow surface temperature, Q_h is positive or an energy gain in the context of an energy budget.

Note that with a higher roughness length of $z_0 = 0.005$ m and $C_{hn} = 0.00446$, C_h becomes 0.00386 after the stability correction and $Q_h = 31.5$ W m^{-2}. In this example, stability conditions, roughness length, wind speed, and the temperature difference all controlled the computed sensible heat flux density.

Rugged terrain, discontinuous snow cover with adjacent bare ground, and presence of trees as roughness elements in the airflow increase the errors in applying one-dimensional vertical exchange theory presented here. In situations where horizontal gradients of roughness, surface temperature, and/or moisture become important, advection or lateral exchange of heat and vapor also becomes important. Advection of heat from bare ground to adjacent snow patches can greatly increase local sensible heat supplied to the snowpack (Liston, 1995; Neumann and Marsh, 1997, 1998). In forests, warm tree boles and branches and exposed ground can also enhance local advection of sensible heat. DeWalle and Meiman (1971) found that sensible heat convection and advection to late-lying snow patches in Colorado subalpine conifer forest contributed 85 W m^{-2} to snowmelt during a two-day period in June with day-time wind speeds in excess of 4 m s^{-1} and maximum air temperatures in excess of 22 °C. Energy exchange due to sensible and latent heat exchange with patchy snow cover and other conditions causing advection, presents special challenges to estimating heat exchange.

6.5 Latent heat convection

Water vapor exchange between the snowpack and atmosphere also occurs due to turbulent mixing in the boundary layer. Transfer of vapor from the snowpack to the atmosphere constitutes a loss of latent heat of vaporization if the snowpack has liquid water present and a loss of latent heat of sublimation if subfreezing temperatures prevail. Conversely, sublimation or condensation of vapor onto the snowpack constitutes an energy gain due to liberation of the latent heat of sublimation or vaporization, respectively.

Generally, during the cold winter season a snowpack loses water due to evaporation and sublimation. For example, Schmidt *et al.* (1998) estimated that 20% of the peak snowpack water equivalent or about 78 mm water equivalent was lost due to sublimation from the snowpack during a 40-d accumulation period in Colorado subalpine forest. However, during melting periods, and during periods with rainfall, the vapor pressure in the atmosphere often exceeds that at the snowpack surface, since vapor pressure at the snowpack surface cannot exceed the saturation vapor pressure at 0 °C. Condensation/sublimation onto the snowpack occurs under these conditions (Leydecker and Melack, 1999). The process of latent heat convection is generally believed to be similar to that for sensible heat exchange and subject to the same errors, with the difference in vapor pressure between snowpack and atmosphere, roughness length, wind speed, and stability again playing major roles.

Some investigators have suggested that $C_e < C_h$ (Granger and Male, 1978), but $C_e = C_h$ is assumed here.

The bulk aerodynamic equation for computing latent heat exchange can be written as Kustas *et al.* (1994):

$$Q_e = (\rho_a\ 0.622\ L/P_a)C_e\ u_a\ (e_a - e_0) \qquad (6.20)$$
$$C_{en} = C_{hn} = k^2\ [\ln(z_a/z_0)]^{-2} \qquad (6.21)$$

where:

ρ_a = density of air, kg m^{-3}
L = latent heat of vaporization or sublimation, J kg^{-1}
P_a = total atmospheric pressure, pascals (Pa)
C_e = bulk transfer coefficient for vapor exchange
C_{en} = bulk transfer coefficient for vapor exchange under neutral stability
e_a = atmospheric vapor pressure at height z, Pa
e_0 = vapor pressure at the snowpack surface, Pa
k = von Karman's constant, 0.4
u_a = wind speed at height z above ground, m s^{-1}
z_0 = roughness length, m

The vapor pressure at the snowpack surface (e_0) is generally assumed to be the saturation vapor pressure (e_s) at the snowpack temperature; computed as e_s over ice for $T < 0\,°C$ and e_s over liquid water for $T = 0\,°C$. Thus, the maximum vapor pressure at the snowpack surface during melting at $0\,°C$ is 610.78 Pa $= 6.1078$ mb. At other times, measurement or computation of snowpack temperature is required to obtain e_0. Refer to Appendix A1 for saturation vapor pressure and latent heat vs. temperature relationships. Measurement of atmospheric vapor pressure (e_a) can be especially problematical in cold winter conditions. Where humidity data are lacking, but air temperatures are available, a useful substitution is to use saturation vapor pressure at the minimum daily air temperature, since the minimum air temperature and dewpoint temperature often are quite similar. The use of Equations (6.20)–(6.21) and the correction for stability effects follows the discussion in the previous section.

An example calculation for vapor transport between a snowpack and the atmosphere will follow the example for sensible heat exchange in the previous section. In addition to a wind speed of 3.2 m s^{-1} measured at 2 m height, it is assumed that vapor pressure in the air at that height is equal to 50% relative humidity. Since the saturation vapor pressure at an air temperature of $2\,°C$ is 7.05 mb and the relative humidity is 50%, the vapor pressure in the air is $(0.5)(7.05) = 3.53$ mb. This is less than the saturation vapor pressure at the melting snow surface of 6.11 mb, so the direction of vapor flow is from the surface to the atmosphere, that is, evaporation is occurring. Evaporation is assumed to be occurring, rather than sublimation, since the snowpack is at $0\,°C$ and liquid water is assumed to be continuously present at the surface. Since $C_e = C_h$ is assumed, the same bulk transfer coefficient of 0.00201 can be used from the previous example based upon a $z_0 = 0.0005$ m. Given a total atmospheric pressure of 1000 mb, air density $\rho_a = 1.27$ kg m^{-3} and $L_v = 2.496 \times 10^6$ J kg^{-1} (see Chapter 1 and Appendix A1), the flux density of latent heat exchange can be computed with Equation (6.20) as:

$$Q_e = (1.27 \text{ kg m}^{-3})(0.622)(2.496 \times 10^6 \text{ J kg}^{-1})$$
$$\times (1/10^3 \text{ mb})(0.00201)(3.2 \text{ m s}^{-1})(3.53 - 6.11 \text{ mb})$$
$$= -32.8 \text{ J s}^{-1} \text{ m}^{-2} = -32.8 \text{ W m}^{-2}$$

If the value of $C_e = C_h = 0.00386$ based upon a roughness length of $z_0 = 0.005$ m were used, the computed Q_e would increase to -63.0 W m^{-2}.

The flux density of latent heat can be converted into flux density of liquid water by dividing by the latent heat of vaporization or sublimation (L) and the density of liquid water ($\rho_w = 1 \times 10^3$ kg m^{-3}), where:

$$E = Q_e/[L\ \rho_w]$$
$$E = (-32.8 \text{ J s}^{-1} \text{ m}^{-2})/[(2.496 \times 10^6 \text{ J kg}^{-1})(1 \times 10^3 \text{ kg m}^{-3})]$$
$$= -1.31 \times 10^{-8} \text{ m s}^{-1} = -0.047 \text{ mm h}^{-1} \qquad (6.22)$$

Although this appears to be a very small evaporation loss, over a 30-day month this rate would amount to a very significant 3.34 cm of water loss. It is also interesting to note that in our example the sensible heat flux density was positive, representing an energy gain by the snowpack, while the latent heat exchange is negative representing an energy and mass loss from the snowpack.

The ratio of sensible heat exchange to latent heat exchange, called the Bowen Ratio, is a useful indicator of snowpack convective exchange. The Bowen Ratio (β) is defined as the ratio of sensible to latent heat convection and can be computed as:

$$\beta = Q_h/Q_e = [\rho_a\ c_p\ C_h\ u_a(T_a - T_s)]/[(\rho_a\ 0.622\ L/P_a)C_e\ u_a\ (e_a - e_0)]$$
$$= [(c_p\ P_a)/(0.622\ L)][(T_a - T_s)/(e_a - e_0)] \qquad (6.23)$$

The assumption of equality of bulk transfer coefficients, $C_e = C_h$, helps to simplify the expression for the Bowen Ratio. The first term in brackets is a slowly varying function of temperature which for the conditions in our example has a value of 0.647 mb $°C^{-1}$ and the second term in brackets is simply the ratio of the temperature difference and the vapor pressure difference over the same height interval. In our example calculation, the Bowen Ratio was:

$$\beta = [0.647][(2 - 0)/(3.53 - 6.11)] = -0.50$$

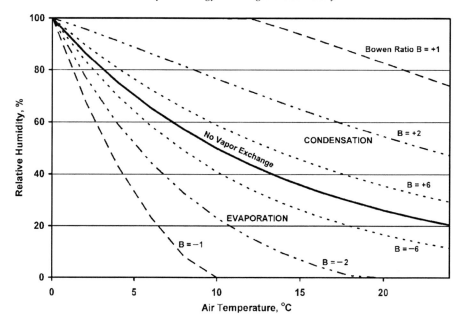

Figure 6.6 Regimes of latent heat exchange over melting snow defined by air temperature and relative humidity. The line of zero vapor pressure difference between the air and snow surface defines two regimes: one with high humidity and condensation energy gains on the snowpack surface and another with low humidity and evaporative heat loss from the snowpack surface. Temperature and humidity conditions defining varying Bowen Ratios ($\beta = Q_h/Q_e$) are also shown above (+) and below (−) the zero vapor exchange line. Computations assume a total atmospheric pressure of 1000 mb.

which indicates the sensible heat flux density was only half as large as that for latent heat exchange and of opposite sign. In other words, the sensible heat gain could only be responsible for about 50% of the latent heat loss due to evaporation/ sublimation.

As illustrated in Figure 6.6, Bowen Ratios can be simply predicted by air temperature and relative humidity conditions over melting snow, where surface temperature ($T = 0\,^{\circ}C$) and vapor pressure ($e_0 = 6.11$ mb) are fixed. Relative humidity is the ratio of actual vapor pressure (e) to saturation vapor pressure (e_s) at that temperature expressed as a percentage [$RH = 100(e/e_s)$]. Temperature and relative humidity conditions exist where evaporative heat losses exactly offset the sensible heat gains to the snowpack (Bowen Ratio = −1). Under such conditions, sensible heat and latent heat exchange offset each other and melt would be primarily dependent upon radiation exchange. Conditions also exist where both sensible and latent heat gains are positive and equal ($\beta = +1$) and energy supply for snowmelt is derived from both net radiation and convection. At any given air temperature, as relative humidity approaches the zero vapor exchange line in Figure 6.6 from either direction, vapor exchange approaches zero, sensible heat exchange becomes relatively important and the Bowen Ratio approaches infinitely large or small values.

When air temperature is less than snow surface temperature, only evaporation or sublimation of vapor from the snowpack can occur, since it is impossible for atmospheric vapor pressure to exceed snow surface vapor pressure even at 100% relative humidity. When air temperature is less than snow temperature, both vapor exchange and sensible heat exchange represent energy losses from the snowpack ($\beta > 0$); a condition common in winter. The specific meteorological conditions encountered and the balance between sensible heat and evaporative/sublimation energy exchange can obviously greatly affect the snowpack energy budget and rates of snowmelt.

The snowpack surface has been likened to a wet-bulb thermometer where latent heat losses are balanced by sensible heat gains (Geiger, 1965). A wet-bulb thermometer when exposed to air with a relative humidity less than 100% will evaporate/sublimate and cool below the current air temperature (dry-bulb temperature). Since cooling of the wet bulb creates a temperature difference between the wet-bulb surface and the air, the wet bulb will gain sensible heat from the surrounding air. When the wet bulb cools to a temperature where the latent heat lost by evaporation/sublimation is matched by the sensible heat gain from the air, an equilibrium exists and the wet-bulb temperature will have been reached.

The wet-bulb analogy with a snowpack surface is correct as long as there are no additional sources of energy to drive snowpack evaporation or sublimation. In the absence of significant radiant energy, such as in winter or at night, a snowpack exposed to relatively dry air can cool by sublimation and approach a balance between latent heat loss and sensible heat gain. Some investigators in areas with relatively dry air have noted the approximate balance between snowpack evaporation/sublimation energy losses and sensible heat gains which leads to a zero net turbulent energy exchange (Marks and Dozier, 1992). However, during periods with significant radiant energy supply, the snowpack surface does not cool below $0\,^\circ\text{C}$, melting will utilize some of the available energy, and the wet-bulb analogy fails.

6.6 Rainfall energy

Rainfall can influence the energy budget of snowpacks in three ways:

- sensible heat additions due to the heat added by a volume of relatively warm rain,
- release of the latent heat of fusion if rainfall freezes on a sub-zero snowpack, and
- condensation on the snowpack due to high humidity associated with rainy weather.

The first type of energy input that pertains when the snowpack is at $0\,^{\circ}$C can be simply computed as:

$$Q_{r1} = P_r\,\rho_w\,c_w\,(T_r - T_s) \qquad (6.24)$$

where:

Q_{r1} = sensible heat exchange due to rainfall on snow, W m^{-2}
P_r = rainfall intensity, m s^{-1}
ρ_w = density of liquid water, liquid water = 1×10^3 kg m^{-3}
c_w = specific heat of liquid water = 4.1876×10^3 J kg$^{-1}\,^{\circ}$C^{-1}
T_r = temperature of rain, $^{\circ}$C, (assumed equal to air temperature during the event)
T_s = temperature of snow, $0\,^{\circ}$C

The energy exchange shows that sensible heat inputs from rain are generally relatively modest, since this example is for a rather large and very warm rain storm. However, this computation is a bit misleading since the time base for rainfall is one day. If all that rain occurred in one hour, then the energy input due to the rain would be huge (e.g. 291 W m^{-2}) relative to other sources for that hour.

An example for a major rainfall event with an intensity of 2.5 cm depth in a day $(P_r = (0.025$ m$)(24$ h $\times 3600$ s h$^{-1})^{-1} = 2.894 \times 10^{-7}$ m s$^{-1})$ and a temperature of $10\,^{\circ}$C can help to show the importance of this energy input:

$$Q_{r1} = (2.89 \times 10^{-7}\text{ m s}^{-1})(1 \times 10^3\text{ kg m}^{-3})(4.1876 \times 10^3\text{ J kg}^{-1}\,^{\circ}\text{C}^{-1})(10 - 0\,^{\circ}\text{C})$$
$$= 12.1\text{ J s}^{-1}\text{ m}^{-2} = 12.1\text{ W m}^{-2}$$

A second possible source of energy input from rain is the release of latent heat of fusion when rain freezes on a subfreezing snowpack. The heat flux due to this component can be computed as:

$$Q_{r2} = P_r\,\rho_w\,L_f \quad T_s < 0 \qquad (6.25)$$

where:

Q_{r2} = flux density of latent heat released due to rainfall freezing, W m^{-2}
L_f = latent heat of fusion, J kg^{-1} (see Chapter 1).

Other terms have been previously defined.

Using the same rainfall intensity as in the above example $(P_r = 2.89 \times 10^{-7}$ m s$^{-1})$ and $L_f = 0.3275 \times 10^6$ J kg^{-1} representing snow temperature of about $-3\,^{\circ}$C, the latent heat flux due to refreezing is:

$$Q_{r2} = (2.89 \times 10^{-7}\text{ m s}^{-1})(1 \times 10^3\text{ kg m}^{-3})(0.3275 \times 10^6\text{ J kg}^{-1})$$
$$= 94.5\text{ J s}^{-1}\text{ m}^{-2} = 94.5\text{ W m}^{-2}$$

Rainfall onto a subzero snowpack would initially give up its sensible heat. If that was not sufficient to warm the snow to $0\,°C$, then the rain water would refreeze and release latent heat of fusion. Thus, the total energy released by a rainfall event for the given conditions would be 12 W m^{-2} due to sensible heating and 94.5 W m^{-2} due to latent heat release for a total of 106.6 W m^{-2}. Computation of this energy flux depends upon knowing the snowpack and rainfall temperatures which are not routinely available. Warming of the snowpack due to sensible and latent gains from rain would be detected in accounting for the heat storage changes in the snowpack discussed in Section 6.8.

Finally, high humidity associated with rainfall could also cause condensation/sublimation onto the snowpack which could add considerable energy. Condensation/sublimation is predicted with Equations (6.20) and (6.21). Each kilogram of vapor that sublimated onto the cold snowpack in our example ($-3\,°C$) would release 2.4943 MJ of energy, while each kilogram that refroze would release 0.3275 MJ. Thus, sublimation releases 7.6 more energy per unit mass of water added to the snowpack than freezing and can be a very important energy source due to rainy weather.

6.7 Ground heat conduction

Ground heat conduction to the base of the snowpack generally represents a very minor energy source for melt. Factors contributing to the relatively small contributions of soil heat conduction are the fact that soil in general is a poor conductor of heat and that the temperatures in the soil are often quite low due to the lack of solar warming beneath deep snowpacks. During the winter snow accumulation season, heat conducted from the soil can contribute to gradual ripening and slow melting of snowpack basal layers, but this contribution gradually diminishes into spring. When snowpack depth decreases to levels that allow transmission of shortwave radiation to the ground in the late spring, 20–25 cm depth based upon Equation (6.8), ground warming and heat conduction become relatively more important.

The heat conduction flux density to the base of the snowpack can be computed as:

$$Q_{\mathrm{g}} = k_{\mathrm{g}}\,dT_{\mathrm{g}}/dz \cong k_{\mathrm{g}}(T_{\mathrm{g}} - T_{\mathrm{sb}})/(z_2 - z_1) \tag{6.26}$$

where:

$Q_{\mathrm{g}} =$ soil heat conduction flux density, W m^{-2}
$k_{\mathrm{g}} =$ thermal conductivity of soil, W m^{-1} °C^{-1}
$z =$ soil depth, m
$T_{\mathrm{g}} =$ soil temperature at depth z_2, °C
$T_{\mathrm{sb}} =$ temperature at base of snowpack at depth $z_1 = 0$ m, °C

An example of computation for ground heat conduction using Equation (6.26) will illustrate the basic units and magnitudes. Assuming a soil temperature of 1 °C at a soil depth of 0.5 m, 0 °C temperature at the snowpack base, and a soil thermal conductivity of 2 W m^{-1} °C^{-1} gives:

$$Q_g = (2 \text{ W m}^{-1} \text{ °C}^{-1})(1 - 0 \text{ °C})(0.5 - 0 \text{ m})^{-1} = 4 \text{ W m}^{-2}$$

This equation was simplified to the expression on the right-hand side by assuming thermal conductivity was constant with depth and integrating. The equation was also written without a negative sign, customarily used to indicate that heat flow is toward lower temperatures, to avoid confusion about whether Q_g is adding energy to or subtracting energy from the snowpack. Thermal conductivity for soil varies widely with moisture and organic matter content and ranges from about 0.2 to 2 W m^{-1} °C^{-1} (Oke, 1987). Temperature at the base of the snowpack is often around 0 °C and soil beneath snow gradually approaches this temperature during the winter-spring (US Army Corps of Engineers, 1956; Marks et al., 1992).

Obviously soil heat conduction flux density is small compared with other energy budget components discussed so far. Soil heat conduction to the snowpack based upon measurements or calculations from field data ranging from about 0 to 10 W m^{-2} have been reported (Cline, 1997; Marks and Dozier, 1992; Male and Gray, 1981), so the calculation above is within the range of observed data. Granger and Male (1978) found that for relatively shallow snowpacks on the Canadian prairie, soil heat conduction varied diurnally and lagged solar radiation when the snowpack was dry, but heat conduction in the soil was relatively constant beneath melting snow. Given the small value of Q_g some investigators have assumed Q_g was negligible or assumed a small positive constant value for heat flow to the snowpack base. Others have employed a model to compute soil heat conduction where supporting snow and soil temperature data were available (Marks and Dozier, 1992; Cline, 1997).

6.8 Internal snowpack energy storage

Changes in the amount of sensible and latent heat stored in the snowpack (Q_i) can at times represent a significant snowpack energy budget component. Sensible heat changes are represented by changes in snowpack temperature and mass over time integrated over the depth of the snowpack. Mass increases due to subfreezing snow falling on the snowpack can also significantly add to snowpack cold content that must be satisfied before liquid water is yielded at the base of the snowpack. Warming of deep, dense snowpacks to 0 °C can represent a significant energy budget component in winter when other energy budget components are quite small. Once the snowpack is isothermal at 0 °C and melting begins, the importance of this energy budget component diminishes drastically. However, even after general snowpack ripening has occurred in spring, surface layers of the snowpack at exposed sites can cool and refreeze overnight creating small energy deficits that must be satisfied before melt can occur on the following day.

Latent heat of fusion storage changes within the snowpack are seldom considered, but would be represented by depth integrated changes in the amount of liquid water derived from melt within the snowpack. If surface melting occurs and the liquid water does not drain completely from the snowpack and does not refreeze, then changes in liquid-water content would represent a form of latent heat of fusion storage. If water from melt at the surface refreezes deeper within the snowpack releasing the latent heat of fusion, then no net energy exchange between the snowpack and its environment has taken place. Of course, if liquid water due to melting drains from the snowpack, that represents a loss of latent heat of fusion energy from the snowpack. Estimation of latent heat of fusion storage changes would require detailed evaluations of snowpack liquid-water content. Only sensible heat storage changes are described quantitatively below.

The general relationship to be used to assess the sensible heat change in snowpack energy storage is:

$$\Delta Q_i = \int_{z=0}^{z=d} (\rho_s \, c_i \, \delta T_s / \delta t) \, dz \tag{6.27}$$

where:

ΔQ_i = change in internal snowpack energy storage, W m^{-2}
ρ_s = snowpack density, kg m^{-3}
c_i = specific heat of ice, J kg^{-1} $^\circ$C^{-1}
T_s = snowpack temperature, $^\circ$C
t = time, s
z = height above ground, $z = 0$ at ground, $z = d$ at snowpack surface, m
d = snowpack depth, m

The integration is performed from ground level to the snowpack surface and requires that measurements or model predictions of snowpack depth, density, and temperature profiles over time are available.

The computation of Q_i for a hypothetical winter day when the snowpack is not yet ripe is given in Table 6.4 for a snowpack with three layers. It is assumed that density and depth of the layers are constant between days although precipitation and sublimation could cause mass changes in a real case. The change in energy storage for each layer then becomes:

$$\Delta Q_i = (\rho_s \, c_i)(\Delta T_s / \Delta t)\Delta z$$

Table 6.4 *Computation of daily change in internal snowpack energy storage*

Depth Interval (m)	Δz (m)	c_i (J kg^{-1} °C^{-1})	ρ_s (kg m^{-3})	ΔT_s (°C)	Δt (s)	ΔQ_i (W m^{-2})
1.5–1.2	0.3	2100	120	3.5	86,400	3.06
1.2–0.5	0.7	2100	150	2	86,400	5.10
0.5–0.0	0.5	2100	200	1	86,400	2.40
sum						10.56

During longer time intervals, the assumption of constant snowpack density and depth cannot be made due to gradual snowpack metamorphosis and addition of new precipitation.

Using the specific heat of ice as $c_i = 2.1 \times 10^3$ J kg^{-1} °C^{-1} and a time step of 86 400 s d^{-1}, the change in energy storage for the uppermost layer shown in Table 6.4 is computed as:

$$\Delta Q_i = -(120 \text{ kg m}^{-3})(2.1 \times 10^3 \text{ J kg}^{-1}\text{ °C}^{-1})(+3.5 \text{ °C}/86\,400 \text{ s})(0.3 \text{ m})$$
$$= -3.06 \text{ J s}^{-1}\text{ m}^{-2} = 3.06 \text{ W m}^{-2}$$

The change in energy storage for other layers, computed in similar manner, and the sum for all layers for the day (10.56 W m^{-2}) is relatively small compared to radiation and convective energy exchange considered above. Regardless, such internal energy storage changes could be significant for days when other energy budget components are small. Note that the middle layer of snowpack in Table 6.4 had the greatest mass and the largest change in energy storage, even though the temperature change was less than in the surface layer. Obviously the depth, density, and temperature changes occurring in the snowpack will dictate the relative importance of this energy exchange component.

6.9 Melt energy

Melt energy (Q_m) represents a loss of latent heat of fusion when liquid water drains from the snowpack. In snowmelt prediction schemes, Q_m is generally solved for as a residual in Equation (6.1). Once the sum of energy available for melt is obtained, the mass flux density of meltwater (M) can be computed as:

$$M = Q_m(\rho_w \, L_f \, B)^{-1} \tag{6.28}$$

where:

M = melt rate, m s^{-1}

Q_m = flux density of melt energy, J s^{-1} m^{-2} or W m^{-2}

ρ_w = density of liquid water, kg m^{-3}

L_f = latent heat of fusion, J kg^{-1}

B = thermal quality of the snowpack, dimensionless

Thermal quality of snow (B) has been previously described in Chapter 3 and for ripe snow B represents the fraction of the snowpack mass that is ice rather than liquid water. For example, a snowpack liquid-water content of 0.03 on a mass basis is often assumed for melting snow giving a thermal quality of 0.97. In other words, 3% more liquid water will be yielded during melting than expected from the available melt energy (Q_m) due to the fraction of snowpack mass that is already liquid water.

If the algebraic sum of average flux density energy available for melt for a day (Q_m) was 25.2 W m^{-2} (= 25.2 J s^{-1} m^{-2}) with $B = 0.97$, $\rho_w = 1 \times 10^3$ kg m^{-3}, and $L_f = 0.334 \times 10^6$ J kg^{-1} at 0 °C (see Appendix A1), then the melt rate would be:

$$M = (25.2 \text{ J s}^{-1} \text{ m}^{-2})[(1 \times 10^3 \text{ kg m}^{-3})(0.334 \times 10^6 \text{ J kg}^{-1})(0.97)]^{-1}$$
$$= 7.78 \times 10^{-8} \text{ m s}^{-1}(10^2 \text{ cm m}^{-1})(8.64 \times 10^4 \text{ s d}^{-1}) = 0.67 \text{ cm d}^{-1}$$

Melt rate can be considered equivalent to outflow of liquid water from the snowpack as long as the transmission time in the snowpack is less than the time step for the computation. For example, with a daily time step and shallow ripe snow, lag times for meltwater outflow can generally be ignored. Transmission and storage of liquid water in snowpacks was also discussed in Chapter 3.

The situation is more complicated for subfreezing snowpacks. In subfreezing snowpacks energy must first be used to warm the snowpack to an isothermal 0 °C condition and satisfy the liquid-water holding capacity of the snow. A liquid-water holding capacity for snow of 3% on a mass basis is generally assumed necessary before liquid water can drain from the snowpack. If the cold content of the snowpack (see Chapter 3) is greater than the energy supply available, then the snowpack is only partially ripened and no melt would occur. If the available energy is greater than the cold content, energy will be first used to warm the snow to 0 °C and then to raise the liquid-water content to 3% on a mass basis. Deep, cold winter snowpacks may require days to weeks to completely ripen, while shallow fresh snowpacks may ripen and completely melt in one day or less. Accounting for changes in internal snowpack energy storage (ΔQ_i) during the ripening process is one of the challenges to accurate modelling of streamflow from snowmelt.

6.10 Energy budget examples and applications

The energy budget of seasonal snowpacks and glaciers has been studied many times generally with the objective of computing rates of snowmelt and/or evaporation/sublimation losses. Collectively the results of these studies have verified the utility of the energy budget approach. A dominant theme of these studies is that net allwave radiation is the major source of energy transported to the snowpack both in the open and forest. Male and Granger (1981) compared results of 15 studies that showed net allwave radiation explained about 59% of total energy exchange on average. The importance of energy exchange due to convection of sensible heat and latent heat of evaporation/sublimation is generally less; representing 35% and −10% of total energy exchange, respectively, in the studies reviewed by Male and Granger (1981). Obviously, the greatest emphasis should be placed on measuring or estimating net allwave radiation exchange. However, variations in the significance of energy budget components occur due to the influence of vegetation, differences in regional climates, and changing climatic patterns in the transition from winter accumulation to spring and summer melt seasons.

Marks and Dozier (1992) showed how snowpack energy exchange varied during the seasons from winter accumulation to summer melt in a Sierra Nevada alpine basin (Figure 6.7). Basically, net longwave radiation exchange and latent heat exchange represented energy losses from the snowpack throughout the November–July period, while net shortwave radiation exchange and sensible heat convection represented energy gains. Net shortwave energy gains in winter were low due to high snowpack albedo, large zenith angles, and short day lengths. In spring and early summer, day length increased and albedo and zenith angles decreased which caused much greater net shortwave radiation energy gains. The net longwave balance was negative in all months due to snowpack exposure to the relatively clear, dry, and cold alpine atmosphere and the relatively warm snowpack once melting began. Turbulent exchange of sensible heat represented a small positive energy supply even in winter and, as the air warmed to average melt-season temperatures of 10 °C, increased to levels equal to or slightly greater than net shortwave radiation exchange. Turbulent exchange of latent heat due to sublimation and evaporation was negative from November to July due to relatively low atmospheric vapor pressures (3.5–4 mb) and high wind speeds at the site (Marks *et al.*, 1992). Thus, the major positive inputs of energy to the snowpack were net shortwave radiation exchange and sensible heat convection balanced against losses due to net longwave exchange and latent heat convection.

Small amounts of energy were added at the snowpack base by soil heat conduction (not shown), but these represented an average monthly flux density of only 7 W m^{-2} in November and diminished to 2 W m^{-2} in July. Precipitation energy

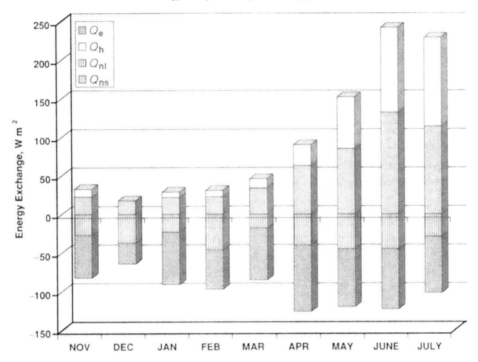

Figure 6.7 Seasonal variation in major snowpack energy exchange components at a Sierra Nevada alpine ridge site in 1986 (data from Marks and Dozier, 1992). (Q_{ns} = net solar radiation exchange, Q_{nl} = net longwave radiation exchange, Q_e = latent heat convection, and Q_h = sensible heat convection).

inputs were zero in all but two months and reached a maximum of only 2 W m^{-2} in February. Net snowpack energy supply in May, June, and July was positive and computed total melt for these months, including small energy inputs due to soil heat conduction, were 26, 94, and 102 cm, respectively. Computed snowpack sublimation and evaporation losses from November to July varied from 2.5 to 8.3 cm of water per month with the peak loss occurring in April. The results of this study clearly show how seasonal changes can affect the snowpack energy balance.

Many other studies of snowpack energy exchange have been conducted since the Male and Granger (1981) review. Several studies have illustrated snowpack energy balances at alpine and high-latitude open sites (Cline, 1995 and 1997; Hood et al., 1999; Pluss and Mazzoni, 1994; Ohno et al., 1992; Hong et al., 1992; Harding and Lloyd, 1998). Recent advances in GIS methods have also lead to the development of distributed models of snowpack energy exchange and snowmelt including, in some cases, the combined effects of forest cover and topography (Link and Marks, 1999; Marks et al., 1999; Tarboton et al., 1995).

One ultimate goal in snow hydrology is to couple energy-budget models with hydrologic models to predict streamflow from snowmelt. Although prediction of streamflow from snowmelt is a subject of a later chapter, mention of a few studies is made here. Anderson's (1968, 1973, 1976) classic work that combined an energy-budget approach to prediction of snowmelt with the US National Weather System River Forecast System was an early successful attempt at such synergy. Kustas *et al.* (1994) developed an energy-budget model that was combined with the Snowmelt Runoff Model (Martinec *et al.*, 1983) and then used to simulate snowmelt runoff from a lysimeter and watershed. Other models using snowpack energy budgets to compute streamflow from snowmelt have been developed by Ujihashi *et al.* (1994), Kuchment and Gelfan (1996) and Ishii and Fukushima (1994). Data availability is a chief limitation to employing the full energy-budget approach in a model representing a complete watershed, but in the final analysis any simplifications must be based upon knowledge of the physics of melt and evaporation/sublimation. In the following chapter, specific effects of topography and forest cover on snowpack energy exchange are given.

6.11 References

Anderson, E. A. (1968). Development and testing of snow pack energy balance equation. *Water Resour. Res.*, **4**(1), 19–37.

Anderson, E. A. (1973). *National Weather Service River Forecast System – Snow Accumulation and Ablation Model.* NOAA Tech. Memo, NWS HYDRO-17. US Dept. Commerce, National Oceanic Atmos. Admin.

Anderson, E. A. (1976). *A Point Energy and Mass Balance Model of a Snow Cover.* NOAA Tech. Rpt., NWS 19. US Dept. Commerce, National Oceanic Atmos. Admin.

Arya, S. P. S. (1988). *Introduction to Micrometeorology*, vol. 42. Internat. Geophysics Series. San Diego, CA: Academic Press.

Bellocchi, G., Acutis, M., Fila, G. and Donatelli, M. (2002). An indicator of solar radiation model performance based on a fuzzy expert system. *Agron. J.*, **94**, 1222–33.

Bristow, K. L. and Campbell, G. S. (1984). On the relationship between incoming solar radiation and daily maximum and minimum temperature. *Agric. and Forest Meteorol.*, **31**, 159–66.

Brutsaert, W. (1975). On a derivable formula for long-wave radiation from clear skies. *Water Resour. Res.*, **11**, 742–4.

Brutsaert, W. (1982). *Evaporation into the Atmosphere: Theory, History, and Applications.* Dordrecht: D. Reidel Publ. Co.

Cline, D. (1995). Snow surface energy exchanges and snowmelt at a continental alpine site. In *Biogeochemistry of Seasonally Snow-Covered Catchments*, Publ. No. 228, ed. K. A. Tonnessen, M. W. Williams, and M. Tranter. International Association of Hydrological Sciences, pp. 157–66.

Cline, D. W. (1997). Snow surface energy exchanges and snowmelt at a continental, midlatitude alpine site. *Water Resour. Res.*, **33**(4), 689–701.

DeWalle, D. R. and Meiman, J. R. (1971). Energy exchange and late season snowmelt in a small opening in Colorado subalpine forest. *Water Resour. Res.*, **7**(1), 184–8.

Donatelli, M., Bellocchi, G., and Fontana, F. (2003). RadEst 3.00: Software to estimate daily radiation data from commonly available meteorological variables. *Eur. J. Agron.*, **18**, 363–7.

Frank, E. C. and Lee, R. (1966). Potential solar beam irradiation on slopes: tables for 30° to 50° latitude, Res. Paper RM-18. Rocky Mtn. For. Range Exp. Sta.: US Dept. Agric., Forest Service.

Geiger, R. (1965). *The Climate Near the Ground*, Revised Edition. Cambridge, MA: Harvard University Press.

Granger, R. J. and D. H. Male. (1978). Melting of a prarie snowpack. *J. Appl. Meteorol.*, **17**, 1833–42.

Harding, R. J. and Lloyd, C. R. (1998). Fluxes of water and energy from three high latitude tundra sites in Svalbard. *Nordic Hydrol.*, **29**(4/5), 267–84.

Hogstrom, U. (1988). Non-dimensional wind and temperature profiles in the atmospheric surface layer: a re-evaluation. *Boundary-Layer Meteorol.*, **42**, 55–78.

Hong, M., Zongchao, L., and Yifeng, L. (1992). Energy balance of a snow cover and simulation of snowmelt in the western Tien Shan mountains, China. *Annals of Glaciol.*, **16**, 73–8.

Hood, E., Williams, M., and Cline, D. (1999). Sublimation from a seasonal snowpack at a continental mid-latitude alpine site. *Hydrol. Process.*, **13**, 1781–97.

Ishii, T. and Fukushima, Y. (1994). Effects of forest coverage on snowmelt runoff. In *Snow and Ice Covers: Interactions with the Atmosphere and Ecosystems*, IAHS Publ. No. 223, ed. H. G. Jones, T. D. Davies, A. Ohmura, and E. M. Morris. International Association of Hydrological Sciences, pp. 237–45.

Koh, G. and Jordan, R. (1995). Sub-surface melting in a seasonal snow cover. *J. Glaciol.*, **41**(139), 474–82.

Kondrat'yev, K. Y. (1965). *Radiative Heat Exchange in the Atmosphere*. Oxford: Pergamon Press.

Kuchment, L. S. and Gelfan, A. N. (1996). The determination of the snowmelt rate and the meltwater outflow from a snowpack for modelling river runoff generation. *J. Hydrol.*, **179**, 23–36.

Kustas, W. P., Rango, A., and Uijlenhoet, R. (1994). A simple energy budget algorithm for the snowmelt runoff model. *Water Resour. Res.*, **30**(5), 1515–27.

Kuz'min, P. P. (1961). *Protsess tayaniya shezhnogo pokrova (Melting of Snow Cover)*. *Glavnoe Upravlenie Gidrometeorologicheskoi Sluzhby Pri Sovete Ministrov SSSR* Gosudarstvennyi Gidrologicheskii Institut. Main Admin. Hydrometeorol. Service, USSR Council Ministers, State Hydrol. Institute. Translated by Israel Program for Scientific Translations. Avail from US Dept. Commerce, National Tech. Inform. Service, 1971, TT 71–50095.

Lee, R. (1963). *Evaluation of Solar Beam Irradiation as a Climatic Parameter of Mountain Watersheds*, vol. 2. Ft. Collin, CO: Colorado State University.

Leydecker, A. and Melack, J. M. (1999). Evaporation from snow in the Central Sierra Nevada of California. *Nordic Hydrol.*, **30**(2), 81–108.

Linacre, E. (1992). *Climate Data and Resources*. London: Routledge.

Link, T. and Marks, D. (1999). Distributed simulation of snowcover mass- and energy-balance in the boreal forest. *Hydrol. Processes*, **13**, 2439–2452.

Liston, G. E. (1995). Local advection of momentum, heat, and moisture during the melt of patchy snow covers. *J. Appl. Meterol.*, **34**, 1705–15.

Male, D. H. and Granger, R. J. (1981). Snow surface energy exchange. *Water Resour. Res.*, **17**(3), 609–27.

Male, D. H. and Gray, D. M. (1981). Chapter 9. Snowcover ablation and runoff. In *Handbook of Snow: Principles, Processes, Management and Use*, ed. D. M. Gray and D. H. Male. Toronto: Pergamon Press, pp. 360–436.

Marks, D. and Dozier, J. (1992). Climate and energy exchange at the snow surface in the alpine region of the Sierra Nevada, 2. snow cover energy balance. *Water Resour. Res.*, **28**(11), 3043–54.

Marks, D., Dozier, J., and Davis, R. E. (1992). Climate and energy exchange at the snow surface in the alpine region of the Sierra Nevada, 1. metrological measurements and monitoring. *Water Resour. Res.*, **28**(11), 3029–42.

Marks, D., Domingo, J., Susong, D., Link, T., and Garen, D. (1999). A spatially distributed energy balance snowmelt model for application in mountain basins. *Hydrol. Processes*, **13**, 1935–59.

Martinec, J., Rango, A., and Major, E. (1983). *The Snowmelt-Runoff Model (SRM) User's Model*, NASA Ref. Publ. 1100. US National Aeronautics and Space Administration.

Moore, R. D. (1983). On the use of bulk aerodynamic formulae over melting snow. *Nordic Hydrol.*, **14**(4), 193–206.

Morris, E. M. (1989). Turbulent transfer over snow and ice. *J. Hydrol.*, **105**, 205–23.

Nakawo, M., Hayakawa, N., and Goodrich, L. E. (eds.) (1998). Chapter 6: Heat budget of a snow pack. In *Snow and Ice Science in Hydrology*, Intern. Hydrol. Prog., 7th Training Course on Snow Hydrology. Nagoya Univ. and UNESCO, pp. 69–87.

Neumann, N. and Marsh, P. (1997). Local advection of sensible heat during snowmelt. In *Proceedings of the 54th Annual Meeting Western Snow Conference*, Banff, Alberta, Canada May 4–8, 1997, pp. 175–85.

Neumann, N. and Marsh, P. (1998). Local advection of sensible heat in the snowmelt landscape of Arctic tundra. *Hydrol. Processes*, **12**, 1547–60.

Oke, T. R. (1987). *Boundary Layer Climates*, 2nd edn. London: Methuen, Inc.

Ohno, H., Ohata, T., and Higuchi, K. (1992). The influence of humidity on the ablation of continental-type glaciers. *Annals Glaciol.*, **16**, 107–14.

Paeschke, W. (1937). Experimentelle untersuchungen zum rauhigkeits- und stabilitatsproblem in der bodennahen luftschicht. *Beitrage z. Phys. d. freien Atmos.*, **24**, 163–89.

Pluss, C. and Mazzoni, R. (1994). The role of turbulent heat fluxes in the energy budget of high alpine snow cover. *Nordic Hydrol.*, **25**(1994), 25–38.

Rohrer, M. B. and Braun, L. N. (1994). Long-term records of snow cover water equivalent in the Swiss Alps, 2. Simulation. *Nordic Hydrol.*, **25**, 65–78.

Schmidt, R. A., C. A. Troendle, and J. R. Meiman. (1998). Sublimation of snowpacks in subalpine conifer forests. *Can. J. For. Res.*, **28**, 501–13.

Shutov, V. A. (1993). Calculation of snowmelt. *Russian Meteorol. Hydrol.*, **4**, 14–20.

Shutov, V. A. (1997). Radiation factors of snowmelt. *Russian Meteorol. Hydrol.*, **9**, 60–7.

Singh, P. and Singh, V. P. (2001). *Snow and Glacier Hydrology*. Dordrecht: Kluwer Academic Publishers.

Tarboton, D. G., Chowdhury, T. G., and Jackson, T. H. (1995). A spatially distributed energy balance snowmelt model. In *Biogeochemistry of Seasonally Snow-Covered Catchments*, IAHS Publ. No. 228, ed. K. A. Tonnessen, M. W. Williams, and M. Tranter. International Association of Hydrological Sciences, Proc. Boulder Symp., pp. 141–55.

Thornton, P. E. and Running, S. W. (1999). An improved algorithm for estimating incident daily solar radiation from measurements of temperature, humidity and precipitation. *Agric. and Forest Meteorol.*, **93**(4), 211–28.

Ujihashi, Y., Takase, N., Ishida, H. and Hibobe, E. (1994). Distributed snow cover model for a mountainous basin. In *Snow and Ice Covers: Interactions with the Atmosphere and Ecosystems*, IAHS Publ. No. 223, ed. H. G. Jones, T. D. Davies, A. Ohmura, and E. M. Morris. International Association of Hydrological Sciences, pp. 153–62.

US Army Corps of Engineers. (1956). *Snow Hydrology, Summary Report of the Snow Investigations*. Portland, OR: US Army, Corps of Engineers, N. Pacific Div.

Warren, S. G. (1982). Optical properties of snow. *Rev. Geophys. Space Physics*, **20**(1), 67–89.

Weiss, A., Hays, C. J., Hu, Q. and Easterling, W. E. (1993). Incorporating bias error in calculating solar irradiance. *Agron. J.*, **93**, 1321–6.

Winther, J. G. (1993). Short- and long-term variability of snow albedo. *Nordic Hydrol.*, **24**, 199–212.

7

Snowpack energy exchange: topographic and forest effects

7.1 Introduction

Recent innovations in computing techniques and the need for modelling snow-pack energy exchange processes for large diverse basins have lead to consideration of topographic and forest effects. Both topography and forests can have a major impact on snowpack radiation energy exchange. Effects on convective exchange are less well known. Methods for adjusting the basic theory of snowpack energy exchange for the effects of topography and forests are considered in this chapter. Topographic effects on precipitation and snow accumulation and forest effects on snow interception were previously discussed in Chapter 3.

7.2 Topographic influences

Topography controls the elevation, slope and aspect, and exposure of snowpack surfaces that can significantly influence energy exchange and melt across a watershed. Since many watersheds with significant snowpacks are quite mountainous, correction for the effects of topography becomes important when trying to develop spatially accurate models of snowmelt-runoff processes. Slope inclination and aspect angles modify the exchange of direct-beam and diffuse shortwave radiation and longwave radiation. Topographic relief can also lead to shading of surfaces by adjacent terrain. Precipitation amounts and types, wind speeds, temperatures, and humidities that are used in computing the energy budget for snowpack surfaces are also influenced by elevation and slope orientation. Snowpack exposure to prevailing winds and typical storm patterns can be influenced by topography and can greatly affect snowpack energy exchange. Lapse rates can be used to adjust temperature data for effects of elevation, but corrections for topographic effects on other meteorological parameters important to energy exchange are more problematical. Ideally, a network of monitoring sites would be available to provide representative meteorological data for landscape types within large watersheds.

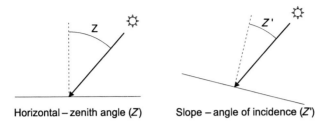

Horizontal – zenith angle (Z)   Slope – angle of incidence (Z')

Figure 7.1 Relationship between zenith angle on horizontal surface (Z) and angle
of incidence on sloping surface (Z').

7.2.1 Shortwave radiation on slopes

Shortwave radiation received by a snowpack on a slope can be broken into two
components: a direct-beam component and a diffuse component. The direct-beam
component varies from that on the horizontal because beam radiation on the slope is
received at a different angle of incidence than a horizontal surface (see Figure 7.1).
Diffuse shortwave radiation on a slope differs from that on a horizontal surface
because the slope is tilted towards surrounding terrain and the view from the slope
to the atmosphere is restricted. Thus, a slope receives a portion of diffuse shortwave
radiation that is reflected from the surrounding terrain and a portion of diffuse
radiation scattered from the atmosphere overhead.

The incoming flux density of shortwave radiation on a slope can be computed
as:

$$K_s\downarrow = I_b \cos Z' + D_s \qquad (7.1)$$

where:

$K_s\downarrow$ = incoming shortwave radiation flux density on the slope
I_b = incoming direct-beam shortwave radiation measured at normal incidence
Z' = angle between solar beam and a perpendicular to the slope
D_s = incoming diffuse shortwave radiation flux density to slope

Net shortwave (Q'_{ns}) radiation exchange on a slope, following earlier derivations in
Chapter 6, can be computed as:

$$Q'_{ns} = K_s\downarrow - \alpha K_s\downarrow = (1 - \alpha)K_s\downarrow \qquad (7.2)$$

where the snowpack albedo (α) remains the same as for the horizontal case. Derivations of the direct-beam and diffuse shortwave components are given separately
below.

Direct-beam shortwave radiation on slopes

Potential solar irradiation theory can be used to illustrate the patterns of beam
radiation received on slopes with varying slope steepness and aspect (Appendix B).

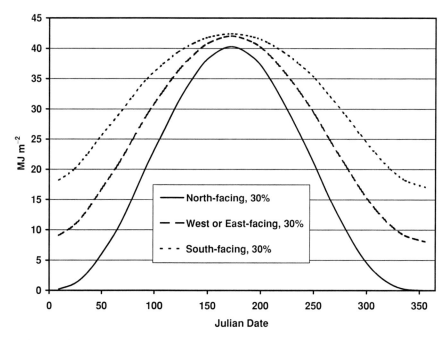

Figure 7.2 Daily potential solar irradiation for 30% slopes facing N, S and E–W at various times of the year and a latitude of 50 °N (Frank and Lee, 1966; see Appendix B).

Tables given by Frank and Lee (1966) show potential solar beam radiation for a wide range of slope steepness and slope aspects for latitudes varying from 30° to 50°. Unfortunately, too many possible combinations of slope steepness, aspect, time of year, and latitude exist to show solutions graphically, but potential solar irradiation of slopes for a latitude = 50° and three different slope orientations (N, S, and E–W) for a slope steepness of 30% (k_s = arctan (30/100) = 16.7°) are shown in Figure 7.2 for illustrative purposes. Computed daily potential solar irradiation received on E- and W-facing slope are the same, only with times of sunrise and sunset reversed. In summer, relatively small differences in potential solar irradiation exist at higher latitudes like 50° N. The largest differences in incidence angle and potential solar irradiation among slope orientations occur in winter which generally is a time of relative importance to snow hydrologists.

   These patterns of potential irradiation apply strictly to beam radiation and mimic conditions likely to occur when beam radiation is dominant such as during clear weather. However, in climates with cloudy weather, such as coastal regions, or at high-latitude sites with relatively large zenith angles, diffuse radiation may dominate and factors discussed in the section on "diffuse shortwave radiation on slopes" will become relatively more important. Understanding actual solar radiation received on slopes must begin with an understanding of how atmospheric conditions affect diffuse and beam radiation fraction.

An empirical relationship given by Linacre (1992) inter-relates the daily diffuse shortwave flux density on a horizontal surface (D), the total incoming shortwave flux density on the horizontal ($K\downarrow$) and the daily potential solar radiation on a horizontal surface (I_q) as:

$$D/K\downarrow = 1 - 1.2(K\downarrow/I_q) + 0.13(K\downarrow/I_q)^2 \tag{7.3}$$

and can be used to compute the fraction of incoming solar radiation that is diffuse. This relation was derived by Linacre (1992) from several empirical studies over the range of $K\downarrow/I_q$ from 0.3 to 0.7 and should be used with caution. Other approaches to obtaining instantaneous diffuse and direct-beam shortwave radiation for clear and cloudy conditions that involve numerical integration to obtain daily totals have been employed (Kustas *et al.*, 1994). This relationship suggests that with only 30% of potential solar radiation reaching the ground ($K\downarrow/I_q = 0.30$), which is equivalent to heavy cloud cover, the diffuse fraction of incoming shortwave radiation ($D/K\downarrow$) is 0.65. Thus, by difference, 35% is direct-beam shortwave radiation. In contrast, for mostly clear skies when incoming shortwave is 70% of potential ($K\downarrow/I_q = 0.70$), Equation (7.3) indicates that about 22% is diffuse radiation and 78% is direct beam.

The direct-beam shortwave radiation received on a slope depends upon the flux density of beam radiation at normal incidence (I_b) being transmitted through the atmosphere and the angle of incidence Z' of the beam radiation on the slope, as described in Appendix B (see Equation (7.1)). The angle of incidence can be computed either for an instant in time or for an entire solar day or part thereof. Beam radiation can either be measured with radiometers (see Chapter 6 and Figure 6.1) at a reference station or approximated using empirical relationships and potential solar irradiation theory. Given the fraction $D/K\downarrow$ using Equation (7.3) and an independent measurement of $K\downarrow$ on a horizontal surface from radiometer measurements or other computations (see Chapter 6), I_b can be found and used to compute beam radiation received by a variety of slope orientations.

To illustrate atmospheric effects on diffuse and direct beam radiation and I_b, we can complete our example calculations given in Chapter 6 for solar radiation received at a latitude of 55° N on Feb 7. For an assumed cloud cover fraction of 0.3, the estimated global radiation flux density was 78.7 W m^{-2} for a potential solar flux density of 111 W m^{-2} on a horizontal surface (Figure 6.3). By Equation (7.3) above, the diffuse fraction would be:

$$D/K\downarrow = 1 - 1.2(78.7/111) + 0.13(78.7/111)^2 = 0.214$$

Thus, 21.4% of the 78.7 W m^{-2} of estimated incoming shortwave radiation on this example day, or 16.8 W m^{-2}, would be diffuse shortwave radiation scattered by the atmosphere and by difference the direct-beam flux density would be $78.7 - 16.8 = 61.9$ W m^{-2}. Using the average zenith angle for the horizontal surface on this day of $77.5°$, derived using potential solar irradiation theory in Appendix B, Section B.1, the daily total beam radiation at normal incidence would be:

$$I_b = (61.9 \text{ W m}^{-2})/(\cos 77.5°) = 286 \text{ W m}^{-2}$$

This value for I_b then can be used to compute direct-beam radiation received by slopes ($I_b \cos Z'$) with other incidence angles, computed using potential solar radiation theory as described in Appendix B.

As a further illustration of the use of $I_b = 286$ W m^{-2} for a sloping surface, an example of beam radiation calculation can be given for a SE aspect slope with an inclination angle of $30°$ for the same day. For February 7 at latitude $55°$ N, the effective angle of incidence for beam radiation during the solar day on this slope would be $Z' = 51°$ (see Appendix section B.2.2). Thus, the beam radiation received by the slope on this day would be approximately:

$$I_b \cos Z' = (286 \text{ W m}^{-2})(\cos 51°) = 180 \text{ W m}^{-2}$$

which is much greater than the 61.9 W m^{-2} received by the horizontal surface on the same day. These computations are based upon an empirical relationship used to obtain $D/K{\downarrow}$ and thus are only approximate. Ideally, beam solar radiation can be routinely measured with a pyrheliometer or obtained by difference between measurements of $K{\downarrow}$ and D with unshaded and shaded pyranometers, respectively (see Chapter 6; Figure 6.1).

Computation of the length of the solar day for sloping surfaces does not account for reductions in receipt of direct-beam shortwave radiation due to shading by surrounding topography. Shading by surrounding topography on steep mountain basins or circumpolar basins with low solar altitudes can change the times that direct-beam shortwave radiation can actually be received on a particular watershed subunit between times of theoretical sunrise and sunset computed above. Digital terrain models can be used to compare solar altitude for all solar azimuths between theoretical sunrise and sunset with the altitude angle to surrounding topography for those same solar azimuths (Dozier and Frew, 1990). Solar azimuth can be computed for any time of day using formulas given in many references (List, 1968; Oke, 1987). Once actual sunrise and sunset times corrected for topographic shading are obtained for a watershed subunit, the potential solar equations can be integrated as shown in Appendix B to obtain corrected mean angle of incidence and potential solar irradiation.

An illustration of the potential effects of topographic shading on receipt of potential solar irradiation is given in Figure 7.3. The reduction in potential solar irradiation

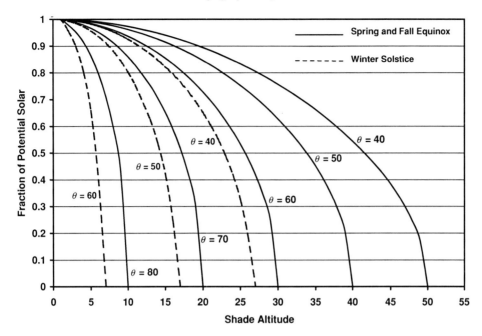

Figure 7.3 Reduction in daily potential solar irradiation received by a horizontal surface due to simulated topographic shading at varying shade altitudes. Curves are given for several latitudes (θ) at times of the winter solstice and the spring and fall equinoxes.

that would reach a horizontal surface due to shading by surrounding topography is given for several latitudes and times of the equinoxes and winter solstice and a range of shade altitudes. The computations assume symmetry of topographic shading around the horizon throughout each day. Clearly, potential solar irradiation and thus beam radiation receipt can be greatly reduced or eliminated by topographic shading for commonly occurring shade altitudes. On high-latitude watersheds or on steep mountain basins, some snowpack surfaces may receive only diffuse solar radiation due to topographic shading.

Diffuse shortwave radiation on slopes

Diffuse shortwave radiation flux density incident on a slope (D_s) is derived from shortwave radiation scattered from the atmosphere and reflected to the slope from the surrounding terrain. The flux density can be approximately computed as:

$$D_s = D(\cos^2 k_s/2) + (1 - \cos^2 k_s/2)\alpha_t\ K_t{\downarrow} \tag{7.4}$$

where:

D_s = flux density of diffuse shortwave radiation on the slope
D = flux density of diffuse shortwave radiation on horizontal surface

Table 7.1 *View factors from slope to sky and surrounding terrain*

Slope inclination angle ($k_s°$)	View factor to sky ($\cos^2 k_s/2$)	View factor to surrounding terrain ($1 - \cos^2 k_s/2$)
0	1.000	0.000
5	0.998	0.002
10	0.992	0.008
15	0.983	0.017
20	0.970	0.030
30	0.933	0.067
45	0.854	0.146
90	0.500	0.500

k_s = slope inclination angle
α_t = albedo of surrounding terrain
$K_t\downarrow$ = total shortwave flux density on surrounding terrain

The first term in Equation (7.4), called a sky-view factor, corrects for the fraction of atmospheric diffuse shortwave radiation received by the slope. As shown in Table 7.1, $\cos^2 k_s/2$ has a maximum value of unity when $k_s = 0°$ and a minimum value of 0.5 when $k_s = 90°$. That is, as k_s gradually increases, less and less sky is viewed by the slope. In the extreme case, a vertical slope, only one-half of the sky radiation is received.

The second term in Equation (7.4) represents the shortwave radiation reflected by the surrounding terrain that is received by the slope. The term $(1 - \cos^2 k_s/2)$ is the fraction of the view from the slope occupied by the surrounding terrain, while $\alpha_t K_t\downarrow$ is a rough approximation of shortwave reflected from the facing terrain. For example, at a slope inclination angle of 30°, 93.3% of atmospheric diffuse shortwave radiation and 6.7% of shortwave diffuse reflected from the facing terrain is received by the slope. $K_t\downarrow$ is seldom known directly, since this surface may also be sloping and shaded by topography, and for convenience the radiation received on the horizontal ($K\downarrow$) can be substituted as a first approximation.

Equation (7.4) assumes that the atmosphere is a diffuse radiator, that is, scattered radiation from the atmosphere varies only with the cosine of the angle from the zenith, and further assumes that reflection from the surrounding terrain is also non-directional and diffuse. In reality neither of these assumptions is strictly correct, especially on clear days, but the relatively small view factors for slope inclinations up to about 20–30°, suggests that errors due to these assumptions are generally acceptable. In specialized applications or for very steep slopes, the distribution of diffuse shortwave radiation from the atmosphere (see discussion in Robinson, 1966) can be accounted for using an approach given by Perez *et al.* (1990) developed for solar energy collectors.

An example of the correction for diffuse shortwave radiation on a slope can be given based upon assumptions in previous examples. Earlier we derived a daily diffuse shortwave flux density on a horizontal surface of 16.4 W m^{-2} for February 7 at a latitude of 55° N and a fractional cloud cover of 0.3. Combining this estimate of D for a horizontal surface with a snowpack albedo of 0.6, $Kt\!\downarrow = 78.7$ W m^{-2} (assumed) and a 30° slope inclination, gives the following solution for diffuse solar radiation on a slope using Equation (7.4):

$$D_s = (16.4 \text{ W m}^{-2})(\cos^2 30°/2) + (1 - \cos^2 30°/2)(0.6)(78.7 \text{ Wm}^{-2})$$
$$= 18.5 \text{ W m}^{-2}$$

In this case, the diffuse solar on the slope is about 13% greater than on a horizontal surface, primarily due to the relatively high albedo of $\alpha_t = 0.6$ assumed for landscape facing the slope. Topographic effects on diffuse shortwave radiation are much less than for shortwave beam radiation; therefore, conditions where diffuse shortwave radiation dominates, e.g. cloudy weather and large zenith angles, produce smaller differences in shortwave radiation received among slopes.

Completing the computation of $K_s\!\downarrow$ for the slope by combining the diffuse and direct beam components in our example gives:

$$K_s\!\downarrow = I_b \cos Z' + D_s = 180 \text{ W m}^{-2} + 18.5 \text{ W m}^{-2} = 198 \text{ W m}^{-2}$$

where the major increase in total shortwave radiation on the slope compared to the horizontal is due to increased beam shortwave radiation rather a correction for diffuse shortwave. Net shortwave radiation exchange on the slope for an albedo of 0.6 would thus be:

$$Q'_{ns} = (1 - \alpha)K_s\!\downarrow = (1 - 0.6)(198 \text{ W m}^{-2}) = 79.2 \text{ W m}^{-2}$$

The comparable net shortwave radiation flux density found previously for the horizontal surface ($Q_{ns} = 31.5$ W m^{-2}) shows that net shortwave exchange is essentially doubled on the slope. Again, most of this difference is due to the direct-beam component.

Effects of surrounding topography on view factors of slopes to the atmosphere can also be significant on steep mountain basins. On certain shaded subunits of steep mountain basins, the fraction of the sky viewed can actually be much less than that computed for the surface slope inclination and azimuth angle as given in Table 7.1. For example, a sloping snowpack surface in a deeply incised basin may receive reflected shortwave radiation from the facing valley slopes and slopes on the opposite side of the valley. View factors for several different basin geometries have been derived that could be used to approximate shading effects of surrounding topography on incoming diffuse shortwave radiation (Oke, 1987) following proce- dures given above. These same view factors can also be used to correct incoming longwave radiation.

7.2.2 Longwave radiation on slopes

Longwave radiation received by a slope comes partially from the atmosphere and partially from the terrain facing the slope; assumed here to be snow covered. View factors previously used for diffuse shortwave are again applied along with the assumption that longwave radiation from the sky and facing terrain are isotropic or perfectly diffuse. Outgoing longwave radiation from the slope is the same as for the horizontal case. Assuming reflected longwave radiation to be negligible, the net longwave radiation exchange for the slope (Q_{nl}') can be approximately written as:

$$Q_{nl}' = \varepsilon_s L_s\!\downarrow - \varepsilon_s\, \sigma T_s^4 \tag{7.5}$$

$$L_s\!\downarrow = L\!\downarrow(\cos^2 k_s/2) + (1 - \cos^2 k_s/2)\varepsilon_t\, \sigma T_t^4 \tag{7.6}$$

where:

$L_s\!\downarrow$ = incoming longwave radiation flux density to slope
$L\!\downarrow$ = incoming longwave radiation flux density from sky to a horizontal surface
ε_s = snowpack emissivity
σ = Stefan–Boltzmann constant, 5.67×10^{-8} W m^{-2} K^{-4}
T_s = snowpack surface temperature
k_s = slope inclination angle
ε_t = emissivity of the facing terrain
T_t = surface temperature of facing terrain, K

In effect, the longwave radiation emitted by the atmosphere that reaches the slope is reduced by the factor $\cos^2 k_s/2$ (see Table 7.1) and a fraction $(1 - \cos^2 k_s/2)$ of the longwave radiation emitted by the facing terrain is added. If the facing terrain is snow covered, then we are substituting near-black body radiation from snow for lower effective emissivity radiation from a fraction of the sky. If melting snow with $\varepsilon_s = 0.98$ is assumed to be on the surface of the slope and facing terrain ($L\!\uparrow = 309$ W m^{-2}), then net longwave radiation exchange for our 30° slope with previously computed incoming longwave radiation from the atmosphere with 0.3 cumulus cloud cover on the horizontal surface ($L\!\downarrow = 228$ W m^{-2}) becomes:

$$Q_{nl}' = (228 \text{ W m}^{-2})(\cos^2 30°/2) + (1 - \cos^2 30°/2)(309 \text{ W m}^{-2}) - 309 \text{ W m}^{-2}$$
$$= -75 \text{ W m}^{-2}$$

Table 7.2 *Comparison of radiation exchange (W m^{-2}) on exposed horizontal and sloping snowpack surfaces computed for a latitude of 55° N on February 7*[a]

Radiation Term	Horizontal	30° SE Slope
Incoming beam shortwave	62	180
Incoming diffuse shortwave	16	18
Total incoming shortwave	78	198
Outgoing shortwave	−47	−119
Net shortwave	**31**	**79**
Incoming longwave	228	234
Outgoing longwave	−309	−309
Net longwave	**−81**	**−75**
Net allwave	**−50**	**4**

[a] Assumptions: $C = 0.3$ cumulus clouds, $\alpha = \alpha_t = 0.6$, $T_a = 2\,°C$, RH $= 50\%$, $T_s = 0\,°C$, $\varepsilon_s = \varepsilon_t = 0.98$, no longwave reflection, see text.

The net longwave exchange on the slope is thus only about 7% greater than the −81 W m^{-2} for a horizontal surface for conditions of relatively clear skies, 50% relative humidity and air temperature of 2 °C that were assumed. The correction would be even less under cloudy skies for our given assumptions of melting snow on the facing terrain. Non-isotropic longwave emission from the atmosphere could modify these results, but as with the correction for diffuse shortwave radiation, the effects would be generally small given the rather small view factors to facing terrain typically found in nature.

Differences in the radiation budget calculations for sloping and horizontal snow-pack surfaces in the example are summarized in Table 7.2. By far the largest difference due to topography is the direct-beam shortwave radiation and the net shortwave components resulting from the SE orientation of the sloping surface towards the Sun. The rather high snowpack albedo tends to minimize this difference in the overall radiation balance however. Differences in incoming diffuse shortwave and incoming longwave radiation due to the 30° slope inclination angle are relatively small. Regardless, the differences in absorbed direct-beam shortwave radiation produced a positive net radiation flux density on the slope, while a relatively large negative net radiation would occur on a horizontal snowpack surface for the same conditions.

The net allwave radiation flux density for snowpacks on an entire watershed can be approached in various ways. Lee (1963) introduced the "watershed lid" concept that represents an entire watershed as a single plane fit to the watershed topographic boundaries. Once the slope and aspect of the plane representing the watershed lid is found, net allwave radiation can be computed for this plane to represent the net exchange of radiant energy for the entire watershed. Another approach to computing net allwave radiation exchange across watersheds is to divide the watershed area into subareas with fixed slope inclination and aspect characteristics and compute the net radiant exchange for each subarea. Several models employing this latter method have been developed with the advent of geographic information system methods (Ohta, 1994; Matsui and Ohta, 2003). Also see discussion of modelling in Chapter 10.

7.2.3 Topography and convection

As distributed models of snowpack energy exchange advance, greater attention will be given to adjusting parameters such as wind speed, air temperature, and humidity for the effects of topography. It is common practice to adjust air temperature data in degree-day applications for the effects of elevation using lapse rates. Humidity is often assumed invariant across a basin and wind-speed adjustments are problematical and the subject of considerable research. Marks *et al.* (1992) reported essentially no difference in vapor pressure but a nearly 50% difference in wind speed between a ridge top and lake site in the alpine in California. Susong *et al.* (1999) developed methods to estimate precipitation, air temperature, vapor pressure, wind speeds, radiation fluxes, snowpack properties, and soil temperature across complex terrain for use in distributed models of snowmelt.

Correction of air temperatures for the effect of elevation usually takes the form:

$$\Delta T = \gamma (h_{st} - h)/100 \tag{7.7}$$

where:

ΔT = change in temperature with elevation, °C
γ = temperature lapse rate, °C 100 m^{-1}
h_{st} = elevation of temperature index station, m
h = elevation of station to be estimated, m

The temperature lapse rates are best derived from local climatic data. Values used in snow studies range between the dry adiabatic lapse rate of -0.98 °C 100 m^{-1} up to about -0.2 °C 100 m^{-1}. Values of about -0.5 to -0.65 °C 100 m^{-1} are typically employed (Verdhen and Prasad, 1993; Braun *et al.*, 1993; Kumar *et al.*, 1991; Zhenniang *et al.*, 1991 and 1994; Ujihashi *et al.*, 1994; Martinec *et al.*, 1998). Lapse rates also vary with time of year (Zhenniang *et al.*, 1994; Martinec *et al.*, 1998). A simple example calculation of the use of a temperature lapse rate follows. Given a lapse rate of -0.6 °C 100 m^{-1}, if the air temperature at a station at 3500 m elevation was needed, the ΔT needed to correct temperatures at an index station at 2700 m elevation would be:

$$\Delta T = (-0.6\,°C/100\ m)(3500\ m - 2700\ m)/100 = -4.8°C$$

The proper elevation (h) of the station that temperature is being estimated for is an issue within watersheds with many terrain subunits. For subunits delineated strictly as elevation zones, the hypsometric mean elevation, or elevation above and below which 50% of the area in that zone occurs, is often used.

Variation of wind speed with topography controls not only the accumulation and sublimation of snow as described in Chapter 3, but also the convective heat exchange of the snowpack. Protection from or exposure to winds by the landscape can vary with wind direction and the need to model wind speed across large areas will require routine observation of wind direction. Hartman *et al.* (1999) adapted a snow transport model developed by Liston and Sturm (1998) to generate a wind-speed field for an alpine/subalpine basin in Colorado for typical westerly winds. The model was able to predict increased wind speeds on ridges and windward slopes and reduced speeds in valleys and lee slopes. Nakayama and Hasegawa (1994) used a shear flow model to elucidate the effects of topography on wind speed and snowfall around Ishikari Bay in Japan. In the future, more applications like these will undoubtedly be used to predict distributed wind-speed maps for large watersheds that will help simulate convection of energy to and from the snowpack.

7.3 Forest influences

Forests have a major influence on the energy budget of snowpacks, especially radiative and convective heat exchange (Figure 7.4). Unfortunately, climatic data such as radiation, wind speed, air temperature, and humidity are generally not available for snowpacks in forests and adjustments to data for open conditions are needed. The one-dimensional equations given previously in this chapter do not strictly apply to snowpack surfaces in forests, where air flow may be de-coupled from that above the forest and turbulence is generated by air moving past tree boles and understory vegetation. Forest effects also vary due to varying species composition, age, and density of trees and with the nature of understory vegetation. At the least, broad distinctions need to be made between the effects of coniferous and leafless deciduous forest. Finally, since plant canopies inter-cept precipitation, the impact of rainfall on snowpacks in the forest is likely to be altered.

7.3.1 Shortwave radiation in forests

The flux density of shortwave radiation that reaches the snowpack surface beneath a forest canopy is the net result of complex multiple reflection, transmission and absorption occurrences with foliage, woody plant tissue, and the ground surface (Ross, 1981). Shortwave beam radiation is transmitted through varying path lengths

Figure 7.4 Prediction of snowmelt in forests requires accounting for canopy effects on snowpack energy exchange, especially radiation exchange. Shadow patterns suggest varying patterns of canopy shading between coniferous forests with foliage (top) and deciduous forests without foliage (bottom) in these pictures from mid-latitude Pennsylvania, USA sites. See also color plate.

in the forest during each day due to differences in the solar zenith angle. Larger path lengths in the forest increase the chances for extinction by vegetation. The fraction of beam radiation that reaches the snowpack surface will generally decrease with increasing zenith angles or path lengths. In contrast, a relatively constant fraction of shortwave diffuse radiation, coming more uniformly from the entire atmosphere above, is transmitted by the forest to the snowpack. Varying beam and diffuse shortwave fractions due to variation in cloud cover, zenith angle, and other factors cause differences in the fraction of shortwave radiation transmitted by the forest.

The nature and distribution of the vegetation elements in the forest obviously also play a major role in the degree of interaction with incoming shortwave radiation. The architecture and spectral properties of conifer forests and leafless deciduous forest that are common to snow-covered landscapes can have an important influence on melt rates. A small fraction (10%) of total shortwave radiation is generally transmitted by dense conifer forest, but even fully stocked, leafless decid-uous forest only transmits about 50% of total shortwave radiation. Boreal forests are often quite sparse, but still can significantly influence radiation exchange at the snowpack surface. The albedo of the snowpack even plays a role, since a por-tion of shortwave radiation reflected from the snowpack can be re-reflected by the forest back to the ground. Snow intercepted on the foliage and branches will likely enhance these multiple reflections between the snowpack and forest. Ulti-mately a method is needed to correct the flux density of incoming shortwave radi-ation computed or measured for open conditions for the complex interactions with forests.

Models of shortwave radiation transmission in forests range widely from simple fractional transmission based upon field studies (Kittredge, 1948) to more complex models of atmosphere–forest–ground interactions (Li *et al.*, 1995). In early studies radiation transmission in relatively closed forests was simply related to crown clo-sure or stem density (Reifsnyder and Lull, 1965). Beer's law model, similar to the model for shortwave radiation transmission in snow, has been used with success to describe radiation transmission of beam and diffuse shortwave radiation in leafless deciduous forest canopies by Federer (1971) and Link and Marks (1999). However, Reifsnyder *et al.* (1971/72) found that Beer's law did not represent the transmis-sion of shortwave beam radiation by deciduous forest in full leaf, but appeared to fit data for conifer forest. Models developed to describe radiation within closed crop canopies that consider the distribution of leaf area and leaf angle distribution within the plant canopy (Ross, 1975) have been applied to forests (Jarvis *et al.*, 1976; Rauner, 1976). In relatively sparse boreal or subarctic forests, theoretical models (GORT) that assume the forest to be randomly distributed tree crowns of ellipsoidal shape have been developed and tested (Li *et al.*, 1995; Ni *et al.*, 1997; Hardy *et al.*, 1997). Woo and Giesbrecht (2000) modelled shortwave radiation transmission in a low-density subarctic spruce forest by assuming trees were represented by isolated vertical cylinders. The approach taken to modelling shortwave radiation transmis-sion should vary with the nature of the forest and the data needs to implement the model.

A simple semi-empirical exponential model that considers transmission of total shortwave radiation can be used to approximate forest effects (Monsi and Saeki, 1953). In this model, transmission of total solar radiation through a forest canopy is related to a measure of vegetation density termed the leaf-area index (LAI) which is the projected surface area of leaves and/or needles per unit ground area inte-grated from canopy top to ground level. In leafless deciduous forest the appro-priate measure becomes the plant-area index or the integrated areas of twigs, branches, boles as well as leaves when present. Leaf-area index varies with tree density, age, and species composition. The fraction of shortwave radiation that penetrates the forest canopy to the snowpack $(K\downarrow_f/K\downarrow)$ can be simply written as:

$$K\downarrow_f/K\downarrow = \exp(-\kappa \, LAI) \tag{7.8}$$

where κ is a dimensionless extinction coefficient for shortwave radiation in forest. Incoming shortwave radiation above a forest stand is assumed equal to $K\downarrow$ computed for a horizontal or sloping snowpack surface in the open. The extinction coefficient varies between the visible and near-infrared wavelength regions and also with the zenith angle, the ratio of diffuse to beam shortwave radiation received, and with depth within the forest canopy (Ross, 1981; Jarvis *et al.*, 1976), but some approximate values are given in Table 7.3 for use in estimating daily totals of radiation. These values should be used with caution since most were derived for conditions without snow on the ground or snow intercepted on the canopy, but they give reasonable results. However, Pomeroy and Dion (1996) found that intercepted snow had very little effect on extinction of shortwave radiation in a pine canopy.

If values for pine forest from Table 7.3 are used with $LAI = 4$ and $\kappa = 0.4$, then the fraction of shortwave radiation transmitted would be:

$$K\downarrow_f/K\downarrow = \exp(-0.4(4)) = 0.20$$

A correction for leafless deciduous forest with $LAI = 0.5$ and $\kappa = 1$ from Table 7.3, gives $K\downarrow_f/K\downarrow = 0.61$. In either case, conifer or deciduous hardwoods, the correction for shortwave transmission is not trivial, but determining the leaf-area index and the appropriate extinction coefficient are still problematical. LAI can be estimated by a variety of ground-based methods (Chen *et al.*, 1997b) and from remote-sensing data (Chen and Cihlar, 1996).

Another physically based and relatively simple model for shortwave radiation transmission in forests is to assume that a Beer's-law type extinction process applies to direct-beam shortwave radiation and that a constant transmission fraction can be used for diffuse shortwave radiation (Federer, 1971; Link and Marks, 1999). With this model, the radiation penetrating to the snowpack in forests would be:

$$K\downarrow_f = \tau_d D + I_b \cos Z \exp(-\mu \, ht \, \sec Z) \tag{7.9}$$

where:

$K\downarrow_f =$ incoming shortwave flux density in forest, W m^{-2}
$\tau_d =$ transmissivity for diffuse shortwave radiation, dimensionless
$D =$ diffuse shortwave radiation flux density, W m^{-2}
$I_b =$ direct beam shortwave radiation flux density at normal incidence, W m^{-2}
$Z =$ zenith angle of the Sun (or incidence angle (Z') for sloping surfaces)
$\mu =$ extinction coefficient for direct beam shortwave radiation, m^{-1}
$ht =$ average forest height, m

Table 7.3 *Extinction coefficients for incoming shortwave radiation in forests for Equation (7.8).*

Forest Type		Tree Height (m)	Leaf- or Plant-Area Index	κ	Source[a]
Pine (*Pinus* spp.)	mid-day	15.5, 22	2.6–4.3	0.28–0.57	(1) two stands
Spruce (*Picea* spp.)	mid-day	11.5, 27.5	8.4–9.8	0.28–0.58	(1) two stands
Oak (*Quercus* spp.)	full leaf	–	5.5*	0.42	(2) and (3)
	leafless		0.5*	1.00	
Oak-Hickory (*Q.* and *Carya* spp.)		17–26			(3)
	full leaf		4.9	0.579	
	leafless		0.5*	1.12	
Aspen (*Populus tremuloides*)		21.5			(4)
	full leaf		2.1	0.60	
	leafless		0.2*	0.72	
Birch (*Betula pubescens* spp. *tortuosa*)		5–7			derived with data from (5)
	full leaf		2.0	0.39	
	leafless		0.36*	1.03	

[a] (1) Jarvis *et al.*, 1976, (2) Rauner, 1976, (3) Baldocchi *et al.*, 1984, (4) Chen *et al.*, 1997a, and (5) Ovhed and Holmgren, 1995.
*plant-area index includes leaves, if present, and woody material.

Table 7.4 *Forest shortwave radiation transmission parameters for Equation (7.9)*

Forest Type	Direct beam shortwave μ, m^{-1}	Diffuse shortwave τ_d	Source[a]
Deciduous	0.025	0.44	(1)
Mixed conifer/deciduous	0.033	0.30	(1)
Medium conifer	0.040	0.20	(1)
Dense conifer	0.074	0.16	(1)
Mixed deciduous	0.019–0.027	–	(2) Canopy space only

[a] (1) Link and Marks, 1999; (2) Federer, 1971.

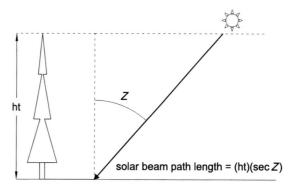

Figure 7.5 Solar beam path length in forest in relation to tree height and zenith angle.

The term "ht sec Z" in Equation (7.9) represents the path length for direct beam radiation in the forest (see Figure 7.5). Extinction coefficients (μ) for beam radiation and average integrated transmissivities for diffuse radiation (τ_d) derived from experimental data are given for several forest types in Table 7.4. Pomeroy and Dion (1996) applied the same model to total incoming shortwave radiation in a boreal pine forest and found that the extinction coefficient varied from zero to 0.14 m^{-1} with solar altitudes from 0 to about 40°, respectively. It is important to realize that τ_d is linked to a specific value of μ, since τ_d represents, theoretically at least, an average value of the direct-beam transmission function integrated over zenith angles of 0 to 90° to represent diffuse short-wave transmission from the entire hemisphere. Neither parameter should be varied independently.

Given the values for $D = 16.4$ W m^{-2}, $I_b \cos \underline{Z} = 62.2$ W m^{-2} and mean zenith angle $\underline{Z} = 77.5°$ from previous calculations, the flux density of incoming shortwave radiation in a leafless deciduous forest with an average height of 15 m, $\mu = 0.02$ m^{-1} and $\tau_d = 0.4$ would be:

$$K\downarrow_f = (0.4)(16.4 \text{ W m}^{-2}) + (62.2 \text{ W m}^{-2}) \exp[-(0.02 \text{ m}^{-1})(15 \text{ m})(\sec 77.5°)]$$
$$= 6.56 \text{ W m}^{-2} + (62.2 \text{ W m}^{-2})(0.25) = 22.1 \text{ W m}^{-2}$$

Thus only 22.1 W m^{-2} or 28% of the total 78.7 W m^{-2} incoming shortwave radiation above the canopy would reach the snowpack for this forest condition. After incoming shortwave radiation at the snowpack surface in the forest is estimated, the reflected shortwave radiation at the snowpack surface in the forest can be simply estimated as $K\uparrow_f = \alpha \ K\downarrow_f$. Computation of net allwave radiation for snowpacks in forests also requires consideration of longwave exchange that is described in the next section.

7.3.2 Longwave radiation in forests

Incoming longwave radiation at the snowpack surface in forests comes from the atmosphere through the gaps and openings and from the forest canopy overhead. Since nearly all longwave radiation received by the forest canopy is absorbed, and little is transmitted or reflected, the computation of longwave exchange beneath the canopy is simple relative to shortwave radiation exchange. If the view factor from the snowpack to the overhead canopy is known (F_{s-f}) then the fraction of the view open to the atmosphere is $1 - F_{s-f}$. With these simple ideas as a model, the net longwave radiation beneath the forest canopy can be written as:

$$Q_{nlf} = L\downarrow(1 - F_{s-f}) + \sigma \ \varepsilon_f \ T_c^4(F_{s-f}) - \sigma \ \varepsilon_s \ T_s^4 \tag{7.10}$$

where:

Q_{nlf} = net longwave radiation flux density beneath a forest canopy, W m^{-2}
$L\downarrow$ = incoming longwave radiation flux density in open (horizontal or sloping surface), W m^{-2}
F_{s-f} = view factor from snowpack surface to overhead forest, dimensionless
σ = Stefan–Boltzmann constant, 5.67×10^{-8} W m^{-2} K^{-4}
ε_f = emissivity of forest canopy, dimensionless
T_c = temperature of forest canopy, K
ε_s = emissivity of snowpack surface
T_s = temperature of snowpack surface, K

Emissivity of the forest canopy elements is quite high and ranges from 0.97 to 0.99. The radiating temperature of the forest canopy overhead is seldom known and is a deficiency in computing net longwave radiation exchange below the canopy. A first approximation would be to assume that canopy temperature was equal to the air temperature within the forest; however, temperatures of sunlit portions of the canopy and tree boles can be heated well above air temperature during the day and increase incoming longwave radiation beneath the canopy. View factors from a plane surface to the overhead canopy are also not easily obtained from forest characteristics, but canopy closure, the projection of overhead canopy and bole area onto a horizontal plane (Storck and Lettenmaier, 1999; US Army Corps of Engineers, 1956) or one minus transmissivity for diffuse shortwave radiation (τ_d) in Table 7.4 can be substituted (Link and Marks, 1999).

The effects of longwave radiation from the forest canopy on net longwave exchange can be computed using Equation (7.10). In previous examples for partially cloudy conditions and melting snow, the incoming longwave radiation from the atmosphere at 2 °C, a $C = 0.3$ cumulus cloud cover and 50% relative humidity was 228 W m^{-2} ($L\downarrow$) and the longwave radiation loss from the melting snow was 309 W m^{-2} ($\sigma \, \varepsilon_s \, T_s^4$ without considering reflected longwave radiation). Net longwave exchange for these conditions was 228–309 W m^{-2} or a relatively large energy loss of −81 W m^{-2}. Assuming a view factor from snow to forest of 0.8, or roughly 80% canopy closure equivalent to a dense conifer canopy, an effective canopy radiating temperature of 3 °C (1 °C above air temperature) and $\varepsilon_f = 0.98$, the snowpack net longwave exchange would be:

$$Q_{nlf} = (228 \text{ W m}^{-2})(1 - 0.8)$$
$$+ (5.67 \times 10^{-8} \text{ W m}^{-2} \text{ K}^{-4})(0.98)(276.16 \text{ K})^4(0.8)$$
$$- 309 \text{ W m}^{-2} = -4.9 \text{ W m}^{-2}$$

The forest canopy can increase the incoming longwave to the snowpack (Hashimoto *et al.*, 1994) and in this example for a simulated coniferous forest produces a small negative net longwave exchange at the snowpack surface. Positive net longwave radiation exchange commonly occurs in forests during the melting season and development of snowpack cold content overnight is often prevented by the overhead canopy.

Theoretical analysis of view factors from the snowpack to individual tree trunks and canopies has been used to help understand forest effects on longwave radiation exchange. Analysis by Bohren and Thorud (1973) suggests that nearly all longwave radiation received by a snowpack from the forest comes from tree boles and crowns within 2–3 times the average crown radius away. Woo and Geisbrecht (2000) presented a theoretical analysis attributed to J. A. Davies that showed that the view factor from the snow surface to an individual tree trunk dropped below 0.01 or 1% of the total view when the point was farther than about one-third the trunk height away. Both of these studies suggest that most longwave radiation received by the snowpack from the forest would be received from nearby trees.

The net allwave radiation at the snowpack surface in a forest can be computed from the sum of net shortwave and net longwave radiation exchange as in above

Table 7.5 *Comparison of snowpack radiation exchange (W m^{-2}) for open, leafless deciduous forest, and dense coniferous forest sites computed for a latitude of 55° N on February 7[a]*

Radiation Term	Open	Leafless Deciduous Forest[b]	Dense Coniferous Forest[c]
Incoming beam shortwave	62	16	1
Incoming diffuse shortwave	16	7	3
Incoming total shortwave	78	23	4
Outgoing shortwave	−47	−14	−2
Net shortwave	**31**	**9**	**2**
Incoming longwave	228	285	304
Outgoing longwave	−309	−309	−309
Net longwave	**−81**	**−24**	**−5**
Net allwave	**−50**	**−15**	**−3**

[a] ht = 15 m, $\varepsilon_f = 0.98$, $T_c = 3\,°C$, other assumptions same as in Table 6.6
[b] leafless deciduous forest: $t_d = 1 - F_{s-f} = 0.4$, $\mu = 0.02$ m^{-1}
[c] coniferous forest: $t_d = 1 - F_{s-f} = 0.2$, $\mu = 0.07$ m^{-1}

examples (Table 7.5). In Table 7.5 a complete example for leafless deciduous forest and coniferous forest is shown using Equations (7.9) and (7.10) and parameters given in Table 7.4. The large reduction in incoming diffuse shortwave and beam shortwave radiation, especially in the coniferous forest, is seen to be partially off-set by the increased incoming longwave radiation from the forest canopy. In this example the forest canopy was assumed to be 1 °C warmer than air temperature for the entire day. The net allwave radiation under dense conifers would be slightly negative under these conditions, while a large negative net allwave radiation sum occurred in the open. The leafless deciduous forest had an intermediate, but nega-tive, net allwave exchange for the example conditions. This comparison generally illustrates the differences between deciduous and coniferous forests on snowpack radiant energy exchange relative to open conditions. Obviously, the choice of dif-fuse and beam radiation transmission parameters and the assumed temperature of the forest canopy and the radiation conditions in the open determine the forest effects using this approach.

Other procedures for estimating snowpack net radiation in forests have been employed. Price (1987) estimated net allwave radiation at the forest floor using a regression analysis with incoming shortwave radiation and air temperature as predictor variables. This procedure could be employed after calibration for the spe-cific forest conditions of interest. Parrott (1974) showed relationships between above and below canopy net allwave radiation in leafless deciduous forest in

Pennsylvania during periods with snow cover. Peters (1980) showed that snow-pack net radiation in deciduous forest was only half that in an adjacent clear-cut opening on days with complete snow cover. Extinction coefficients for predicting net allwave radiation in the forest similar to those used for incoming shortwave radiation (Equation (7.8) and Table 7.3) have also been derived (Jarvis *et al.*, 1976), but introduce the additional requirement that net radiation above the forest canopy be estimated.

7.3.3 Forest effects on wind speed and convection

Convective exchange of sensible and latent heat between the snowpack and air in the forest is poorly understood. Due to the very low wind speeds often observed above snowpacks in forests, convective sensible and latent heat exchange computed using Equations (6.14) and (6.20), respectively, would be quite low (Price, 1988; Ohta *et al.*, 1993; Barry *et al.*, 1990). Sauter and McDonnell (1992) state that during passage of frontal storms convective exchange in forests could be significant on a diurnal basis. In addition, equations derived for uniform, open snowpack conditions do not strictly apply within forests. Jarvis *et al.* (1976) indicate that for conifer forest, wind speed near ground level is often controlled by horizontal pressure gradients that negate simple one-dimensional analysis of heat, mass, and momentum beneath the canopy. Turbulence within the forest is generated more by air flow around tree boles and through understory and lower canopy vegetation, rather than by air flow over the snowpack surface. Not surprisingly then, results of some studies have suggested that the bulk transfer coefficients (C_h and C_e) necessary to account for observed convective exchange at ground level in forests would have to be several to many times greater than those based upon the roughness length at the snowpack surface for open conditions (Parrott, 1974; Ishii and Fukushima, 1994; Tanaka, 1997; Suzuki and Ohta, 2003). Kuz'min (1961) also suggested use of adjustment factors for bulk transfer coefficients in forests that ranged up to 2 for conifers and 1.6 for deciduous forests, but indicated these had not been experimentally verified.

   Estimation of wind speeds at the snowpack surface in forests from open mea-surements is also not a straightforward procedure without site-specific data. Ratios of wind speed in the forest to that in the open without consideration of meteoro-logical, topographic, and forest characteristics do not appear to produce consistent results. In addition, wind speeds below the canopy are often quite low (<1 m s^{-1}) and difficult to measure accurately without sensitive instruments. A wide range of ratios between wind speed in deciduous forest and that in the open or above the canopy are given in the literature: 0.4 for mixed oak (*Quercus* spp.) forest on sloping ground (Parrott, 1974), 0.9 for maple (*Acer mono*) in mountainous terrain in Japan (Ishii and Fukushima, 1994), 0.51 and 0. 77 for two other deciduous stands in Japan

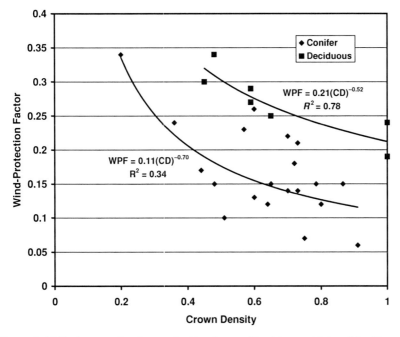

Figure 7.6 Wind-protection factors for Russian conifer (pine and fir) and deciduous forest as a function of crown density (redrawn from Kuz′min, 1961).

(Hashimoto *et al.*, 1994), 0.0–0.22 in aspen (*Populus tremuloides*) computed from relationships given by Hardy *et al.* (1997) for windspeeds up to 5 m s^{-1}. The ratios of wind speed in forests to open conditions for conifer stands are equally variable: 0.12 to 0.51 for conifer stands reviewed by Kittredge (1948), 0.8 for two different evergreen (*Crytomeria japonica* and *Abies sachalinensis*) forests in Japan (Ishii and Fukushima, 1994), 0.0 to 0.06 in black spruce (*Picea mariana*) forest based upon relationships given by Hardy *et al.* (1997) for windspeeds up to 5 m s^{-1}, and 0.76 for evergreens in Japan (Hashimoto *et al.*, 1994). Woo and Giesbrecht (2000) found that the ratio of wind speed beneath an isolated subarctic spruce tree to that in the open varied exponentially from near zero at open wind speeds below 0.4 m s^{-1} to a ratio above 0.8 for open wind speeds greater than about 1.5–2.0 m s^{-1}. These studies suggest a wide range of wind-speed reductions can occur in forests depending on site conditions and that a more general method related to canopy characteristics is desirable.

An alternative to using simple wind-speed ratios, are curves for deciduous and conifer (pine and fir) forest that consider canopy density based upon data given by Kuz′min (1961) for Russian forests (Figure 7.6). Wind-protection factors are analogous to ratios of wind speed in the forest to wind speed in the open and can be simply multiplied by measured open wind speeds to correct for forest effects.

Although the scatter of data is large, especially for conifer stands, curves fit to these data show that for a given crown density wind speeds in conifers are reduced more than in deciduous forest. These curves should obviously not be used outside the range of data used to derive them. The term crown density used in the Russian literature is related to canopy closure, but the terms may not be strictly equivalent.

Another quantitative approach is to use plant- or leaf-area index (LAI or PAI) to estimate wind speed in forests as (Rauner, 1976):

$$u_z/u_h = \exp(-n\,\text{PAI}) \tag{7.11}$$

where:

u_z = wind speed near ground level in the forest, m s^{-1}
u_h = wind speed at the top of the canopy, m s^{-1}
PAI = plant-area index, dimensionless
n = wind-speed extinction coefficient, dimensionless

Rauner (1976) found that the wind-extinction coefficient "n" ranges from 0.25 to 0.35 for several deciduous forests in full leaf for wind speeds of 1–3 m s^{-1}. At higher wind speeds the value of n decreases, decreasing by 50% at 5 m s^{-1}. Plant-area index is the area of leaves or needles, twigs, branches, and boles integrated from the top of the canopy to the ground and expressed per unit ground area. Using this equation, given a plant-area index for deciduous forest of 2 and $n = 0.3$, the ratio of below canopy to above canopy wind speed would be about 0.55. It is not certain whether this range of extinction coefficients applies to winds in leafless deciduous and conifer stands at different plant-area index values. To use Equation (7.11) to estimate wind speed in forests, wind speed measured above the canopy must be available or assumed equal to that in the open.

Air temperature and humidity above the snowpack surface are also modified somewhat by forest cover. Generally, maximum daily air temperatures are reduced due to canopy shading, while minimum daily air temperatures are increased due to enhanced longwave exchange beneath the canopy relative to an open area. The net result is generally a significant reduction in the range of daily temperatures and a decrease in mean daily air temperatures beneath the canopy. Forest effects on air temperature would be greater in coniferous than leafless deciduous forest. Reductions in daily mean air temperatures relative to an open site for conifer stands are about 0.5–1.5 °C in winter (Kittredge, 1948), but negligible in leafless deciduous forest (Parrott, 1974; Peters, 1980). Peters (1980) showed maximum and minimum daily air temperatures in leafless deciduous forest with snowcover were 0.9 °C lower and 0.6 °C higher, respectively, than in an adjacent clear-cut.

Water vapor pressure in the air within the forest can also be slightly elevated compared to the open, due to reduced convection and evaporation/sublimation at the snowpack surface and to transpiration by the overhead canopy late in the melt season. Differences in relative humidity between open and forest areas are quite large due to temperature differences, but in absolute terms vapor pressure elevations of <0.5 mb occur in forest relative to open areas during snowmelt. Corrections for forest effects on air temperature and humidity are often small during snowmelt (Kuz'min, 1961; Ishii and Fukushima, 1994), but for some applications and locations they could be important. DeWalle and Meiman (1971) showed that vapor pressure gradients above late-season melting snowpacks in Colorado conifer forest varied diurnally leading to evaporation at night and condensation during daytime.

Atmospheric stability conditions over snowpacks may also be affected by forests, following the computation scheme given previously. Reductions in wind speed and possible changes in the temperature difference between snowpack surface and air by forests could affect the magnitude of the bulk Richardson number (Ri_B) and stability corrections for convective transport. Since reductions in wind speeds will probably dominate any changes in temperature within the forest, Ri_B will likely increase further into the stable range. Thus, the bulk transfer coefficients C_h and C_e given by Equation (6.22b) would also likely be less than in the open due to stability effects.

7.3.4 Forest energy budget examples

Studies of snowpack energy budgets in forest have enhanced our understanding of forest effects on radiative and convective exchange (Ishii and Fukushima, 1994; Hashimoto *et al.*, 1994; Yamazaki and Kondo, 1992; Davis *et al.*, 1997; Hardy *et al.*, 1997; Fitzgibbon and Dunne, 1983; Metcalfe and Buttle, 1997; Barry *et al.*, 1990). In a study near Mt. Iwate in Japan, Ohta *et al.* (1993) evaluated the snowpack energy budget at an open site and an adjacent leafless, deciduous forest site both on level terrain (Figure 7.7). Incoming shortwave radiation was reduced by about 38% and wind speed was reduced by 71% in the forest relative to the open site, but air temperatures and vapor pressures were similar between sites. Latent heat convection represented an energy gain to the snowpack at both sites on most days, due to the generally high vapor pressures in the air in the region. Very high latent heat convection occurred on March 20 due to high wind speeds (8.6 m s^{-1}) and very high humidity (9.6 mb) due to the occurrence of a small rainfall event. Sensible heat convection was also positive on all days at both sites due to warm daytime air temperatures. Average daily air temperatures of 7.6 °C combined with high wind speeds on March 20 also produced the highest sensible heat convection to the snowpack. Latent heat and sensible heat convection were much lower in the forest than in the open due to reduced wind speeds. Net radiation was the dominant snowpack energy budget component at both the open and forest sites. Net radiation

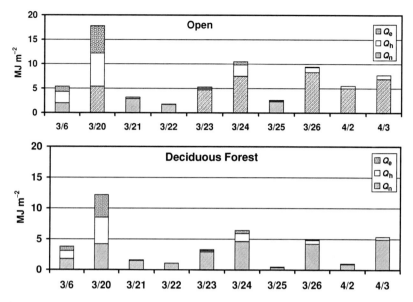

Figure 7.7 Comparison of snowpack energy budgets during melt at open and deciduous forest sites in Japan (redrawn from Ohta *et al.*, 1993). (Q_e = latent heat convection, Q_h = sensible heat convection, and Q_n = net allwave radiation).

was reduced by about 20% by the forest relative to the open, based upon regression analysis of hourly data. Because of overall differences in net radiation and total energy supply to the snowpack, daily snowmelt rates in the forest were reduced by 15–60% compared to melt at the open site. Results of this study illustrate how forest cover can reduce the snowpack energy supply and how more humid conditions can increase the importance of latent heat convection as a source of melt energy.

7.4 References

Baldocchi, D. B., Matt, D. R., Hutchison, B. A., and McMillen, R. T. (1984). Solar radiation within an oak-hickory forest: an evaluation of the extinction coefficients for several radiation components during fully-leafed and leafless periods. *Agric. For. Meteorol.*, **32**, 307–22.

Barry, R., Prevost, M., Stein, J., and Plamondon, A. P. (1990). Application of a snow cover energy and mass balance model in a balsam fir forest. *Water Resour. Res.*, **26**(5), 1079–92.

Bohren, C. F. and Thorud, D. B. (1973). Two theoretical models of radiation heat transfer between forest trees and snowpacks. *Agric. Meteorol.*, **11**(1973), 3–16.

Braun, L. N., Grabs, W., and Rana, B. (1993). Application of a conceptual precipitation-runoff model in the Langtang Khola Basin, Nepal Himalaya. In *Snow and Glacier Hydrology*, Publ. 218, ed. G. J. Young. International Association of Hydrological Sciences, pp. 221–37.

Chen, J. M. and Cihlar, J. (1996). Retrieving leaf area index of boreal conifer forest using LANDSAT TM images. *Remote Sens. Environ.*, **55**(2), 153–62.

Chen, J. M., Blanken, P. D., Black, T. A., Guilbeault, M., and Chen, S. (1997a). Radiation regime and canopy architecture in a boreal aspen forest. *Agric. For. Meteorol.*, **86** (1997), 107–25.

Chen, J. M., Rich, P. M., Gower, S. T., Norman, J. M., and Plummer, S. (1997b). Leaf area index of boreal forests: theory, techniques, and measurements. *J. Geophys. Res.*, **102**(D24), 29 429–43.

Davis, R. E., Hardy, J. P., Ni, W., Woodcock, C., McKenzie, J. C., Jordan, R., and Li., X. (1997). Variation of snow cover ablation in the boreal forest: a sensitivity study on the effects of conifer canopy. *J. Geophys. Res.*, **102**(D24), 29 389–95.

DeWalle, D. R. and Meiman, S. R. (1971). Energy exchange and late season snowmelt in a small opening in Colorado subalpine forest. *Water Resour. Res.* **7**(1), 184–8.

Dozier, J. and Frew, J. (1990). Rapid calculation of terrain parameters for radiation modelling from digital elevation data. *IEEE Trans. Geosci. Remote Sens.*, **28**(5), 963–9.

Federer, C. A. (1971). Solar radiation absorption by leafless hardwood forests. *Agric. Meteorol.*, **9**(1971/1972), 3–20.

FitzGibbon, J. E. and Dunne, T. (1983). Influence of subarctic vegetation cover on snowmelt. *Phys. Geogr.*, **4**(1), 61–70.

Frank, E. C. and Lee, R. (1966). *Potential Solar Beam Irradiation on Slopes: Tables for 30° to 50° Latitude*, Res. Paper RM-18, Rocky Mtn. For. Range Exp. Sta.: US Dept. Agric., Forest Service.

Hardy, J. P., Davis, R. E., Jordan, R., Ni, W., and Woodcock, C. (1997). Snow ablation modelling in conifer and deciduous stands of the boreal forest. In *Proceedings of the 54th Annual Meeting Western Snow Conference*, Banff, Alberta, Canada May 4–8, 1997, pp. 114–24.

Hartman, M. D., Baron, J. S., Lammers, R. B., Cline, D. W., Band, L. E., Liston, G. E., and Tague, C. (1999). Simulations of snow distribution and hydrology in a mountain basin. *Water Resour. Res.*, **35**(5), 1587–603.

Hashimoto, T., Ohta, T., Fukushima, Y., and Ishii, T. (1994). Heat balance analysis of forest effects on surface snowmelt rates. In *Snow and Ice Covers: Interactions with the Atmosphere and Ecosystems*, IAHS Publ. No. 223, ed. H. G. Jones, T. D. Davies, A. Ohmura, and E. M. Morris. International Association of Hydrological Sciences, pp. 247–58.

Ishii, T. and Fukushima, Y. (1994). Effects of forest coverage on snowmelt runoff. In *Snow and Ice Covers: Interactions with the Atmosphere and Ecosystems*, IAHS Publ. No. 223, ed H. G. Jones, T. D. Davies, A. Ohmura, and E. M. Morris. International Association of Hydrological Sciences, pp. 237–45.

Jarvis, P. G., James, G. B. and Landsberg, J. J. (1976). Chapter 7: coniferous forest. *Vegetation and the Atmosphere Case Studies*, vol. 2, ed. J. L. Monteith. London: Academic Press, pp. 171–240.

Kittredge, J. (1948). *Forest Influences*. New York: McGraw-Hill.

Kumar, V. S., Haefner, H., and Seidel, K. (1991). Satellite snow cover mapping and snowmelt runoff modelling in Beas basin. In *Snow Hydrology and Forests in High Alpine Areas*, Publ. No. 205, ed. H. Bergmann, H. Lang, W. Frey, D. Issler and B. Salm. International Association of Hydrological Sciences, pp. 101–9.

Kustas, W. P., Rango, A., and Uijlenhoet, R. (1994). A simple energy budget algorithm for the snowmelt runoff model. *Water Resour. Res.*, **30**(5), 1515–27.

Kuz'min, P. P. (1961). *Protsess tayaniya shezhnogo pokrova (Melting of Snow Cover).* *Glavnoe Upravlenie Gidrometeorologischeskoi Sluzhby Pri Sovete Ministrov SSSR* Gosudarstvennyi Gidrologischeskii Institut. Main Admin. Hydrometeorol. Service, USSR Council Ministers, State Hydrol. Institute. Translated by Israel Program for Scientific Translations. Avail from US Dept. Commerce, National Tech. Inform. Service, 1971, TT 71–50095.

Lee, R. (1963). *Evaluation of Solar Beam Irradiation as a Climatic Parameter of Mountain Watersheds*, Hydrol. Paper No. 2, Ft. Collin, CO: Colorado State University.

Li, X., Strahler, A. H., and Woodcock, C. E. (1995). A hybrid geometric optical-radiative transfer approach for modelling albedo and directional reflectance of discontinuous canopies. *IEEE Trans. Geos. Remote Sens.*, **33**, 466–80.

Linacre, E. (1992). *Climate Data and Resources.* London: Routledge.

Link, T. and Marks, D. (1999). Distributed simulation of snowcover mass- and energy-balance in the boreal forest. *Hydrol. Processes*, **13**, 2439–52.

List, R. J. (1968). *Smithsonian Meteorological Tables*, 6th edn., vol. 114, Smithsonian Publ. 4014. Washington, DC: Smithsonian Institution Press.

Liston, G. E. and Sturm, M. (1998). A snow-transport model for complex terrain. *J. Glaciol.*, **44**(148), 498–516.

Marks, D., Dozier, J., and Davis, R. E. (1992). Climate and energy exchange at the snow surface in the alpine region of the Sierra Nevada, 1. metrological measurements and monitoring. *Water Resour. Res.*, **28**(11), 3029–42.

Martinec, J., Rango, A. and Roberts, R. (1998). Snowmelt Runoff Model (SRM) user's manual, updated edn. 1998, ver. 4.0. In *Geographica Bernensia*, ed. M. F. Baumgartner. Berne: Department of Geography, University of Berne.

Matsui, K. and Ohta, T. (2003). Estimating the snow distribution in a subalpine region using a distributed snowmelt model. In *Water Resources Systems: Water Availability and Global Change*, IAHS Publ. 280, ed. S. Franks, G. Bloschl, M. Kumagai, K. Musiake, and D. Rosbjerg. International Association of Hydrological Sciences, pp. 282–91.

Metcalfe, R. A. and Buttle, J. M. (1997). Spatially-distributed snowmelt rates in a boreal forest basin. In *Proceedings of the 54th Annual Meeting Western Snow Conference*, Banff, Alberta, Canada May 4–8, 1997, pp. 163–74.

Monsi, M. and Saeki, T. (1953). Uber den lichtfaktor in den pflanzengesellschaften und seine bedeutung fur die stoffproduktion. *Japan. J. Bot.*, **14**, 22–52.

Nakayama, K. and Hasegawa, K. (1994). Analysis of wind fields in winter by using AMeDAS and considering topography. In *Snow and Ice Covers: Interactions with the Atmosphere and Ecosystems*, Publ. 223, ed. H. G. Jones, T. D. Davies, A. Ohmura, and E. M. Morris. International Association of Hydrological Science, pp. 177–86.

Ni, W., Xiaowen, L., Woodcock, C. E., Roujean, J. L., and Davis, R. E. (1997). Transmission of solar radiation in boreal conifer forests: measurements and models. *J. Geophys. Res.*, **102**(D24), 29 555–66.

Ohta, T. (1994). A distributed snowmelt prediction model in mountain areas based on an energy balance method. *Annals Glaciol.* **19**, 107–13.

Ohta, T., Hashimoto, T., and Ishibashi, H. (1993). Energy budget comparison of snowmelt rates in a deciduous forest and an open site. *Annals Glaciol.*, **18**, 53–9.

Oke, T. R. (1987). *Boundary Layer Climates*, 2nd edn. London: Methuen, Inc.

Ovhed, M. and Holmgren, B. (1995). Spectral quality and absorption of solar radiation in a mountain birch forest, Abisko, Sweden. *Arctic and Alpine Res.*, **27**(4), 380–8.

Parrott, H. A. (1974). Radiation exchange and snowmelt characteristics at deciduous forest and clearcut sites in central Pennsylvania. Unpublished MS Thesis, The Pennsylvania State University.

Peters, J. G. (1980). The effects of clearcutting deciduous forest on net radiation exchange and snowmelt in central Pennsylvania. Unpublished MS Thesis, The Pennsylvania State University.

Perez, R., Ineichen, P., Seals, R., Michalsky, J. and Stewart, R. (1990). Modelling daylight availability and irradiance components from direct and global irradiance. *Sol. Energy*, **44**(5), 271–89.

Pomeroy, J. W. and Dion, K. (1996). Winter radiation extinction and reflection in a boreal pine canopy: measurements and modelling. *Hydrol. Processes*, **10**, 1591–608.

Price, A. G. (1987). Modelling of snowmelt rates in a deciduous forest. In *Seasonal Snowcovers: Physics, Chemistry, Hydrology*, NATO ASI Series, Series C: Mathematical and Physical Sciences, vol. 211, ed. H. G. Jones and W. J. Orville-Thomas. Dordrecht: D. Reidel Publishing Company, pp. 151–65.

Price, A. G. (1988). Prediction of snowmelt rates in a deciduous forest. *J. Hydrol.*, **101**, 145–57.

Rauner, J. L. (1976). Chapter 8: deciduous forests. In *Vegetation and the Atmosphere Case Studies*, vol. 2, ed. J. L. Monteith. London: Academic Press, pp. 241–64.

Reifsnyder, W. E. and Lull, H. W. (1965). *Radiant Energy in Relation to Forests*. Tech. Bull. 1344. US Department of Agriculture.

Reifsnyder, W. E., Furnival, G. M., and Horowitz, J. L. (1971/72). Spatial and temporal distribution of solar radiation beneath forest canopies. *Agric. Meteorol.*, **9**, 21–7.

Robinson, N. (Ed.). (1966). *Solar Radiation*. Amsterdam: Elsevier.

Ross, J. (1975). Chapter 2: radiative transfer in plant communities. In *Vegetation and the Atmosphere Principles*, vol. 1, ed. J. L. Monteith. London: Academic Press, pp. 13–55.

Ross, J. (1981). *The Radiation Regime and Architecture of Plant Stands*. The Hague: Dr W. Junk Publishers.

Sauter, K. A. and McDonnell, J. J. (1992). Prediction of snowmelt rates at a forested alpine site in northern Utah. In *Proceedings of the Western Snow Conference*, Jackson Hole, WY April 14–16, 1992, pp. 95–102.

Storck, P. and Lettenmaier, D. P. (1999). Predicting the effect of a forest canopy on ground snow pack accumulation and ablation in maritime climates. *Proceedings of the 67th Annual Western Snow Conference*, Lake Tahoe, CA April 19–22, 1999, pp. 1–12.

Susong, D., Marks, D., and Garen, D. (1999). Methods for developing time-series climate surfaces to drive topographically distributed energy- and water-balance models. *Hydrol. Processes*, **13**, 2003–21.

Suzuki, K. and Ohta, T. (2003). Effect of larch forest density on snow surface energy balance. *J. Hydrometeorol.* **4**, 1181–93.

Tanaka, Y. (1997). Evaporation and bulk transfer coefficients on a forest floor during times of leaf-shedding. *J. Agric. Meteorol.* (Japan), **53**(2), 119–29.

Ujihashi, Y., Takase, N., Ishida, H., and Hibobe, E. (1994). Distributed snow cover model for a mountainous basin. In *Snow and Ice Covers: Interactions with the Atmosphere and Ecosystems*, Publ. 223, ed. H. G. Jones, T. D. Davies, A. Ohmura, and E. M. Morris., International Association of Hydrological Sciences, pp. 153–62.

US Army Corps of Engineers. (1956). *Snow Hydrology, Summary Report of the Snow Investigations*. Portland, OR: US Army, Corps of Engineers, N. Pacific Div.

Verdhen, A. and Prasad, T. (1993). Snowmelt runoff simulation models and their suitability in Himalayan conditions. In *Snow and Glacier Hydrology*, Publ. 218, ed. G. J. Young. International Association of Hydrological Sciences, pp. 239–48.

Woo, M. and Giesbrecht, M. A. (2000). Simulation of snowmelt in a subarctic spruce woodlands: 1. tree model. *Water Resour. Res.*, **36**(8), 2275–85.

Yamazaki, T. and Kondo, J. (1992). The snowmelt and heat balance in snow-covered forested areas. *J. Appl. Meteorol.*, **31**, 1322–27.

Zhenniang, Y., Zhihuai, W., and Qiang, W. (1991). Characteristics of hydrological processes in a small high mountain basin. In *Snow Hydrology and Forests in High Alpine Areas*, Publ. 205, ed. H. Bergmann, H. Lang, W. Frey, D. Issler, and B. Salm, International Association of Hydrological Sciences, pp. 229–36.

Zhenniang, Y., L. Fungxian, Y. Zhihuai, W., and Qiang, W. (1994). The effect of water and heat on hydrological processes of a high alpine permafrost area. In *Snow and Ice Covers: Interactions with the Atmosphere and Ecosystems*, Publ. 223, ed. H. G. Jones, T. D. Davies, A.Ohmura, and E. M. Morris. International Association of Hydrological Sciences, pp. 259–68.

8

Snowfall, snowpack, and meltwater chemistry

8.1 Introduction

Biogeochemical cycling of nutrients and pollutants in the environment is significantly affected by the occurrence of snowfall and snowpacks. Snowpacks can be viewed as reservoirs of chemicals that, unlike substances dissolved in rainfall, can be largely stored for significant periods of time during winter until melting occurs. Snowpacks reflect the chemical nature of the original snowfall events or wet deposition that accumulated to create them as well as the dry deposition of chemicals occurring as aerosol droplets, particles, and gases in the atmosphere deposited on the snowpack surface during non-precipitation periods. Both naturally cycling chemicals and pollutants in the environment end up in the snowpack in this manner. Snowpack chemistry may also be affected by interactions with plant canopies during interception and the activity of organisms that find a home in the snow. Once melting or rain-on-snow events occur, chemicals are redistributed and released from the snowpack non-uniformly due to fractionation processes leading to relative enrichment of the initial liquid-water releases. In polluted environments, early spring fish kills have been attributed to effects of these initial concentrated acidic snowmelt-runoff events that caused toxic levels of dissolved aluminum to occur in streams. Studies have also shown that some ions are preferentially eluted or leached from the snowpack in different ratios than found in the snow itself. Stable isotope fractionation also occurs in snow during winter and can be used to study phase changes important in snow hydrology.

8.2 Snowfall and wet deposition

The chemical nature of precipitation occurring as snow is influenced by both naturally occurring and anthropogenically derived gases and aerosol particles and droplets in the atmosphere. Aerosols here refer to droplets and solid particles that

Table 8.1 *Comparison of rainfall and snowfall chemistry at Brookhaven*
National Laboratory at Upton, NY over five winters (Raynor and Hayes, 1983)

Precipitation type	Concentrations (ueq L^{-1})					
	H$^+$	SO$_4{}^{2-}$	NO$_3{}^-$ + NO$_2{}^-$	NH$_4{}^+$	Na$^+$	Cl$^-$
Rain	20	40	16	11	23	31
Snow	25	57	46	20	56	62

are small enough to be borne aloft by atmospheric turbulence. Gases that react with atmospheric water are derived from natural biological processes, volcanoes, and wild fires. Natural aerosols are derived from sea salt, soil dust, wild fires, and vegetation. Human-related processes such as combustion of fuels, incineration of waste, high-temperature industrial processes, and photochemical reactions in the atmosphere also contribute gases and aerosols. Aerosol particles initially influence snowfall chemistry by acting as condensation nuclei for initial ice crystal formation. Once ice crystals are formed in clouds and snowfall begins, the gases and aerosols can be scavenged by snow falling within the clouds (washout) or by snow falling below the cloud base to the ground (rainout).

Riming of ice crystals within clouds due to impact with super-cooled water droplets can significantly affect snowfall chemistry. Super-cooled droplets are often highly concentrated due to interactions between gases and aerosols within clouds. Many high-elevation sites are shrouded by clouds during major snowfall events, thus the depth of the atmosphere for scavenging is limited, but riming of intercepted snow or the snowpack surface can also occur.

Snowflakes are more efficient at scavenging due to their large surface area to mass ratio and thus snowfall is generally more highly concentrated than rainfall under the same atmospheric conditions. Table 8.1 shows the concentrations of substances in snow relative to rain from a study in New York state (Raynor and Hayes, 1983). Higher concentrations in snowfall than rain were particularly noticeable for nitrate plus nitrite. Particulates in the atmosphere during winter snowfalls are thus quite likely to be scavenged by snowfall and accumulated in the snowpack.

Scavenging of pollutants causes some interesting variations in snowfall chemistry in the Cairngorm mountains of Scotland. Higher concentrations of acidic pollutants in fresh snowfalls are found at lower altitudes in this region (Davies *et al.*, 1988) due to greater scavenging by snowflakes falling through increased depths of the atmosphere. It is also in this region where highly acidic "black snowfall" events are experienced due to scavenging of carbon aerosols and other pollutants. Back-trajectory analysis of these storm events from Scotland reveals air-mass origins from eastern Europe and the former USSR (Davies *et al.*, 1984; 1988).

Chemical interactions of snowfall with forest canopies are not as well understood as that for rainfall. Enrichment of dissolved solids in throughfall and stemflow during rainfall commonly occurs due to washoff of intercepted dry deposition, leaf and bark tissue exchange, surficial biological activity and evaporation loss (see review by Schaefer and Reiners, 1990). Reductions in nitrogen concentrations in throughfall compared to rainfall in the open have also been commonly observed due to assimilation by canopy organisms. The extent of enrichment with intercepted snow at subfreezing temperatures in the forest canopy is not well known. Studies by Stottlemyer and Troendle (1995) show that a conifer canopy in Colorado caused reductions in snowfall NH_4^+ and NO_3^- concentrations, likely due to uptake by lichens in the forest canopy. In contrast, Pomeroy *et al.* (1999) reported that SO_4^{2-} and Cl^- concentrations in intercepted snow were 5–6 times greater than snow away from trees in a boreal forest at Inuvik, Northwest Territories, Canada, apparently due to enhanced dry deposition on intercepted snow from high levels of atmospheric aerosols found at that site.

Regional differences in snowfall chemistry are quite pronounced in the United States and serve to illustrate differences that can occur due to pollution and natural factors. Regions like the western United States with relatively low levels of atmospheric pollutants give rise to precipitation with relatively low dissolved solids and a pH around 5.6 due to dissolution of CO_2. In contrast, air pollution in regions such as the northeastern United States greatly increases concentrations of dissolved solids and acidity in precipitation with pHs often approaching 4.0. Wet-deposition data collected as part of the National Atmospheric Deposition Program, http://nadp.sws.uiuc.edu/ in regions where snowfall dominates in winter precipitation provide a useful comparison of a polluted site in New Hampshire (Hubbard Brook) and a relatively unpolluted site in California (Yosemite National Park). Wet deposition data for January, 2000 in Figure 8.1 show a 20 times greater total dissolved solids due primarily to sulfate and nitrate ions at the more polluted New Hampshire site. Average pH of wet deposition was 5.4 at the California site and 4.5 at the New Hampshire site during January 2000. At coastal sites it is customary for sodium and chloride concentrations to be relatively high in wet deposition, but this is difficult to see at the Yosemite site due to the very low concentrations of all constituents.

The rate of transfer of a chemical represented by wet deposition of snow is often computed for use in biogeochemical cycle studies. Just as for energy exchange the mass rate of transfer can be expressed as a mass flux (mass transferred per unit time) or preferably for watershed studies as the mass flux density (mass transferred per unit watershed area per unit time). The flux density in precipitation (F_p) can be computed as the product of snowfall intensity expressed as the volumetric water equivalent of snowfall per unit of ground area and per unit of time (P) and the

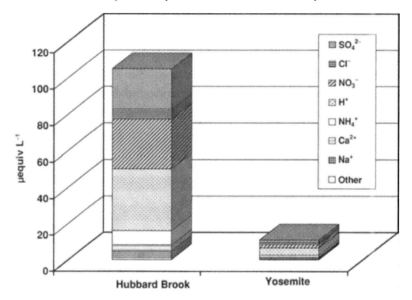

Figure 8.1 Chemistry of winter wet deposition at Hubbard Brook, New Hampshire and Yosemite, California USA during January, 2000 (data from National Atmospheric Deposition Program, http://nadp.sws.uiuc.edu/).

concentration of the ion of interest in the snowfall as mass per unit volume (c_{sp}) as:

$$F_p = Pc_{sp} \qquad (8.1)$$

For example, given that the precipitation for January, 2000 at Hubbard Brook, NH had a water equivalent of 0.102 m³ m⁻² (= 0.102 m or 10.2 cm of water) and a sulfate concentration of 1.01 mg L⁻¹ (equal to 21 μeq L⁻¹ shown in Figure 8.1), we can compute the flux density of sulfate during that month using Equation (8.1):

$$F_p = (0.102 \text{ m mo}^{-1})(1.01 \text{ mg L}^{-1})(10^3 \text{ L m}^{-3}) = 103.0 \text{ mg SO}_4^{2-} \text{ m}^{-2} \text{ mo}^{-1}$$
$$F_p = (103.0 \text{ mg SO}_4^{2-} \text{ m}^{-2} \text{ mo}^{-1})(32 \text{ mg S}/96 \text{ mg SO}_4^{2-}) = 34.3 \text{ mg S m}^{-2} \text{ mo}^{-1}$$

The flux density of sulfate was also expressed in terms of mass of sulfur by multiplying by the ratio of mass of S per unit mass of $SO_4{}^{2-}$ ion to make it comparable with flux densities of SO_2 as dry deposition discussed later. Annual values of flux density are often expressed in kg ha⁻¹ y⁻¹, where 1 hectare (ha) = 10⁴ m². Interestingly, even though the corresponding sulfate concentration at Yosemite, CA for the same month was only 0.005 mg L⁻¹ or about 0.5% of that at the New Hampshire site, the mass flux was 2.24 mg m⁻² mo⁻¹ or about 2% of that at Hubbard Brook due to the relatively large precipitation amount at Yosemite (0.448 m mo⁻¹).

This comparison shows that concentrations as well as precipitation intensity can be influential in controlling mass fluxes in precipitation or snowfall at a site.

In order to assess wet-deposition fluxes for a site, it may be possible to find representative wet-deposition chemistry data through networks, such as the National Atmospheric Deposition Program (NADP) Internet site for the United States, http://nadp.sws.uiuc.edu/. Other monitoring networks and experimental sites exist that can provide similar data, for example wet deposition data for Canada are available through the Internet, www.msc-smc.ec.gc.ca/NatChem/.

8.3 Dry deposition on snowpacks

Dry deposition of gases and aerosols on snow surfaces can also significantly affect chemical inputs to the landscape. Unlike wet deposition that occurs only during precipitation events, dry deposition occurs continuously. The rate of dry deposition is controlled by a large number of factors (Davidson and Wu, 1990):

- atmospheric conditions (wind speed, turbulence, stability conditions, etc.)
- the nature of the snow surface (roughness, chemistry, terrain features, etc.)
- the properties of the depositing gas or aerosol (particle sizes and shapes, aerosol reactivity, etc.).

Transport of dry deposits to the snow surface occurs in three stages: aerodynamic transport to the top of the boundary sublayer by turbulent motion, transport through the quasi-laminar boundary layer by diffusion, impaction, sedimentation or interception; and finally by direct surface interactions such as dissolution, adhesion, adsorption, or chemical reaction. Dry deposition on porous snowpacks can also take place below the surface due to natural and forced convection of gases and aerosols in the snowpack pore space. Any of these three stages can limit or control the dry deposition rate. To further complicate matters, surface gaseous reactions are sometimes reversible and/or particulates can rebound and re-suspend in the turbulent air stream. Blowing surface snow that is sublimated to vapor can also concentrate substances in the remaining snow or re-suspend gases and aerosols in the atmosphere. These processes are very complex and readers are referred to papers by Davidson and Wu (1990), Waddington *et al.* (1996), Davidson *et al.* (1996), and Barrie (1985), for more detailed discussions.

Flux density as dry deposition (F_d) can be computed as the product of the concentration of the chemical species in the atmosphere (c_c) given as mass per unit volume and the dry-deposition velocity (v_d) which has units of length per unit time:

$$F_d = v_d c_c \qquad (8.2)$$

Table 8.2 *Average wintertime atmospheric concentrations and computed dry deposition velocities to the snowpack for several chemical species at a rural Michigan site (Cadle and Dasch, 1987b)*

Species	Concentrations ($\mu g\ m^{-3}$)	Dry Deposition Velocity ($m\ s^{-1} \times 10^{2}$)
SO_2	7.6	0.093 ± 0.093
$SO_4{}^{2-}$	2.7	0.10^{a}
HNO_3	0.69	1.4 ± 1.0
$NO_3{}^{-}$	1.2	0.10^{a}
NO_2	3.8	0.02^{a}
$NH_4{}^{+}$	0.78	0.10 ± 0.11
NH_3	0.03	1.0^{a}
Ca^{2+}	0.15	2.1 ± 1.8
Mg^{2+}	0.04	1.5 ± 1.3
Na^{+}	0.10	0.44 ± 0.48
K^{+}	0.08	0.51 ± 0.60

[a] assumed values, others computed from experimental data

Concentration data are often available from air quality monitoring stations, but dry-deposition velocities vary for each gaseous or aerosol chemical species and for the specific atmospheric and surface conditions being considered. Given detailed site and meteorological data, v_d may be computed (Davidson *et al.*, 1996), or published values for v_d tabulated from experimental snow studies can be employed (see reviews of deposition velocity data by Davidson and Wu, 1990 and DeWalle, 1987).

Average wintertime dry-deposition velocities in Table 8.2 from a study in Michigan by Cadle and Dasch (1987b) illustrate typical values for snow surfaces for major chemical species. Note that the error bars around these v_d estimates are quite wide. In general the uncertainty in estimating dry deposition is much larger than for wet deposition. To illustrate the use of deposition velocities from the study by Cadle and Dasch, using an average atmospheric SO_2 concentration of 7.6 $\mu g\ m^{-3}$ and a representative deposition velocity from Table 8.2 of about 0.00093 $m\ s^{-1}$ gives a dry-deposition rate of:

$$F_d = (7.6\ \mu g\ SO_2\ m^{-3})(0.00093\ m\ s^{-1})(2.592 \times 10^{6}\ s\ mo^{-1})(10^{-3}\ mg\ \mu g^{-1})$$
$$= 18.3\ mg\ SO_2\ m^{-2}\ mo^{-1}$$
$$F_d = (18.3\ mg\ SO_2\ m^{-2}\ mo^{-1})\ (32\ mg\ S/64\ mg\ SO_2) = 9.16\ mg\ S\ m^{-2}\ mo^{-1}$$

Flux density here was converted to units for a 30-day month and also expressed as mass of S by multiplying by the ratio of mass of S per unit mass of SO_2 (32/64) to be comparable with wet-deposition flux density of S (34.3 mg S m^{-2} mo^{-1}) in the earlier example. Conversions to make flux density of other compounds comparable, such as the varying forms of N in Table 8.3, are made in similar fashion. Cadle and Dasch (1987b) estimated that average wintertime dry deposition to snowpacks at this rural Michigan site supplied as little as 6% of total atmospheric deposition for H^+ to as much as 34% of total atmospheric deposition for Ca^{2+}. However, contributions of dry deposition at a more polluted urban site were much greater for some ions (e.g. 80% for Cl^-). At remote sites dry deposition will generally be a minor contributor to snowpack chemistry compared to wet deposition, while in heavily industrialized and urbanized regions it can be of greater significance.

In the United States there are two networks that provide dry deposition data for selected sites. The first of these is the Clean Air Status and Trends Network (CASTNET) with a network of 70 stations operated by the Environmental Protection Agency. These sites are co-located with NADP wet-deposition sites to assess the effectiveness of air pollution control efforts on total deposition, www.epa.gov/castnet/. The other network, known as AIRMoN for Atmospheric Integrated Monitoring Network, www.arl.noaa.gov/research/projects/airmon_dry.html/, is primarily a research network with 13 stations operated by the National Oceanic and Atmospheric Administration in cooperation with both CASTNET and NADP. Both networks provide data in down-loadable files that give measured atmospheric concentrations, computed v_d, and computed dry-deposition fluxes for weekly, seasonal, and annual time periods and several chemical species.

Early studies of atmospheric deposition employed measurement of bulk deposition. Bulk deposition represents the flux of wet deposition and dry deposition in continuously open buckets or tubs. Dry deposition to an open bucket or tub is not necessarily equal to that on the surrounding landscape due to bucket aerodynamics and differences between physical-chemical reactions between plastic bucket and natural surfaces, but this approach was used to provide a relative index and to avoid measurement of atmospheric concentrations and meteorological data at a site needed to more accurately estimate dry deposition. Thus, bulk-deposition measurements are not necessarily directly comparable to the sum of wet plus dry deposition given by Equations (8.1) and (8.2).

8.4 Snowpack chemistry

Seasonal snowpacks accumulate and lose chemical substances through several mechanisms summarized in Table 8.3. The major pathways for chemical gains are new snow and rain events and dry deposition to the snowpack. Chemically distinct layers or horizons often form in snowpacks from each major wet-deposition event and these layers often persist with identifiable chemical characteristics until they come in contact with liquid water from rain or melt. A true loss of chemical

Table 8.3 *Factors affecting snowpack chemistry*

Gains/Losses
Gains wet deposition of rain, snow, sleet, fog droplets dry deposition of gases and aerosols on or within snowpack deposition of organic debris in forests blowing snow-deposition vapor and capillary liquid transport from underlying soil
Losses meltwater and rain water flux out of snowpack blowing snow-scouring volatilization and vapor transport to atmosphere

from the snowpack in liquid water only occurs when liquid water carrying solutes drains from the snowpack. Liquid water that refreezes or is stored internally within the snowpack only serves to redistribute the chemical element burden. Snowpacks in windy locations can also gain or lose significant mass and chemical load due to drifting or scouring action by the wind, respectively. In contrast, forest snowpacks are generally protected from the wind, but can receive large amounts of organic debris from the overhead canopy.

Blowing-snow particles can become significantly enriched due to sublimation losses of pure water that concentrate ions in the remaining mass, scavenging of gases and aerosols and other processes. Pomeroy and Jones (1996) reported concentrations of ions in wind-deposited, partially sublimated snow in the western Canadian Arctic that were 1.7 (NO_3^-) to 7.6 (Cl^-) times greater than sites without wind effects. Thus, deposition of blowing snow can significantly affect snowpack ion concentrations. If the blowing-snow particles are completely vaporized, then gases and aerosols can be re-released to the atmosphere. Snow crystals at the snowpack surface can be similarly sublimated without wind transport.

Other pathways for chemical gains or losses due to vapor transport and micro-biological processing within the snowpack also exist, but their significance is less well understood. Results from Cragin and McGilvary (1995) suggest that the sulfate compounds in snow may also volatilize back into the atmosphere with water vapor. In addition, Pomeroy *et al.* (1999) reported NO_3^- volatilization losses from snow intercepted on a boreal forest canopy of about 62% per unit snow mass sublimated in Northwest Territories, Canada. Such gaseous losses may be viewed as a form of delayed, reverse dry deposition and may need to be added to detailed chemical budgets of the snowpack. Massman *et al.* (1995) found that rates of CO_2

Table 8.4 *Comparison of snowpack chemistry (μeq L^{-1}) from regions with widely varying levels of pollution*

Ion	Tien Shan Mtns. Northwest China (1)*	Sierra Nevada Mtns., NV and CA, USA (2)	Appalachian Mtns., PA, USA (3)	Black Snow, Cairngorm Mtns., Scotland (4)
pH	6.9	5.6	4.3	3.0
H^+	0.13	2.5	50	1,000
Ca^{2+}	53	2.5	28	31
Mg^{2+}	5.9	0.8	4	43
NH_4^+	13	–	15	17
Na^+	9.7	3.9	14	197
K^+	1.2	1.0	–	10
SO_4^{2-}	16	3.5	42	412
NO_3^-	5.7	2.1	53	337
Cl^-	9.9	5.9	9	418

*Sources: (1) May 1990 survey data, Williams *et al.*, 1992, (2) 4 surveys in 1975, Brown and Skau, 1975, (3) 12 survey averages during 1979–81, DeWalle *et al.*, 1983, (4) February 1984 survey, Davies *et al.*, 1984.

movement from soil to atmosphere were augmented, beyond rates of simple molecular diffusion, by "pressure pumping" of snowpack air by the wind. Hogan and Leggett (1995) showed that vapors from synthetic organic compounds used as tracers in the soil, could be detected in snowpack samples after 3–75 day periods. These studies of gaseous movement through the snowpack suggest that gases originating from the soil or within the snowpack could also be transported and affect snowpack chemistry.

Internal processing of the snowpack chemical load by microbes and animals can represent either a source (decomposition) or sink (assimilation by organisms) for dissolved solids (Jones, 1999). Jones (1987) presents evidence that NO_3^- and NH_4^+ are lost from meltwater in forest snowpacks due to assimilation by organisms such as lichens, algae, bacteria, and fungi associated with organic debris from the canopy that accumulates in the snowpack. Such losses due to snowpack biota may be more important for forest snowpacks where a greater supply of organic matter and associated microflora and fauna is found.

Studies of snowpack chemistry throughout the world illustrate a variety of chemical element concentrations that reflect differences in wet-deposition chemistry due to atmospheric pollution and other natural factors. Data from studies of two relatively clean and two polluted snowpacks are shown in Table 8.4 to illustrate the range of conditions that can be encountered. Samples of relatively clean snowpacks collected in the Tien Shan Mountains in northwest China (Williams *et al.*, 1992) show a basic reaction due to the effects of aeolian dust, suggesting a $CaSO_4$ source in the soil. Unpolluted snowpacks in the Sierra Nevada Mountains of Nevada and California show a pH expected due to typical levels of CO_2 in the atmosphere (Brown and Skau, 1975). It is interesting to note that the major cation and anion in these Sierra samples was Na^+ and Cl^-, respectively, which suggests an influence of sea salt. Snowpacks in Pennsylvania (DeWalle *et al.*, 1983) are relatively acidic (pH $= 4.3$) with H^+ the major cation and SO_4^{2-} and NO_3^- the major anions. The relatively high Ca^{2+} concentrations in Pennsylvania snow was attributed to dry deposition of dust from limestone quarries. Finally, data for a fresh "black snow" layer in Scotland (Davies *et al.*, 1984) is an extreme example of a polluted snow-pack with high H^+, SO_4^{2-}, and NO_3^-. Both Cl^- and Na^+ levels are also high in the black snow likely due to sea salt and industrial sources of chloride, such as coal combustion.

The load of a chemical compound or ion stored within a snowpack or a snow layer is often needed to document snowpack chemical conditions at a given time and place. The load of a specific chemical species (W_s, mass per unit ground area) is computed as the product of the water equivalent (SWE, depth of liquid water) of the snow layer and the concentration of the chemical compound or ion in the snow layer (c_{ss}, mass per unit volume) as:

$$W_s = SWE\, c_{ss} \qquad (8.3)$$

For example, if the water equivalent of a snowpack or snowpack layer was 0.056 m of liquid water and the nitrate concentration of the snowpack was 3.2 mg L^{-1}, then the load of nitrate would be:

$$W_s = (0.056\ \text{m})(3.2\ \text{mg L}^{-1})(10^3\ \text{L m}^{-3}) = 179\ \text{mg NO}_3\ \text{m}^{-2}$$

Obviously, computing the load in a snowpack depends upon an accurate mea-surement of concentration as well as water equivalent. Unfortunately, no routine nation-wide measurement network for snowpack chemistry exists in the United States. Snow depth on the ground is measured at selected meteorological stations, but snow density observations are generally not available. A density of 100 kg m^{-3} is often assumed for fresh snow on the ground as a rough approximation, but the snowpack on the ground will quickly densify due to snowpack metamorphism thereafter.

One problem with chemical characterizations of snowpacks is that concentra-tions are measured after the samples have been melted and liquid water has had some time to interact with particulates and organic debris in the snowpack. Of course such solids do come into contact with liquid water during melt and rain for indeterminate lengths of time, but there is no standard way to approximate the length of contact. Another problem is the extremely high variability of snowpack chemistry that causes great difficulty in obtaining representative samples. The best index to snowpack impacts on chemical cycling in the environment and streamflow often is to measure the chemistry of the water leaving the snowpack.

8.5 Meltwater chemistry

Meltwater releases from a snowpack also represent a chemical flux density delivered to the soil surface. The meltwater flux density (F_{mw}) for a chemical species is computed (like the wet deposition flux in Equation (8.1)) as:

$$F_{mw} = M c_{mw} \qquad (8.4)$$

where M is the melt water expressed as a liquid volume per unit ground area and per unit time ($m^3\ m^{-2} = m$) and c_{mw} is the concentration of the species in the meltwater.

Mass fluxes of ions in meltwater can then be compared to mass fluxes in wet and dry deposition. Concentrations of ions in meltwater, however, are not simply a reflection of the average snowpack chemistry due to the effects of fractionation and preferential elution processes to be discussed below. Nakawo *et al.* (1998) have also reviewed the effects of these processes on snowpack meltwater chemistry.

8.5.1 Melt fractionation

Impurities in snowpacks are generally located on the surface of ice grains that allows the early liquid water passing through a snowpack layer to flush out a greater amount of substances than later liquid water passing through that layer. This process is referred to as melt fractionation; a term from chemistry that refers to separation of constituents into parts based upon their different properties, in this case their solubility and availability on the surface of ice grains. During formation of ice crystals the bulk of impurities is excluded from the crystal matrix and migrates to the crystal surface where even at subfreezing temperatures a quasi-liquid layer of water can be found (Davies *et al.*, 1987). Substances scavenged from the atmosphere during snowfall also accumulate on the ice crystal surfaces. Once in the snowpack, metamorphosis of ice grains also tends to cause migration of impurities to the crystal surface. The end result is that the early flushes of water from a snowpack due to melt, or rain-on-snow, tends to wash out more of the impurities relative to average snowpack chemical concentrations than late-season meltwater which is relatively depleted in impurities. Another type of fractionation, isotope fractionation, also occurs in the snowpack that is discussed later in Section 8.7.

Johannessen and Hendriksen (1978) were the first to describe the fractionation process during snowmelt. They showed with both field and controlled laboratory studies that H^+, SO_4^{2-}, NO_3^-, and Pb were present in early meltwater fractions

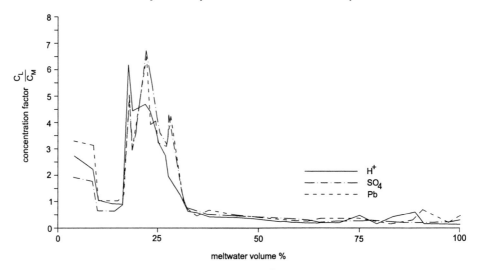

Figure 8.2 Concentration factors for H⁺, SO₄²⁻, and Pb in meltwater illustrating the effects of fractionation on enrichment of early meltwater releases from the snowpack (Johannessen and Henriksen, 1978, © 1978 American Geophysical Union). Concentration factors are given as the ratio of concentration in meltwater to the average concentration in the initial snowpack.

at concentrations three to six times greater than the overall initial snowpack concentration, while late season releases had only a fraction of the concentration of the initial snowpack. They found that the first 30% of meltwater released from the snowpack contained 2–2.5 times higher concentrations than the snowpack average (Figure 8.2). Consequently, streamflow from snowmelt during the initial flush of acid meltwater can lead to stream acidification and high aluminum concentrations toxic to fish.

Since this early study, fractionation of snowpack chemicals into the early meltwater volumes has been studied by many investigators (for example Colbeck, 1981; Cadle and Dasch, 1987a; Williams and Melack, 1991; Davies *et al.*, 1987; Bales *et al.*, 1989). Results from Hudson and Golding (1998) show the effects of fractionation on the distribution of snowpack impurities within snowpack layers at a British Columbia site. Concentrations in the snowpack for several ions were completely shifted from higher concentrations in the surface layers to higher concentrations in deeper layers (Figure 8.3) due to fractionation during a relatively brief April 13–15 melt cycle.

Winter climate and the frequency of melt–freeze events helps to control the magnitude of acidic snowmelt releases. Regions with very cold winters, infrequent thaws, and deep snowpacks will have fractionation effects isolated in a few large

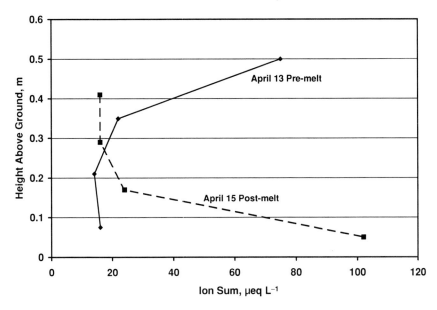

Figure 8.3 Change in the sum of ion concentrations with height above ground in the snowpack during a three-day melt sequence in a British Columbia, Canada snowpack (adapted from Fig 2–3, Hudson and Golding 1998, with permission of copyright holders, International Water Association Publishing). The sum of concentrations represents nitrate, sulfate, chloride, calcium, and sodium concentrations in the snowpack at the 240 Watershed site.

acidic snowmelt events in spring. Davis *et al.* (1995) have found that the highest concentrations in meltwater occur with longer-duration melt events and lower-melt-rate events that allow more time for washout of impurities and for meltwater to reach the base of the snowpack. Melt or rain water that gradually refreezes in deep snowpacks as ripening occurs, will tend to store impurities in ice lenses near the base of the snowpack until the beginning of melt. In regions with shallower snowpacks and frequent winter thaws and rainfall, fractionation tends to gradually reduce snowpack acidity due to frequent liquid-water releases and lessen the severity of snowmelt events.

The mechanism for water delivery from the snowpack can interact with fractionation effects. Suzuki *et al.*, 1994 showed that meltwater released from the snowpack during a daily melt cycle came first from meltwater with high NO_3^- concentration that was stored in the base of the snowpack from prior days' melt. Meltwater release later in the same day came from current melting of surface snow with a low NO_3^- concentration. Thus it appeared that the current day's melt forced the release of stored water from the snowpack base in a type of translatory flow

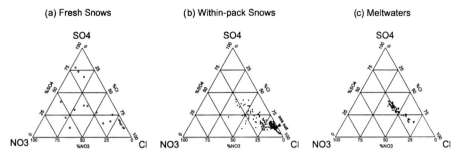

Figure 8.4 Ionic composition of (a) fresh snow, (b) within-pack snow, and (c) meltwater in the Scottish Highlands showing effects of preferential elution (Tranter *et al.*, 1986, © 1986 by permission from Elsevier).

process often ascribed to watershed soils and caused the NO_3^- concentrations in meltwater to vary diurnally from high to low.

8.5.2 *Preferential elution*

More recent studies have shown that in addition to fractionation effects, not all ions are flushed from the snowpack with the same efficiency in the early meltwater (Davies *et al.*, 1982). The tendency for some ions to be washed out before others is referred to as preferential elution. Some conflicting results from field and laboratory experiments occur, but Brimblecombe *et al.* (1985) suggest the following elution sequence from high to low mobility in snowpacks:

$$SO_4^{2-} > NO_3^- > NH_4^+ > K^+ > Ca^{2+} > Mg^{2+} > H^+ > Na^+ > Cl^-.$$

The reasons for preferential elution are not clear, but may be related to the ability of ions to become incorporated into the ice crystal lattice rather than on the surfaces of ice grains (Davies *et al.*, 1987). Since the major anions in snow are generally SO_4^{2-}, NO_3^+, and Cl^-, the shifts in snowpack anion chemistry due to preferential elution can be shown on a triangular diagram as in Figure 8.4 (Tranter *et al.*, 1986). Preferential elution caused fresh snow that had an initial wide range of anion concentrations (Figure 8.4(a)) to be gradually depleted of SO_4^{2-} and NO_3^- and dominated by Cl^- (Figure 8.4(b)). Meltwaters from these snowpacks were relatively enriched in SO_4^{2-} and NO_3^+ compared to within-pack snow (Figure 8.4(c)).

The effects of fractionation and preferential elution combined tend to create more acidic releases of meltwater early in the season. Not only are more impurities in the snowpack released early in the melt season, but the more acidic anions,

SO_4^{2-} and NO_3, and H^+ tend to be released earlier. Thus, where deep acidic snow-packs occur these processes will tend to create an early acidic pulse of meltwater that can influence soil-water and streamflow chemistry.

8.6 Streamflow chemistry and snowmelt

Streamflow during snowmelt represents a mixture of event water from snowmelt or rain-on-snow and pre-event water from subsurface soil and groundwater. Melt-waters can be acidic especially when derived from polluted snowpacks. Pre-event water itself represents a mixture of waters that have been in contact with subsurface organic and inorganic soil and rock material on a watershed for varying times from days to decades. Pre-event water that has been in contact with easily weathered min-erals and/or been in the subsurface for long periods of time may have high levels of dissolved solids and considerable acid neutralizing capacity (ANC). Depend-ing upon the fractional contribution and chemistry of meltwater and the buffering capacity of pre-event water, streamflow events from snowmelt can become quite acidic and cause toxic conditions for fish.

Results from Sharpe *et al.* (1984) given in Figure 8.5 serve as an example of the interaction of meltwater and pre-event water on two adjacent basins in Penn-sylvania, USA. One stream, Wildcat Run, had high ANC pre-event water due to contact of groundwater with easily weathered limestone strata. On the adjacent basin, McGinnis Run, pre-event water had a low ANC because groundwater was largely exposed to weather-resistant sandstone. During a large rain-on-snow event both watersheds registered significant increases in streamflow and drops in pH, but stream pH in Wildcat Run never dropped below 5.6 while pH in McGinnis Run dropped to nearly 4.4 at peak flow. Low pH in McGinnis Run was accompanied by high dissolved-aluminum concentrations toxic to fish. McGinnis Run has no resi-dent trout, while the adjacent Wildcat Run has a healthy, reproducing population of trout. Acidic meltwater obviously can have significant impacts on stream chemistry and fish, but the impacts will vary depending upon the biogeochemical nature of the watershed.

Another example of meltwater effects on stream chemistry is the hydrologic flushing of dissolved organic carbon (DOC) in a Colorado alpine basin dominated by snowmelt (Boyer *et al.*, 1997). In this Colorado basin where flushing of surface soils generally only occurs once annually during snowmelt, stream DOC was observed to increase rapidly early in the melt season, peak before peak streamflow, and then decline rapidly thereafter. Timing of flushing of DOC from soil varied across the basin due to asynchronous snowmelt, but was integrated by streamflow to give an overall flushing time constant of 84 days (time required for DOC to decrease to 1/e or 37% of its initial value) for the basin.

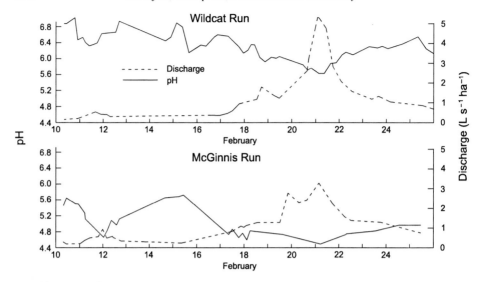

Figure 8.5 Comparison of stream discharge and pH between two adjacent water-sheds during a major rain-on-snow event in southwestern Pennsylvania during 1981 (from Sharpe *et al.*, 1984, © 1984 by permission from American Society of Agronomy). Wildcat Run with well-buffered pre-event water showed a much smaller pH decline during the event than McGinnis Run with poorly buffered pre-event water.

The fraction of streamflow due to current melting can be determined using a simple two-component model and conservative chemical or isotopic tracers:

$$Q_e/Q_t = (c_{stream} - c_{pre\text{-}event})/(c_{event} - c_{pre\text{-}event}) \qquad (8.5)$$

where Q_e is the streamflow rate due to event water from current melt, Q_t is the total streamflow rate, c_{stream} is the concentration of some tracer in the total streamflow, $c_{pre\text{-}event}$ is the concentration of the tracer in the stream baseflow prior to melt, and c_{event} is the concentration in the meltwater or snowpack. An example from work by Sueker (1995) using Na^+ as a tracer in the Rocky Mountains will be used to illustrate this method. Sueker (1995) found that the Na^+ concentration in the bulk snow on Spruce Creek was about 1.5 μeq L^{-1} which was used to represent c_{event}, while the stream at baseflow was 44.2 μeq L^{-1} which was used to represent $c_{pre\text{-}event}$. Spruce Creek at peak streamflow during melt had a Na^+ concentration of about 28 μeq L^{-1}, which gave a fraction of event water at peak flow of:

$$Q_e/Q_t = (28 - 44.2)/(1.5 - 44.2) = 0.38 \text{ or } 38\%$$

Stable isotopes ^2H and ^{18}O in water preferably should be used as conservative tracers in computing the fraction of event water during snowmelt (see Rodhe, 1998). Ideally, meltwater chemistry should be used to index event-water concentrations.

A chemical mass flux density in streamflow can be computed just as for fluxes in precipitation and meltwater exiting the snowpack. If the streamflow rate (Q_t) and ion concentration (c_{stream}) are known, then the mass flux density in streamflow (F_{st}) can be obtained as:

$$F_{st} = Q_t c_{stream} \qquad (8.6)$$

where Q_t is expressed in volume per unit watershed area per unit of time, c_{stream} is given in mass per unit volume and computations follow previous examples for Equation (7.1) and (7.4). It is customary to compare the mass flux densities of ion inputs to watersheds computed with Equation (8.1) and Equation (8.2) with streamflow export computed with Equation (7.4) and Equation (7.6).

Another major, albeit indirect, impact of snowpacks on the biogeochemistry of watersheds and streamflow is control of nitrogen processing in the soil (Brooks and Williams, 1999). The depth and the duration of snowcover can affect soil temperatures and the rates of nitrogen assimilation by soil microbes. Shallow or intermittent snowcover allows surface soils to cool to temperatures below the threshold for soil-heterotrophic microbes that can assimilate nitrogen. Without uptake by soil heterotrophs, relatively large amounts of nitrogen from decomposition of organic matter over the winter will be leached and appear in streamflow. In contrast, soils protected by deep snow throughout much of the winter season will stay warm enough to permit heterotroph activity and consequent low nitrogen export to streamflow. Nitrogen export in streamflow from watersheds in the same region, e.g. alpine vs. subalpine (Heuer *et al.*, 1999), can vary significantly depending upon depth and duration of snow cover. In the growing season uptake by plants generally ties up much of the available nitrogen.

8.7 Snowpack stable isotopes

Isotopes of hydrogen (^1H, ^2H, ^3H) and oxygen (^{16}O, ^{17}O, ^{18}O) in water have played an increasingly important role as conservative tracers and indicators of snowpack processes in snow hydrology research in recent years. Tritium (^3H) is a radioactive isotope with a very low abundance which decays with a half-life of 12.26 years. Other isotopes do not decay or spontaneously disintegrate and are defined as stable isotopes. Water molecules formed from heavier isotopes of hydrogen and oxygen (e.g. $^1H_2{}^{18}O$ or $^2H_2{}^{16}O$) have greater mass than $^1H_2{}^{16}O$ molecules that increases the binding energy and reduces the mobility of the heavier molecules. This can result in shifts in the relative abundances of heavy and light isotopes, during phase changes and chemical and biological reactions, referred to as isotope fractionation. A brief review of stable-isotope geochemistry with applications to snow hydrology is given here. More complete discussions can be found in Kendall and McDonnell (1998, see especially Chapters 4 and 12), Clark and Fritz (1997), and Stichler (1987).

Abundances of heavier isotopes of hydrogen and oxygen in water molecules are small but can be measured quite accurately with a mass spectrometer. Natural hydrogen isotopes in water ^{1}H, ^{2}H (deuterium or D), and ^{3}H (tritium) occur with relative abundances of about 0.99985, 0.00015, and 10^{-16} to 10^{-18}, respectively, while the oxygen isotopes in water ^{16}O, ^{17}O, ^{18}O occur with relative abundances of about 0.99757, 0.00038, and 0.00205, respectively. Abundance of a heavier stable isotope relative to a lighter isotope in a water sample (R_x) is generally described relative to the ratio of the heavier to lighter abundances in a standard (R_s) by:

$$\delta = (R_x/R_s - 1)\,1000 \tag{8.7}$$

where δ or delta is dimensionless and commonly given as per mil or parts per thousand (‰), R_x is the ratio of numbers of heavier atoms to lighter atoms (e.g. ^{18}O/^{16}O) in a sample, and R_s is the same ratio found in a known standard. Vienna Standard Mean Ocean Water or VSMOW is commonly used as a standard for ^{2}H or ^{18}O. For example, if a precipitation sample had an abundance ratio of 0.0019650 for ^{18}O/^{16}O based upon analysis with a mass spectrometer and given the VSMOW reference ratio for ^{18}O/^{16}O was 0.0020052 (Kendall and Caldwell, 1998), then:

$$\delta\,^{18}O = [(0.0019650/0.0020052) - 1]\,1000 = -20.05\,‰$$

This indicates that the water sample was depleted in ^{18}O relative to VSMOW by 20.05 ‰ which is equivalent to depletion by 2.005%. Ocean water would have a relative abundance similar to that of VSMOW and $\delta\,^{18}O \approx 0$.

Deuterium $\delta\,^2$H values in precipitation are related to $\delta\,^{18}$O values due to similarity in fractionation processes for these isotopes as (Craig, 1961):

$$\delta\,^2H\,(‰) = 8\,\delta\,^{18}O + 10 \tag{8.8}$$

This relationship, known as the Global Meteoric Water Line (GMWL), expresses the average relationship found in freshwaters worldwide and is a useful reference for comparisons with waters from various climates and geographic locations. In the above example, the value of $\delta\,^2$H for the precipitation sample corresponding to the value of $\delta\,^{18}$O given would be -150.4 ‰ according to the GMWL, Equation (8.8). VSMOW has a deuterium abundance ratio of 0.00015575 (Kendall and Caldwell, 1998).

Due to differences in rates of evaporation–sublimation and melt–freezing among heavier (^{1}H$_2$^{18}O or ^{2}H$_2$^{16}O) and lighter (^{1}H$_2$^{16}O) water molecules, the ratio of isotopes that appear in liquid, solid, and vapor phases of snowpack water can change over time and space. Fractionation among the phases of water under equilibrium conditions gives:

$$\delta\ ^{18}O\ \text{ice} > \delta\ ^{18}O\ \text{liquid} > \delta\ ^{18}O\ \text{vapor}$$

and

$$\delta\ ^2H\ \text{ice} > \delta\ ^2H\ \text{liquid} > \delta\ ^2H\ \text{vapor}.$$

For example, lighter water molecules $^1H_2{}^{16}O$ have a higher surface vapor pressure causing them to evaporate or sublimate (lose mass) more rapidly than heavier $^1H_2{}^{18}O$ or $^2H_2{}^{16}O$ molecules. Consequently, snowpack ice and liquid surfaces exposed to the atmosphere can become relatively enriched in the heavier $^1H_2{}^{18}O$ or $^2H_2{}^{16}O$ molecules during vapor loss, while vapor becomes enriched with $^1H_2{}^{16}O$ molecules.

Sublimation and vapor migration during crystal metamorphism within and at the surface of snowpacks can cause isotope fractionation during winter. Cooper (1998) showed the results of a classic experiment by Sommerfield *et al.* (1991) (Figure 8.6) that showed how sublimation under subfreezing conditions could alter the stable-isotope distribution in a snow column subjected to temperature gradient metamorphism. Surface as well as deeper snow column layers showed relative enrichment of heavier ^2H (deuterium or D) and ^{18}O isotopes due to fractionation caused by sublimation and vapor migration after 31 days.

During condensation in the atmosphere, the vapor phase becomes relatively enriched in $^1H_2{}^{16}O$, while the condensate (precipitation) becomes relatively enriched in $^1H_2{}^{18}O$ or $^2H_2{}^{16}O$. Thus atmospheric vapor transported from the oceans over land becomes gradually depleted in heavier water molecules due to rain and snowfall occurring along the air-mass trajectory, causing vapor and precipitation to be progressively lighter or more depleted in $^1H_2{}^{18}O$ or $^2H_2{}^{16}O$ with increasing distance from the ocean. Isotopic fractionation is also more pronounced with colder temperatures such that this depletion in $^1H_2{}^{18}O$ or $^2H_2{}^{16}O$ in precipitation is generally more pronounced in winter and at higher elevations and higher latitudes due to temperature. This causes snowfall, snowpack, and meltwater to be relatively depleted in heavier isotopes compared to summer rainfall, soil water, and streamflow. Since air masses have varying trajectories over land, varying amounts of mixing with vapor from terrestrial sources, and differing temperature chronologies, each precipitation event can therefore be somewhat unique in terms of its stable-isotope profile. Individual snowfall events can often be detected from their isotope signatures within the snowpack, but the snowpack isotope profile can also change and homogenize due to crystal metamorphism, vapor migration, wind drifting and scour, and rainfall or intermittent melt events.

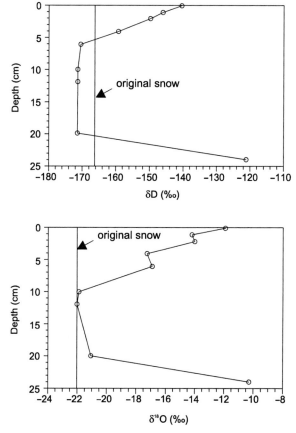

Figure 8.6 Isotopic profiles of $\delta\,^2H$ and $\delta\,^{18}O$ with depth in a snow column after 31 days of temperature gradient metamorphism showing enrichment in surface and bottom layers due to fractionation caused by vapor movement and sublimation (Cooper, 1998 after Sommerfield *et al.*, 1991, © 1991 with permission from Elsevier).

Melting also causes fractionation, whereby $^1H_2^{16}O$ molecules become enriched in meltwater, while heavier $^1H_2^{18}O$ molecules become enriched in the residual snowpack. This fractionation occurs both during melt of surface layers of the snowpack and as water percolates downward through the snowpack and contacts deeper ice layers. Initial meltwater leaving a seasonal snowpack will be relatively depleted and later melt will be relatively enriched in heavier isotopes. Stable-isotope levels in meltwater leaving the snowpack and travelling through or over the soil to streams will thus vary during the melt season (Figure 8.7). Taylor *et al.* (2001) showed that melt rate can also affect the degree of fractionation, with meltwater being more depleted in heavier isotopes during periods of lower flow rates from the

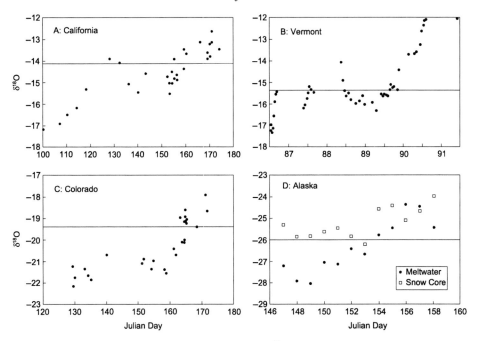

Figure 8.7 Progressive changes in meltwater $\delta^{18}O$ during the melt season at four US sites showing depletion in early melt followed by enrichment in later melt relative to the original snowpack condition (horizontal line) (Taylor *et al.*, 2002, © 2002 with permission from John Wiley and Sons).

snowpack. Overall the message is clear, when doing hydrograph separations using stable isotopes (see Equation (8.5)) it is important to sample the meltwater leaving the snowpack for the period of interest, rather than using an average snowpack sample to characterize the new water input (Taylor *et al.*, 2002).

8.8 References

Bales, R. C., Davis, R. E., and Stanley, D. A. (1989). Ion elution through shallow homogeneous snow. *Water Resour. Res.*, **25**, 1869–77.

Barrie, L. A. (1985). Atmospheric particles: their physical and chemical characteristics, and deposition processes relevant to the chemical composition of glaciers. *Annals Glaciol.*, **7**, 100–8.

Boyer, E. W., G. M. Hornberger, K. E. Bencala, and D. M. McKnight. (1997). Response characteristics of DOC flushing in an alpine catchment. *Hydrol. Processes*, **11**, 1635–47.

Brimblecombe, P., Tranter, M., Abrahams, P. W., Blackwood, I., Davies, T. D., and Vincent, C. E. (1985). Relocation and preferential elution of acidic solute through the snowpack of a small, remote, high-altitude Scottish catchment. *Annals Glaciol.*, **7**, 141–47.

Brooks, P. D. and Williams, M. W. (1999). Snowpack controls on nitrogen cycling and export in seasonally snow-covered catchments. *Hydrol. Processes*, **13**(14–15), 2177–90.

Brown, J. C. and Skau, C. M. (1975). *Chemical composition of snow in the East Central Sierra Nevada*. Coop. Rpt. Series, Publ. AG-1, Reno, NV: Renewable Natural Resour. Div., University of Nevada.

Cadle, S. H. and Dasch, J. M. (1987a). Composition of snowmelt and runoff in northern Michigan. *Environ. Sci. Technol.*, **21**(3), 295–9.

Cadle, S. H. and Dasch, J. M. (1987b). The contribution of dry deposition to snowpack acidity in Michigan. In *Seasonal Snowcovers: Physics, Chemistry, Hydrology*, NATO ASI Series C: Math. and Phys. Sci., vol. 211, ed. H. G. Jones and W. J. Orville-Thomas, pp. 299–320.

Clark, I. D. and Fritz, P. (1997). *Environmental Isotopes in Hydrogeology*. Boca Raton, FL: Lewis Publishers.

Colbeck, S. C. (1981). A simulation of the enrichment of atmospheric pollutants in snow cover runoff. *Water Resour. Res.*, **17**(5), 1383–8.

Cooper, L. W. (1998). Chapter 4: Isotopic Fractionation in Snow Cover. In *Isotope Tracers in Catchment Hydrology*, ed. B. V. Kendall and J. J. McDonnell, Amsterdam: Elsevier, pp. 119–36.

Cragin, J. H. and McGilvary, R. (1995). Can inorganic chemical species volatilize from snow? In *Biogeochemistry of Seasonally Snow-Covered Catchments*, IAHS Publ. 228, ed. K. A. Tonnessen, M. W. Williams, and M. Tranter. Proc. Boulder Symp., July 1995, pp. 11–16.

Craig, H. (1961). Isotopic variations in meteoric waters. *Science*, **133**, 1702–3.

Davidson, C. I. and Wu, Y. L. (1990). Dry deposition of particles and vapors. In *Acidic Precipitation: Sources, Deposition, and Canopy Interactions*, vol. 3, ed. S. E. Lindberg, A. L. Page, and S. A. Norton. New York: Springer-Verlag, pp. 103–216.

Davidson, C. I., Bergin, M. H., and Kuhns, H. D. (1996). The deposition of particles and gases to ice sheets. In *Chemical Exchange between the Atmosphere and Polar Snow*, NATO ASI Series I, Global Environ. Change, vol. 43, ed. E. W. Wolff and R. C. Bales. Berlin: Springer-Verlag, pp. 275–306.

Davies, T. D., Abrahams, P. W., Tranter, M., Blackwood, I., Brimblecombe, P., and Vincent, C. E. (1984). Black acidic snow in the remote Scottish Highlands. *Nature*, **312**, 58–61.

Davies, T. D., Brimblecombe, P., Blackwood, I. L., Tranter, M., and Abrahams, P. W. (1988). Chemical composition of snow in the remote Scottish Highlands. In *Acid Deposition at High Elevation Sites*, NATO ASI Series C, vol. 252, ed. M. H. Unsworth and D. Fowler. Dordrecht: Kluwer Academic, pp. 517–40.

Davies, T. D., Brimblecombe, P., Tranter, M., Tsiouris, S., Vincent, C. E., Abrahams, P. and Blackwood, I. L. (1987). The removal of soluble ions from melting snowpacks. In *Seasonal Snowcovers: Physics, Chemistry, Hydrology*, NATO ASI Series C: Math. and Phys. Sci., vol. 211, ed. H. G. Jones and W. J. Orville-Thomas, pp. 337–92.

Davies, T. D., Vincent, C. E., and Brimblecombe, P. (1982). Preferential elution of strong acids from a Norwegian ice cap. *Nature*, **300**, 161–3.

Davis, R. E., Petersen, C. E., and Bales, R. C. (1995). Ion flux through a shallow snowpack: effects of initial conditions and melt sequences. In *Biogeochemistry of Seasonally Snow-Covered Catchments.*, IAHS Publ. 228, ed. K. A. Tonnessen, M. W. Williams, and M. Tranter, pp. 115–28.

DeWalle, D. R. (1987). Review of snowpack chemistry studies. In *Seasonal Snowcovers: Physics, Chemistry, Hydrology*, NATO ASI Series C: Math. and Phys. Sci., vol. 211, ed. H. G. Jones and W. J. Orville-Thomas, pp. 255–68.

DeWalle, D. R., Sharpe, W. E., Izbicki, J. A., and Wirries, D. L. (1983). Acid snowpack chemistry in Pennsylvania, 1979–81. *Water Resour. Bull.*, **19**(6), 993–1001.

Heuer, K., Brooks, P. D. and Tonnessen, K. A. (1999). Nitrogen dynamics in two high elevations catchments during spring snowmelt. *Hydrol. Processes*, **13**(14–15), 2203–14.

Hogan, A. and Leggett, D. (1995). Soil-to-snow movement of synthetic organic compounds in natural snowpack. In *Biogeochemistry of Seasonally Snow-Covered Catchments*, IAHS Publ. 228, ed. K. A. Tonnessen, M. W. Williams, and M. Tranter, pp. 97–106.

Hudson, R. O. and Golding, D. L. (1998). Snowpack chemistry during snow accumulation and melt in mature subalpine forest and regenerating clear-cut in the southern interior of B. C. *Nordic Hydrol.*, **29**, 221–4.

Johannessen, M. and Henriksen, A. (1978). Chemistry of snow meltwater: changes in concentration during melting. *Water Resour. Res.*, **14**(4), 615–19.

Jones, H. G. (1987). Chemical dynamics of snowcover and snowmelt in a boreal forest. In *Seasonal Snowcovers: Physics, Chemistry, Hydrology*. NATO ASI Series C: Math. and Phys. Sci., vol. 211, ed. H. G. Jones and W. J. Orville-Thomas, pp. 531–74.

Jones, H. G. (1999). The ecology of snow-covered systems: a brief overview of nutrient cycling and life in the cold. *Hydrol. Processes*, **13**, 2135–47.

Kendall, C. and Caldwell, E. A. (1998). Chapter 2: Fundamentals of Isotope Geochemistry. In *Isotope Tracers in Catchment Hydrology*, ed. B. V. Kendall and J. J. McDonnell. Amsterdam: Elsevier, pp. 51–86.

Kendall, C. and McDonnell, J. J. (1998). *Isotope Tracers in Catchment Hydrology*. Amsterdam: Elsevier.

Massman, W., Sommerfeld, R., Zeller, K., Hehn, T., Hudnell, L., and Rochelle, S. (1995). CO2 flux through a Wyoming seasonal snowpack: diffusional and pressure pumping effects. In *Biogeochemistry of Seasonally Snow-Covered Catchments*, ed. K. A. Tonnessen, M. W. Williams, and M. Tranter. IAHS Publ. 228, pp. 71–80.

Nakawo, M., Hayakawa, N., and Goodrich, L. E. (1998). Chapter 9: Changes in snow pack and melt water chemistry during snowmelt. In *Snow and Ice Science in Hydrology. International Hydrological Programme*, 7th Training Course on Snow Hydrology, Nagoya University and UNESCO, pp. 119–32

Pomeroy, J. W. and Jones, H. G. (1996). Wind-blown snow: sublimation, transport and changes to polar-snow. In *Chemical Exchange between the Atmosphere and Polar Snow*, NATO ASI Series I, Global Environ. Change, vol. 43, ed. E. W. Wolff and R. C. Bales. Berlin: Springer-Verlag, pp. 453–89.

Pomeroy, J. W., Davies, T. D., Jones, H. G., Marsh, P., Peters, N. E., and Tranter, M. (1999). Transformations of snow chemistry in the boreal forest: accumulation and volatilization. *Hydrol. Processes*, **13**(14–15), 2257–73.

Raynor, G. S. and Hayes, J. V. (1983). Differential rain and snow scavenging efficiency implied by ionic concentration differences in winter precipitation. In *Precipitation Scavenging, Dry Deposition, and Resuspension, 4th Intern. Conf Proc. Precipitation Scavenging.*, vol. 1, US DOE and EPA, Santa Monica, CA, ed. H. R. Pruppacher, R. G. Semonin and W. G. N. Slinn. New York: Elsevier, pp. 249–62.

Rodhe, A. (1998). Chapter 12: Snowmelt-dominated systems. In *Isotope Tracers in Catchment Hydrology*, ed. Kendall, C. and J. J. McDonnell. Amsterdam: Elsevier, pp. 391–433.

Schaefer, D. A. and Reiners, W. A. (1990). Throughfall chemistry and canopy processing mechanisms. In *Acidic Precipitation: Sources, Deposition, and Canopy Interactions*, vol. 3, ed. S. E. Lindberg, A. L. Page, and S. A. Norton. New York: Springer-Verlag, pp. 241–84.

Sharpe, W. E., DeWalle, D. R., Leibfried, R. T., Dinicola, R. S., Kimmel, W. G., and Sherwin, L. S. (1984). Causes of acidification of four streams on Laurel Hill in southwestern Pennsylvania. *J. Environ. Qual.*, **13**(4), 619–31.

Sommerfield, R. A., Judy, C., and Friedman, I. (1991). Isotopic changes during formation of depth hoar in experimental snowpacks. In *Stable Isotope Geochemistry: A Tribute to Samuel Epstein*, ed. H. P. Taylor, J. R. O'Neil, and I. R. Kaplan. San Antonio, TX: The Geochemical Society, pp. 205–10.

Stichler, W. (1987). Snowcover and snowmelt processes studied by means of environmental isotopes. In *Seasonal Snowcovers: Physics, Chemistry, Hydrology*. NATO ASI Series C: Math. and Phys. Sci., vol. 211, ed. H. G. Jones and W. J. Orville-Thomas, pp. 673–726.

Stottlemyer, R. and C. A. Troendle. (1995). Surface water chemistry and chemical budgets, alpine and subalpine watersheds, Fraser Experimental Forest, Colorado. In *Biogeochemistry of Seasonally Snow-Covered Catchments*. IAHS Publ. 228, ed. K. A. Tonnessen, M. W. Williams, and M. Tranter, pp. 321–8.

Sueker, J. K. (1995). Chemical hydrograph separation during snowmelt for three headwater basins in Rocky Mountain National Park, Colorado. In *Biogeochemistry of Seasonally Snow-Covered Catchments*, IAHS Publ. 228, ed. K. A. Tonnessen, M. W. Williams, and M. Tranter, pp. 271–82.

Suzuki, K., Ishii, Y., Kodama, Y., Kobayashi, D., and Jones, H. G. (1994). Chemical dynamics in a boreal forest snowpack during the snowmelt season. In *Snow and Ice Covers: Interactions with Atmosphare and Ecosystems*, IAHS Publ. 223, ed. H. G. Jones, T. D. Davies, A. Ohmura, and E. M. Morris, pp. 313–22.

Taylor, S., Feng, X., Kirchner, J. W., Osterhuber, R., Klaue, B., and Renshaw, C. E. (2001). Isotopic evolution of a seasonal snowpack and its melt. *Water Resour. Res.*, **37**(3), 759–69.

Taylor, S., Feng, X., Williams, M., and McNamara, J. (2002). How isotopic fractionation of snowmelt affects hydrograph separation. *Hydrol. Processes*, **16**, 3683–90.

Tranter, M., Brimblecombe, P., Davies, T. D., Vincent, C. E., Abrahams, P. W., and Blackwood, I. (1986). The composition of snowfall, snowpack and meltwater in the Scottish Highlands-evidence for preferential elution. *Atmos. Environ.*, **20**(3), 517–25.

Waddington, E. D., Cunningham, J., and Harder, S. (1996). The effects of snow ventilation on chemical concentrations. In *Chemical Exchange between the Atmosphere and Polar Snow*, NATO ASI Series I, Global Environ. Change, vol. 43, ed. E. W. Wolff and R. C. Bales. Berlin: Springer-Verlag, pp. 403–52.

Williams, M. W. and Melack, J. M. (1991). Solute chemistry of snowmelt and runoff in an alpine basin, Sierra Nevada. *Water Resour. Res.*, **27**(7), 1575–88.

Williams, M. W., Tonnessen, K. A., Melack, J. M., and Daqing, Y. (1992). Sources and spatial variation of the chemical composition of snow in the Tien Shan, China. *Annals Glaciol.*, **16**, 25–32.

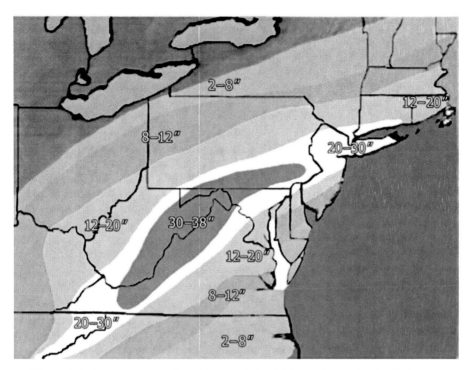

Figure 1.4 Snow depth map from January 6–8, 1996 associated with the "Blizzard of 1996," centered over the middle Atlantic United States (after WRC-TV/NBC 4, Washington, DC analysis).

Figure 1.5 Transportation department snow blower at work clearing Highway 143 between Cedar Breaks and Panguitch, Utah in 2005. (Courtesy R. Julander.)

Figure 1.7 Instrumentation at the Central Sierra Snow Laboratory in Soda Springs, California. (Courtesy R. Osterhuber.)

Figure 1.9 Site of the 239 km^2 Reynolds Creek Experimental Watershed in the Owyhee Mountains about 80 km southwest of Boise, Idaho, operated by the Agriculture Research Service. (Courtesy USDA, Agriculture Research Service; www.ars.usda.gov/is/graphics/photos/.)

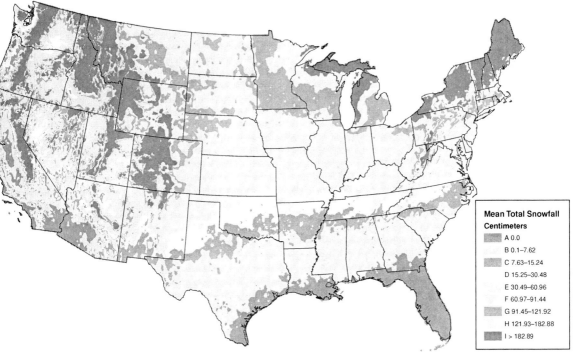

Mean Total Snowfall
Centimeters

A 0.0
B 0.1–7.62
C 7.63–15.24
D 15.25–30.48
E 30.49–60.96
F 60.97–91.44
G 91.45–121.92
H 121.93–182.88
I > 182.89

Data Source: NOAA, National Environmental Satellite, Data and Information Service (NESDIS), National Climatic Data Center

Figure 2.6 Mean annual snowfall in the continental United States (modified from USDC, NOAA, National Climate Data Center, Climate Maps of the United States).

Figure 3.13 Flow fingers for preferential liquid-water movement that formed during initial day's melting of a homogeneous 28-cm depth snowpack resulting from a single snowfall in Pennsylvania (photo D. R. DeWalle). Water mixed with food coloring was uniformly sprayed over the snowpack surface to trace water movement. Also note evidence of meltwater movement downslope to the right in the surface layer and at the soil–snow interface.

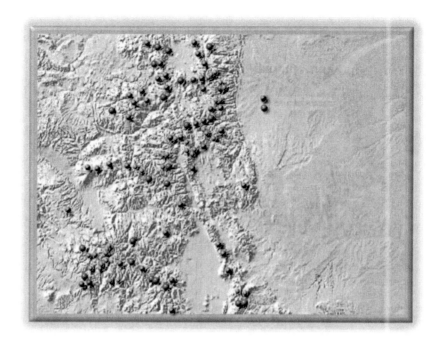

Figure 4.9 Colorado SNOTEL sites (courtesy Natural Resources Conservation Service, US Department of Agriculture).

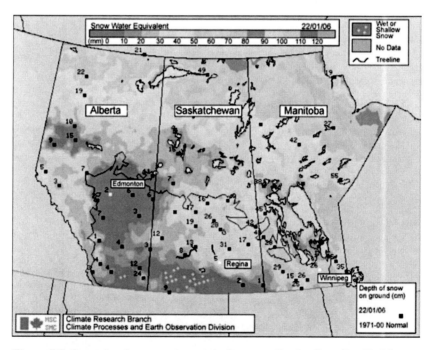

Figure 5.3 Operational snow-water-equivalent map for the Canadian Prairies on January 22, 2006 derived from passive-microwave satellite data. Numbers on the map indicate point measurements of observed snow depth. Maps are distributed by posting on the World Wide Web, www.socc.ca/SWE/snow˙swe.html/. (Reproduced with the permission of the Minister of Public Works and Government Services Canada, 2006.)

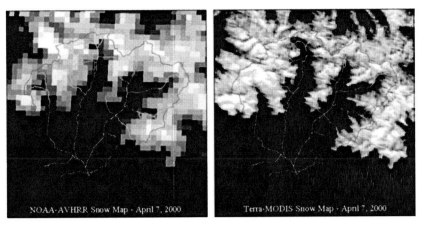

Figure 5.4 Comparison of NOAA-AVHRR and MODIS derived snow cover for the Noguera Ribagorzana Basin (572.9 km^2) in the Central Pyrenees of Spain on April 7, 2000. The different gray levels correspond to different percents of snow cover in each pixel. Snow cover in the Basin totals 181 km^2 from AVHRR and 184 km^2 from MODIS as reported by Rango *et al.* (2003).

Figure 6.1 Radiation instruments at the Penn State, Pennsylvania SURFRAD Network site: (a) pyranometer used to measure incoming shortwave radiation on a horizontal surface, (b) a continuously shaded pyranometer (near) and pyrgeometer (far) used to measure incoming diffuse shortwave and incoming longwave radiation, respectively, (c) tubular pyrheliometer (see arrow) mounted on a solar tracker to measure direct-beam solar radiation at normal incidence, and (d) inverted pyranometer and pyrgeometer on tower to measure reflected shortwave and outgoing longwave radiation from the ground surface (photographs by D. DeWalle).

Figure 6.5 Solar-powered climate stations equipped with data loggers (left) are used to provide data (wind speed, air temperature, humidity) to support computations of snowpack convective exchange. The addition to climate stations of sonic anemometers (top right), fine-wire thermocouples, and high-speed open-path gas analyzers for CO_2 and water vapor (bottom right) allows computation of convective exchange using the eddy covariance method (photographs by D. DeWalle).

Figure 7.4 Prediction of snowmelt in forests requires accounting for canopy effects on snowpack energy exchange, especially radiation exchange. Shadow patterns suggest varying patterns of canopy shading between coniferous forests with foliage (top) and deciduous forests without foliage (bottom) in these pictures from mid-latitude Pennsylvania, USA sites.

Figure 9.4 Rapid delivery of meltwater as overland flow on frozen ground in a pasture swale that has turned the snow into a slush layer.

Figure 12.10 Urban snow requires dedicated areas for snow storage (left) and may lead to accumulation of surface contaminants later in the spring (right) (photos by D. R. DeWalle).

Figure 12.11 Ratnik low-energy modified snow gun (left) and modified fan-type Lenko snow gun (right) being used to augment snowpack for skiing (courtesy of Ratnik Industries, Inc. and Lenko Quality Snow, Inc., respectively).

Figure 12.12 Large slab avalanche in Colorado, USA that began in the alpine startup zone (photo courtesy of Richard Armstrong, US National Snow and Ice Data Center, Boulder, CO).

9

Snowmelt-runoff processes

9.1 Introduction

Understanding the generation of streamflow from snowmelt involves integration of processes that produce meltwater within the snowpack, processes that delay and store liquid water in the snowpack and processes that direct the flow of meltwater through watersheds to stream channels. Computations of liquid-water dynamics in snowpacks and melt rates have been covered in prior chapters. In this chapter these concepts and computations are synthesized with a discussion of the movement of meltwater within watersheds and the resulting hydrographs of snowmelt runoff. Emphasis is given to factors unique to generation of streamflow from snowmelt, such as frozen-ground effects, rain-on-snow flooding and snow damming.

9.2 Snowpack water balance

The snowpack water balance is fundamental to understanding and modelling snowmelt runoff generation. The snowpack water balance can be written as:

$$\Delta SWE = P \pm E - O \qquad (9.1)$$

where:

$\Delta SWE = (SWE_{t2} - SWE_{t1}) =$ change in water equivalent (liquid and solid) of the snowpack over a time interval $t_2 - t_1$
$P =$ net precipitation inputs from rainfall, snowfall, and wind redistribution
$E =$ net vapor exchange between the snowpack and environment by sublimation, evaporation, and condensation
$O =$ outflow of liquid water from the snowpack base

All terms in the Equation (9.1) can be expressed either as a water mass change per unit area per unit time such as kg m^{-2} or m depth per unit time. The water-balance model is linked to the snowpack energy-budget model given in Chapter 6 through the respective outflow (melt), evaporation/sublimation, and precipitation terms.

Outflow of liquid water from a snowpack represents water from melt (M) and rainfall (Pr) and internal changes in liquid-water holding capacity due to metamorphosis. During non-rain periods, nearly all outflow is due to melting. Daily melt rates up to 2 cm d^{-1} occur frequently and maximum daily rates of 7–8 cm d^{-1} have been noted (Kuchment, 1997; Cooley and Palmer, 1997). Hourly melt rates can reach 0.1 to 0.3 cm h^{-1}. Meltwater generated at the surface of the snowpack can refreeze deeper within unripe snowpacks or be temporally stored as liquid water within the snowpack by ice layers. Thus the timing of outflow at the snowpack base can be altered considerably by storage and transmission within the snowpack.

Precipitation inputs to the snowpack in the form of liquid or solid are considered in the mass balance. Both solid and liquid precipitation may be intercepted in forest canopies with significant canopy interception losses before being added to the snowpack (see Chapter 2). Wind transport of snow may also cause scouring or deposition of snow at a particular site (see Chapter 2). Rain falling on a snowpack with subfreezing temperatures can be refrozen in transit and through release of latent heat of fusion can alter the snowpack energy budget (Chapter 6). Rainfall on ripe snow can move quickly to the soil below and rain-on-snow events are commonly associated with annual peak flows and flooding for climates where heavy rain can occur with snowcover.

Net vapor exchange between the snowpack and the atmosphere can represent a small but significant loss in dry, windy climates. Seasonally, evaporation/sublimation loss from snowpacks generally represents <10% of maximum snowpack water equivalent (Leydecker and Melack, 1999; Cline, 1997), although Aizen *et al.* (1995) reported losses of 30% of annual snow accumulation in central Asia. Sublimation losses from blowing snow can equal 15–45% of winter snowfall in grassland prairie settings (Marsh, 1999). Schmidt *et al.* (1998) estimated snowpack sublimation losses equaled 20% of peak snowpack accumulation or 78 mm over a 40-d period in Colorado subalpine forest. Vapor losses from snowpacks on a daily basis are generally much smaller than melt, ranging up to about 0.2–0.3 cm d^{-1} (Leydecker and Melack, 1999). Vapor gains within the snowpack are common when moist air masses come in contact with cold snowpacks; the process becomes more important late in the melt season when humidity increases due to evapotranspiration. Vapor is also exchanged between the snowpack and soil when moisture mainly in the form of vapor, coupled with heat, is forced upward along the temperature gradient. The importance of this form of vapor exchange has not been completely assessed but in settings like boreal forests it can be important. Vapor-exchange computations are described in Chapter 6.

The net change in snowpack water equivalent (SWE) accounts for changes in the amount of liquid and solid phase water stored in the snowpack over time. Meltwater or rainfall that freezes within the snowpack, liquid water temporally detained within the snowpack and snowfall would represent storage gains, while melt and rain water outflows at the snowpack base and evaporation/sublimation losses at the snowpack surfaces represent storage losses.

Melt rates for comparisons with energy-budget or degree-day computations are often computed using a snowpack water balance. By monitoring the changes in snowpack water equivalent and precipitation over time and assuming no change in storage of meltwater or rainfall in the snowpack, melt rates can be estimated. Evaporation/sublimation losses are generally assumed to be negligible for short time periods with this method. For example, if the water equivalent of a ripe snowpack changed from 27.9 cm to 26.7 cm over a single 24-h period with no precipitation inputs and no vapor exchange losses or gains assumed, the outflow and melt rate for the day would be computed as:

$$(26.7 - 27.9) \text{ cm d}^{-1} = -O$$
$$O = M = 1.2 \text{ cm d}^{-1}$$

If all of the liquid water due to melting during the period had not drained from the snowpack or if some of the melt was refrozen due to incomplete snowpack ripeness, then the final water equivalent would include some refrozen liquid water or liquid water still in transit. In this case the true melt rate would be underestimated.

The interpretation is more complicated when precipitation occurs during the time period, because of the added uncertainty in correctly assessing how much precipitation was rain or snow. Following from the above example, if a precipitation input of 0.37 cm of rainfall (Pr) and a small vapor loss of 0.04 cm also occurred during the 24-hour period and all liquid water drained from the snowpack, then the melt rate would be:

$$(26.7 - 27.9) \text{ cm d}^{-1} = +0.37 - 0.04 - O$$
$$O = 1.53 \text{ cm d}^{-1} = M + Pr = M + 0.37$$
$$M = 1.16 \text{ cm d}^{-1}$$

Since measurements of changes in snowpack liquid-water content or the form of precipitation (rain or snow) are generally lacking, use of the snowpack mass balance to estimate true melt rates can lead to errors.

9.3 Snowpack storage and time lags

Storage and time lags for liquid water to travel through a snowpack can be due to refreezing of liquid water in subfreezing snowpacks with a significant cold content, time required to satisfy snowpack liquid-water holding capacity and, for ripe snowpacks, the time for transmission of liquid water through the porous matrix of ice crystals. Each time lag and storage can be simply computed separately and the times summed to estimate the total delay for liquid water in the snowpack (US Army Corps of Engineers, 1956). Here we are primarily dealing with vertical flows in snowpacks, situations where lateral flows commonly occur through snowpacks such as with snow damming on low-gradient watersheds would require other analysis and is discussed later in this chapter.

9.3.1 Cold-content lag times

The storage and time lag due to cold content can represent a significant loss of water and delay to snowpack outflow. As described in Chapter 3, snowpack cold content (CC) expresses the amount (cm) of liquid water that would have to refreeze within the snowpack to warm it to 0 °C. The units are convenient since the cold content can be compared directly with the rates of liquid water input from melt and rain. Liquid water refrozen within the snowpack is lost to current outflow and represents the first type of snowpack storage. The time lag for melt or rainfall to satisfy snowpack cold content can be computed as:

$$t_c = CC/(P_r + M) \qquad (9.2)$$

where:

t_c = time needed to satisfy snowpack cold content, h
CC = snowpack cold content, cm
P_r = rainfall intensity, cm h^{-1}
M = melt rate, cm h^{-1}

The small amount of sensible heat added by warm rainfall is ignored in the above calculation (see Chapter 6 for discussion of melting due to rain). Melt rates are computed using formulas given in Chapter 6. Time lags can be computed for cold contents that develop over the snow-accumulation season as well as cold contents that develop overnight.

Given seasonal snowpack cold content of about 0.94 cm of liquid water, representing a typical mid-winter snowpack with 30 cm water equivalent and average temperature of −5 °C, cold-content lag times can be computed for various liquid-water input rates. With modest melt rates of about 0.5 cm d^{-1}, the time lag needed to satisfy this cold content would be:

$$t_c = (0.94 \text{ cm})/(0.5 \text{ cm}/24 \text{ h}) = 45.1 \text{ h} = 1.88 \text{ d}$$

At a higher melt rate of 2 cm per day, only 11.3 hours would be required to warm the snowpack. In contrast, modest rainfall plus melt during rainy weather of 1.5 cm per day would satisfy this cold content in less than four hours. In all cases an actual storage loss of refrozen rain or melt water (S_c) of 0.94 cm would have occurred and this loss is only regained as the snowpack continues to melt. Even for exceptionally large cold-content conditions, such as a snowpack with 100 cm of water equivalent and average temperature of $-10\,°C$, melt would require 12.5 and 3.1 days to ripen the snowpack at melt rates of 0.5 and 2 cm d^{-1}, respectively. Rainfall of only 0.25 cm h^{-1} plus melt could satisfy even this large cold content in about one day. These examples suggest that typically up to several days to over a week of melt in extreme circumstances may be required to satisfy cold contents, but that rainfall plus melt could exceed cold contents in less than one day under most circumstances.

Time lags to daily snowpack meltwater outflow may also occur due to nocturnal cold contents that develop during the previous night. The magnitude of nocturnal cold contents at exposed snowpack sites is typically 0.25 cm (US Army Corps of Engineers, 1956). Given melt rates of 0.083 and 0.33 cm h^{-1}, obtained by assuming 0.5 and 2 cm d^{-1} melt occurs mainly during six daylight hours, the time required to satisfy nocturnal cold contents would range from about 3 to 0.8 h the following day, respectively. Thus, a small to moderate delay in meltwater appearance at the snowpack base and storage loss may occur each melt day to satisfy nocturnal cold contents at open sites. Shutov (1993) describes a procedure used to compute the depth of snowpack refrozen each night to help account for this lag.

9.3.2 Liquid-water holding capacity lag times

Time lags may also occur due to water withheld by snow against rapid gravity drainage. Whenever first ripened, snowpacks exhibit a liquid-water holding capacity that must be satisfied before liquid water can be released at the snowpack base. Irreducible liquid-water holding capacities range from 3–5% on a mass basis and this lag can also be computed as:

$$t_f = (f)(\text{SWE})/[100(P_r + M)] \tag{9.3}$$

where:

$t_f =$ runoff lag time due to the liquid-water holding capacity, h
$f =$ liquid-water holding capacity $= 100\,\theta_{vi}(\rho_w/\rho_s)$, see Section 3.4.2
SWE $=$ snowpack water equivalent, cm
$P_r =$ rainfall rate, cm h^{-1}
$M =$ melt rate, cm h^{-1}

If the liquid-water holding capacity was 4% for a 30-cm water equivalent snowpack and the melting rate without precipitation was 2 cm d^{-1}, then the lag time would be:

$$t_f = (4\%)(30\ \text{cm})/100(0 + 2\ \text{cm d}^{-1}) = 0.6\ \text{d}$$

Over one-half of the day's melt would be required to satisfy the liquid-water holding capacity, before liquid water could be released from the snowpack. The storage loss in this example would be 4% of 30 cm or 1.2 cm of liquid water (S_l). Water initially stored is gradually released as the snowpack is melted. Once the water holding capacity is satisfied, the liquid-water holding capacity only needs to be replenished if the snowpack is refrozen or if new snowfall is added. Surface snowpack layers refrozen overnight during formation of nocturnal cold contents or during sudden extremely cold weather during the melt season exert a new liquid-water holding capacity. The depth and density of refrozen snowpack layers are needed to compute additional storage losses, although these losses are likely to be quite small.

9.3.3 Lag times in ripe snow

Once cold content and liquid-water holding capacities are satisfied, the remaining lag time is the time required for transmission of liquid water from the ripe snowpack to the ground. The delay for water originating at the surface to exit a ripe snowpack would simply be:

$$t_t = d/v_t \qquad (9.4)$$

where:

t_t = time for liquid-water transmission through snowpack, h
d = snowpack depth, cm
v_t = transmission rate, cm h^{-1}

The US Army Corps of Engineers (1956) observed transmission rates that varied from 2 to 60 cm per minute. For a snowpack with a water equivalent of 0.3 m and a density of 350 kg m^{-3}, the depth would be 0.86 m. If the transmission rate was 10 cm per minute or 6 m per hour, the time lag for water to be transmitted through the snowpack would be:

$$t_t = (0.86 \text{ m})/(6 \text{ m h}^{-1}) = 0.14 \text{ h} \cong 0.01 \text{ d}$$

Lag times for transmission in this example are relatively small in comparison to other time lags, but our ability to predict transmission rates is very uncertain. In many snowmelt runoff models, time for transmission through ripe snow is not considered. For example, Zhang *et al.* (2000) assumed that vertical travel time through snow was negligible compared to the error in computing the time of lateral downslope in an Arctic setting. As discussed in Chapter 3, gravity forces generally dominate liquid-water movement in the snowpack and the permeability of the snowpack changes over time due to metamorphism. Permeability also increases dramatically with the snowpack liquid-water content that is controlled by rainfall and melt rates. Thus, water moves the quickest through the snowpack under the highest liquid-water contents during active melt and rainfall. At night as the snowpack drains liquid-water movement is quite slow.

The total time lag due to cold content, liquid-water holding capacity and transmission in the above example calculations is:

$$t_c + t_f + t_t = 1.88 + 0.60 + 0.01 = 2.49 \text{ d} \tag{9.5}$$

The total storage due to cold content and liquid-water holding capacity is $0.94 + 1.2 = 2.14$ cm, respectively. In this analysis the snowpack is assumed to be homogeneous, but occurrence of preferential flow channels that develop in snowpacks can facilitate rapid drainage of liquid water to the ground before the entire snowpack is ripened as discussed in Chapter 3.

Anderson (1973) developed an operational scheme to lag and route melt water through ripe snow in lieu of computations using Equation (9.5). The scheme was based upon empirical snowmelt lysimeter data and a model with a six-hour computation interval. Liquid water added to the snowpack during a time interval, in excess of snowpack liquid-water holding capacity, was initially lagged by an amount that increased with snowpack water equivalent and decreased with the relative amount of excess liquid water in the snowpack. Anderson's (1973) function for lag time is:

$$\text{LAG} = 5.33\{1 - \exp[(-0.03)(\text{SWE/EXCESS})]\} \tag{9.6}$$

where:

LAG = lag time applied to inputs of liquid water during the current time period, h
SWE = snowpack water equivalent, cm
EXCESS = liquid water in excess of water holding capacity in snowpack, cm/6 h

Lag times for a 30 cm snowpack water equivalent with excess liquid water in transit of 5 cm (e.g. one good day's melt) would give the ratio SWE/EXCESS = $30/5 = 6$ and a lag time of:

$$\text{LAG} = 5.33\{1 - \exp[(-0.03)(6)]\} = 0.88 \text{ h}$$

Increasing water equivalent to 100 cm with the same excess liquid content, would increase the lag time from 0.88 to 2.4 hours. If the water equivalent were 30 cm and the excess water was 7 cm, then the computed lag would be reduced from 0.88 to 0.64 hours. The maximum lag time for deep snowpacks that could be computed with Equation (9.6) for conditions of relatively little water in transit would be 5.33 hours.

In Anderson's (1973) scheme, after the time lag was applied to liquid water inputs for the current time period, outflow from the snowpack was determined by a

Table 9.1 *Example of a lag-and-route procedure used to simulate the transmission of liquid water in ripe snow. A constant lag time of 1 h and routing coefficient of 0.6 h^{-1} was assumed*

Time, h.	1	2	3	4	5	6	7	8	9	10	11	12
Input Melt, mm	2	6	12	6	2	0	0	0	0	0	0	0
Lagged Input, mm	0	2	6	12	6	2	0	0	0	0	0	0
Storage, mm	0	0.8	2.72	5.89	4.76	2.70	1.08	0.43	0.17	0.07	0.03	0.01
Outflow, mm	0	1.2	4.08	8.83	7.13	4.05	1.62	0.65	0.26	0.10	0.04	0.02

routing coefficient. The routing coefficient is defined as the fraction of liquid water in transit (current inflow during the time interval plus liquid left from previous periods) that leaves the snowpack per unit of time. The routing coefficient can be computed for a six-hour interval using Anderson's (1973) function adjusted to metric units as:

$$R = \{(0.5)[\exp(-110)(I/\text{SWE}^{1.3})] + 1\}^{-1} \tag{9.7}$$

where:

R = routing coefficient, $(6\ \text{h})^{-1}$

I = lagged inflow of liquid water during the current time interval, cm $(6\ \text{h})^{-1}$

SWE = water equivalent of snowpack, cm

Given a current inflow of 5 cm $(6\ \text{h})^{-1}$ and a water equivalent of 30 cm, following the above example, the routing coefficient would be:

$$R = \{(0.5)[\exp(-110)(5/30^{1.3})] + 1\}^{-1} = \{(0.5)[\exp(-6.61)] + 1\}^{-1}$$
$$= 0.999(6\ \text{h}^{-1})$$

For these conditions essentially all of the excess liquid water would drain from the snowpack during a six-h period. If the current inflow is relatively small compared to snowpack water equivalent, for example $I = 0.05$ cm $(6\ \text{h})^{-1}$ for SWE $= 30$ cm, then the routing coefficient would reduce to 0.68 $(6\ \text{h})^{-1}$. The minimum value for R using Anderson's function is 0.667 $(6\ \text{h})^{-1}$. These computations suggest that routing of liquid water in ripe snow is relatively rapid and essentially all of a given day's melt input would drain from a ripe snowpack in 24 hours. Simulations for hourly time steps would require greater attention to meltwater routing.

An example of the lag-and-route procedure for an assumed daily melt wave in ripe snow is given in Table 9.1. For simplicity, an hourly time step, constant

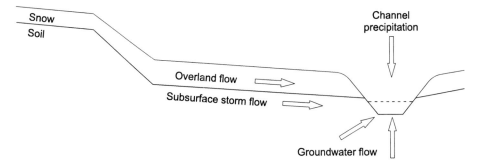

Figure 9.1 Diagram showing four major hydrologic flow paths possible for snow-fall and meltwater.

one-hour lag time and a routing coefficient of 0.6 h^{-1} were assumed. The hourly surface melt would be computed using an energy budget and the lagged input mimics this surface input simply delayed by one hour. Zero snowpack water storage is assumed intially. In the second hour, 0.6 of the hour's lagged melt of 2 mm becomes outflow ($0.6 \times 2 = 1.2$ mm) and the remaining fraction ($0.4 \times 2 = 0.8$ mm) goes into storage as liquid water in transit. In the third hour, outflow equals 0.6 of the current lagged input of 6 mm plus 0.6 of stored water in transit of 0.8 mm, giving:

$$\text{Outflow hour } 3 = (0.6)(6) + (0.6)(0.8) = 4.08 \text{ mm}$$

Storage in hour three becomes 0.4 of the current lagged input of 6 mm plus the remaining 0.4 of the prior hour's storage giving:

$$\text{Storage hour } 3 = (0.4)(6) + (0.4)(0.8) = 2.72 \text{ mm}$$

Each successive hour is handled in a similar manner to predict the hydrograph of outflow from the snowpack. Increases in the routing coefficient reduce the magnitude of storage and vice versa.

9.4 Meltwater flow paths

Once snowpack outflows reach the landscape surface, outflows can often be quickly translated into streamflow due to soil frost, permafrost, or saturated subsurface depending upon the hydrologic flow paths that are active. Hydrologic flow paths involved in the generation of streamflow from snowmelt differ from those for rainfall-generated flows in some respects. Flowpaths can be categorized as: (1) channel precipitation, (2) overland flow, (3) subsurface stormflow, and (4) groundwater flow (Figure 9.1) and each will be discussed relative to meltwater delivery.

Channel precipitation is represented by rain and snow that fall directly into the channel system. Water following this flow path is quickly routed to the basin mouth and contributes the water in the initial stages of streamflow response to precipitation as long as the stream is not frozen. The volume of water delivered by this flow pathway is often limited since the fraction of watershed area covered by channel areas is often quite small (<1% of basin area). However, wetlands associated with the channel system can also become active during these events and greatly expand the area available for receiving atmospheric inputs. Since many basins with hydrology dominated by melt have considerable wetlands, channel precipitation may be a more important pathway for meltwater than frequently realized. Snowfall contributions to flow as channel precipitation are generally separated in time from basin-wide contributions due to delayed melt. When channels and associated wetlands are frozen in winter, snowfall is stored on the stream ice until break-up or melt occurs. In this manner, melt can make significant contributions to flow as channel precipitation. In windy, low-relief environments, additional redistributed snow can find its way into the depressions of the drainage network, thus enhancing early runoff or by causing snow dams delay meltwater delivery downstream.

Overland flow is rain or meltwater delivered to the soil surface at rates in excess of the soil infiltration rate which subsequently flows over the soil surface to a channel. Since rates of rainfall commonly exceed snowmelt rates, rainfall potentially can contribute more overland flow than snowmelt. However, snowmelt onto impermeable, frozen soils can also generate overland flow. Studies by Gibson *et al.* (1993) on a wetland basin with permafrost in the Northwest Territories, Canada and Bengtsson *et al.* (1992) on an agricultural basin in Finland reported overland flow of meltwater on frozen soil. The infiltration capacity of undisturbed soils often exceeds the rates of either rainfall or snowmelt, so that true overland flow is quite rare on undisturbed basins without soil frost or saturated soils. However, seasonal frost on watersheds may limit the infiltration capacity if near surface soil pores are ice-rich, thus producing overland flow (Kane and Stein 1983). On basins with impermeable, paved or compacted surfaces, both rain and melt can create overland flows (Bengtsson and Westerstrom, 1992; Semadeni-Davies, 1998).

Subsurface stormflow or interflow is defined as water from rainfall or melt that infiltrates the soil and moves quickly through shallow soil layers to contribute to streamflow. Infiltrated water can move quickly through shallow soil layers to reach the channel due to the higher permeability near the soil surface as compared to deeper subsurface layers (Kendall *et al.*, 1999) and through soil macropores and natural pipes (Gibson *et al.*, 1993). Bathhurst and Cooley (1996) reported modelling evidence supporting rapid, near-channel subsurface delivery of meltwater to streams in Idaho, USA. Infiltrated meltwater can also displace older stored soil water into the channel through shallow soil layers by a process known as "translatory flow."

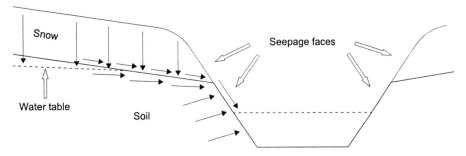

Figure 9.2 Schematic diagram showing flow paths for melt and shallow subsurface
water to reach the stream channel during "saturated overland flow" conditions.

Translatory flow is a concept supported by isotope hydrology studies where much
of streamflow during events has isotope levels similar to water stored in the basin
subsurface prior to the event, rather than isotope levels of event water itself. This
suggests that new water infiltrating a basin acted as a piston to force out stored
water or translate older stored water into streamflow.

Older stored water can exfiltrate the soil at seepage faces on the channel banks or
on adjacent lower slopes where the land surface intersects the water table and flows
over the surface to a channel. This type of flow generation is termed "saturated
overland flow" and commonly occurs during snowmelt. Seepage zones represent
saturated soil areas that not only seep water that had already infiltrated the basin,
but also intercept current snowmelt or rainfall and quickly deliver this water as
overland flow to channels. Figure 9.2 illustrates flow conditions during saturated
overland flow where subsurface water and current melt water flow over the surface
in seepage areas to channels. Small rain events or rainfall on relatively dry soils
often do not generate appreciable subsurface stormflow. However, melt usually
occurs for several consecutive days to weeks at a time and produces very wet soils
that lead to larger contributions as subsurface stormflow.

Groundwater flows reach the channel via pressure gradients and gravity through
saturated porous materials intersected by the channel. Infiltrating melt or rain
water recharge the groundwater as evidenced when the elevation of the water table
increases during and after events. Due to reduced permeabilities common in subsurface rock and till aquifers, recharged deep groundwater may take months to years
before emerging as streamflow. Prolonged periods of rain or melt cause saturated
conditions and significant groundwater recharge that sustains both the baseflow of
streams and groundwater supplies during summer periods after snowpacks have
entirely melted.

The net effect of flow paths followed by rainfall and meltwater can be deduced
from isotope studies of streamflow events. It is common for melt water or rainfall
(event water) to have a different stable-isotope level than baseflow (pre-event water).

This occurs because isotope levels in each precipitation event vary due to atmospheric temperatures during condensation/evaporation and groundwater flow is a mixture of many past rain and melt events. Hydrograph separation of streamflow into event water and pre-event water components can be accomplished using Equation (8.5) previously introduced in Chapter 8:

$$Q_e/Q_t = (c_{\text{stream}} - c_{\text{pre-event}})/(c_{\text{event}} - c_{\text{pre-event}})$$

Using stable isotopes such as ^{2}H and ^{18}O in stream water, the fraction of event water can be computed for rainfall compared to melt events. Results from a large number of studies on small relatively undisturbed basins suggest (Buttle, 1994; Rodhe, 1998) that snowmelt events have a greater event water fraction at peakflows than rainfall events. Such isotope studies indicate the source of the water for streamflow (event vs. pre-event), not how the water got there but, nevertheless, these studies allow some speculation about pathways for greater event water contributions for melt events.

As already implied in the discussion of flow paths, greater event water contributions for melt than rain events would be consistent with a combination of flow paths such as channel precipitation, overland flow, macropore flow and saturated overland flow paths. Greater importance of channel precipitation for melt than rain events could occur early in the melt season on basins with considerable frozen wetland areas with a stored snowpack on the surface. Snow stored in wetland areas connected to the main channel could quickly become streamflow during melt. A greater importance of overland flow for meltwater could also occur because of frozen soils. Macropore flow leading to direct delivery of meltwater to channels could be more important due to the very wet soils that accompany prolonged periods of snowmelt. Finally, greater soil moisture levels during melt may also lead to greater saturated overland flow. Even though much of the water forced out of the soil at seepage faces by infiltrating meltwater would likely be stored water from prior events, the exposed seepage faces create an impermeable surface, like frost, that causes current melt to become overland flow. The exact causes for greater event water contributions probably vary in each individual study but, overall, these flow paths are probably responsible for delivering greater amounts of event meltwater than rain water to streams during events.

9.5 Frozen ground

Snowpack outflow onto frozen, impermeable ground can result in overland flow, surface ponding, and storage and in many cases rapid streamflow response. Rapid delivery of meltwater is a particular problem in arctic basins with permanently frozen ground or permafrost (Woo, 1986). Prévost *et al.* (1990) found that the low

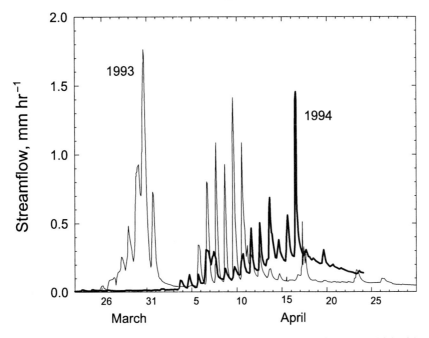

Figure 9.3 Streamflow from snowmelt in 1993 with deep soil frost and 1994 with shallow soil frost on a Vermont agricultural watershed (Shanley and Chalmers, 1999 © 1999 John Wiley & Sons Limited, reproduced with permission).

infiltration capacity of frozen soil caused overland flow and contributed to high peak streamflows during spring snowmelt at Lac Laflamme in Quebec, Canada. In an Arctic watershed, McNamara *et al.* (1997) found from isotope studies that runoff in spring was primarily (90%) due to event melt water from the snowpack, but runoff in summer was typically sustained (70%) by older water released from storage by thawing of the shallow soil layer or active layer above the permafrost. A soil infiltration index was needed to account for frozen soil effects in models for accurate peak-streamflow prediction. Prediction of the depth of soil frost and the infiltration capacity of the frozen ground can be quite important to successful modelling of streamflow from snowmelt.

One of the more dramatic examples of soil-frost effects on the generation of streamflow from snowmelt was presented by Shanley and Chalmers (1999) for a small agricultural watershed (W2) in Vermont, USA (Figure 9.3). In 1993, a year with deep soil frost, daily streamflow from melt occurred in a series of daily runoff events each with a very high peak flow rate followed by a return to about the same relatively low flow rate overnight. At the end of the melt period stream baseflow rates were about the same as prior to melt. In 1994, a year with shallow soil frost, each day's melt event produced smaller peak flows followed by daily minimum flows that gradually increased during the melt period. By the end of the melt period in 1994, baseflow levels had increased dramatically and a period of baseflow recession occurred. These differences suggest that deeper soil frost in 1993 prevented infiltration and caused overland flow that led to rapid streamflow response to melting during each day, but limited any groundwater recharge that could sustain stream baseflow during later periods.

The snowpack, by virtue of its excellent insulating properties, plays a dominant role in the formation and preservation of soil frost despite the many other climatic, vegetative, topographic, and soil factors of importance to frost formation. Soils that are unfrozen prior to snowfall, generally will not freeze appreciably during the winter under snowcover and soils will be maintained in a permeable condition. Snow depths of about 50 cm are generally considered adequate to protect soils from freezing as a rule of thumb in the continental United States. In contrast, soils that are frozen prior to snowfall, either due to seasonal cooling or existence of permafrost, will generally remain frozen throughout the winter when buried deep beneath a snowpack. Once melting begins, water at the snowpack base will refreeze and warm the frost layer to 0 °C, much as meltwater refreezes and reduces snowpack cold contents. Once shallow frost layers are warmed to 0 °C by refreezing meltwater, however, they are protected against further energy gains by the snowpack and can persist until the snow disappears and thawing begins at the surface. In permafrost regions, a layer of soil as shallow as 50 cm above the permafrost zone, called the active layer, thaws and refreezes each year and is primarily responsible for shallow storage of water that influences hydrology on Arctic basins. Regardless, meltwater infiltration into frozen soils may be reduced or completely prevented and relatively rapid flow of liquid water can occur over the soil surface (see Figure 9.4).

Carey and Woo (1998) describe the differences in meltwater delivery between south slopes with seasonal soil frost and north slopes with a thick surface organic mat of mosses, sedge, grass, lichens, and peat overlying permafrost in the clayey till soil. The south slope melted quickly and soil frost did not cause overland flow. When melt began on the north slope the south slope was free of both snow and frost. Melting on the north slope resulted in overland flow through the frozen organic layer and rapid subsurface flow through soil pipes.

Frost depth in watersheds varies with snow depth and many other factors. Snow-depth variations caused by climatic variations, wind re-distribution of snow, inter-ception of snow by vegetation (Hardy and Albert, 1995), and soil properties can all affect the occurrence of soil frost. Climate controls amount of snowfall and the severity and duration of the freezing period which affect frost formation. Stadler *et al.* (1996) also found greater frost penetration and occurrence of overland flow near tree trunks as opposed to canopy gaps in subalpine spruce forest due to effects of canopy interception of snow. Apart from snow-depth effects, greater soil freezing generally also occurs in open terrain where bare mineral soil is exposed, rather than

Figure 9.4 Rapid delivery of meltwater as overland flow on frozen ground in a pasture swale that has turned the snow into a slush layer. See also color plate.

in forests with the protective influence of the overhead canopy (increased longwave radiation from above and insulating organic matter on the soil), or where a thick organic mulch or turf exists on the surface. More rapid soil freezing and slower thawing also occurs on north-facing than south-facing slopes (Zhenniang *et al.*, 1994). Effects of soil properties on soil freezing are difficult to generalize, sandy soils have higher thermal conductivity than silt and clay soils for a given soil density and water content. However, increasing water content of soil, which is also related to texture and other soil properties, tends to reduce the rate of penetration of soil frost because the latent heat released when soils freeze must be conducted to the surface before the freezing front can advance downward into the soil. Overall, all of these factors make the occurrence and depth of soil frost difficult to understand and predict.

9.5.1 Predicting soil frost depth

Analysis of the effects of frozen ground on snowmelt hydrology must begin with prediction of the depth of frost penetration at any given time. Heat transfer during seasonal soil freezing and thawing is very complex and difficult to approximate using limited climatic data. Basically, heat is lost from the soil to the atmosphere or overlying snowpack due to the energy balance at the soil surface. Cooling of the soil surface to freezing temperatures initiates the development of the soil frost layer. The frost layer grows from the surface downward into the soil as heat is conducted through the layer of frozen ground to the surface. The thermal conductivity of the frozen soil layer, which increases with soil density and water content, thus plays a major role in the rate of frost penetration. Cooling of the soil frost layer is also dependent upon the volumetric heat capacity of the soil. Soils with greater density and water content will cool more slowly. The temperature at the base of the growing soil frost layer is maintained at 0 °C as freezing of soil water liberates the latent heat of fusion. Fusion energy released during freezing must also be conducted to the surface before the frost layer can grow further.

The situation is complicated further by heat conduction and migration of water to the base of the frost layer from below due to thermal and pressure gradients. Liquid water moves upward by pressure gradients and water vapor diffuses upward along a temperature-induced vapor pressure gradient in the soil pores. Upward migration of water and freezing at the base of the frost layer often results in a dry soil layer below and a upper layer enriched with ice that can significantly impact infiltration and runoff. Also, solute concentrations in soil water also can cause the migration of water in soil from high to low solute concentrations. Liquid and vapor movement and solute concentrations are generally not considered in soil frost prediction equations currently available, but Flerchinger and Saxton (1989) developed a physically based numerical model to predict depth of soil freezing as affected by tillage and crop residues that includes effects of soil water movement.

One method commonly used for predicting the depth of soil frost is the modified Berggren equation (Aldrich, 1956, Kersten, 1949). This equation considers heat conduction in the frozen soil layer and the unfrozen soil layer below, the volumetric heat capacity of the frozen and unfrozen soil, and the latent heat of the soil per unit volume. It does not consider the effects of liquid or vapor movement in the soil. According to the modified Berggren equation, the depth of soil frost can be computed as:

$$Z_f = \lambda (48 K_m N / L)^{1/2} \tag{9.8}$$

where:

Z_f = frost depth, cm
λ = dimensionless coefficient (see Figure 9.5)
K_m = mean thermal conductivity of the frozen and unfrozen soil layers, J m^{-1}
    h^{-1} °C^{-1}
N = number of degree days in the freezing or thawing period, °C-d
L = latent heat of fusion of water per unit soil volume, J m^{-3}

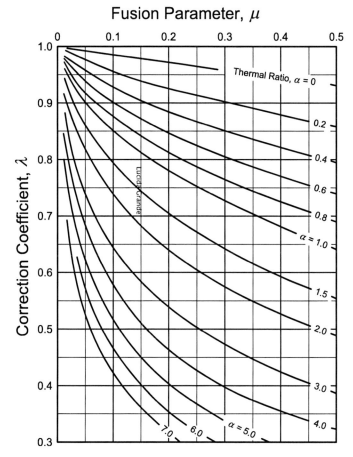

Figure 9.5 Dimensionless parameter λ used in the modified Berggren formula to compute the depth of soil frost (Aldrich, 1956, Transportation Research Board, National Research Council, Washington, DC, Figure 5, p. 133. Reprinted with permission of TRB).

The thermal conductivity of frozen soil is greater than that of unfrozen soil and varies with soil density, texture, and water content (Aldrich, 1956). Values range from about 0.5 to 4 W m^{-1} °C^{-1} or about 1800 to 14 400 J m^{-1} h^{-1} °C^{-1}. The number of degree-days can be estimated from air-temperature data for a brief freezing or thawing period, the spring thawing period or the maximum number of freezing degree-days during an entire winter period at a particular location. The dimensionless coefficient λ was added by Aldrich (1956) to account for the effects of heat capacity of the soil.

The latent heat of fusion per unit soil volume varies with soil density and liquid content according to:

$$L = (L_f)(w/100)(\rho_{ds}) \tag{9.9}$$

where:

L = volumetric latent heat of fusion of water in soil, J m^{-3}
L_f = latent heat of fusion of water, 0.334 MJ kg^{-1} at 0 °C, see Appendix A
w = soil water content, % dry wt.
ρ_{ds} = dry-soil density, kg m^{-3}

The dimensionless coefficient λ can be found in Figure 9.5 as a function of two other dimensionless parameters (α and μ):

$$\mu = \text{fusion parameter} = C(T_s - T_f)/L = C(N/t - T_f)/L \qquad (9.10)$$

where:

C = the mean volumetric heat capacity of the frozen and unfrozen soil layers, J m^{-3} K^{-1}
T_s = temperature of frozen soil layer = N/t, °C
t = duration of the freezing period, d
T_f = freezing temperature of soil water, 0 °C assumed here

The mean soil volumetric heat capacity can be computed as the mean of unfrozen and frozen soil conditions as:

$$C_{\text{unfrozen}} = (\rho_{ds})(712 + 4188w/100) \qquad (9.11)$$

$$C_{\text{frozen}} = (\rho_{ds})(712 + 2094w/100) \qquad (9.12)$$

where the constants 712, 4188, and 2094 represent the specific heats for mineral soil, liquid water, and ice, respectively, in J kg^{-1} K^{-1}. Since the temperature of the frozen soil layer (T_s) is generally not known, the number of freezing degree-days (N) divided by the number of days in the freezing period (t) provides a useful estimate. T_f could be varied to account for solute effects on the freezing point.

The value of the fusion parameter, μ, is the ratio of the sensible heat loss during cooling of the frozen soil layer to the latent heat of fusion released during freezing of soil water. Increasing values of μ reduce the value of λ in Figure 9.5 and thus reduce the depth of frost penetration, other factors held constant. Conditions where the soil cooling is relatively great and the heat capacity is relatively large compared to the latent heat of soil water (e.g. dryer soils) produce such conditions.

A second dimensionless parameter needed to find λ in Figure 9.5 is α where:

$$\alpha = \text{thermal ratio} = (T_0 - T_f)/(T_s - T_f)$$

$$= \text{for annual periods } (T_m - T_f)/(N/t - T_f) \qquad (9.13)$$

where:

T_0 = initial temperature of the soil layer, °C
T_f = freezing temperature of the soil water, °C

T_s = final temperature of the frozen soil layer, °C

T_m = mean annual air temperature, °C

As indicated in Figure 9.5, increases in α reduce λ and the predicted depth of frost penetration. Large values of α occur in climates where cooling of the soil layer from the annual average temperature and the effects of soil volumetric heat capacity are relatively large (more temperate climates) compared to the frozen-soil temperature.

An example computation of maximum depth of frost penetration using the modified Berggren formula (Equation (9.8)) for a region with an average winter $N =$ 833 °C-days below freezing, a duration of freezing period of $t = $ 130 days, and a mean annual air temperature $T_m = $ 4.44 °C follows. These conditions roughly correspond to a location in central Maine in the northeastern United States. A thermal conductivity of 4032 J h^{-1} m^{-1} °C^{-1} (K_m) will be assumed that represents an average for homogeneous silty soil in the frozen and unfrozen state with a 15% water content (w) and a dry density of 1500 kg m^{-3} (ρ_{ds}) based upon curves given by Aldrich (1956). The latent heat of fusion for this soil would be:

$$L = (334000 \text{ J kg}^{-1})(15/100)(1500 \text{ kg m}^{-3}) = 7.52 \times 10^7 \text{ J m}^{-3}$$

The volumetric heat capacity for the frozen and unfrozen soil layers would be:

$$C_{unfrozen} = (1500 \text{ kg m}^{-3})(712 + (4188)(15\%)/100) = 2.01 \text{ MJ m}^{-3} \text{ K}^{-1}$$
$$C_{frozen} = (1,500 \text{ kg m}^{-3})(712 + (2094)(15\%)/100) = 1.54 \text{ MJ m}^{-3} \text{ K}^{-1}$$

giving a mean C of 1.78 MJ m^{-3} K^{-1}. The freezing temperature of soil water is assumed to be 0 °C. The temperature of the frozen soil layer in degrees below freezing is approximated as $N/t = $ 6.4 °C. The dimensionless parameters needed to find λ from Figure 9.5 are:

$$\mu = [(1.78 \times 10^6 \text{ J m}^{-3} \text{ °K}^{-1})(6.4 \text{ °C})/(7.52 \times 10^7 \text{ J m}^{-3})] = 0.15$$

and

$$\alpha = 4.44/6.4 = 0.69$$

The value of λ in Figure 9.5 corresponding to these values of α and μ is about 0.85. The maximum depth of frost penetration would then be:

$$Z_f = (0.85)[48(4,032 \text{ J h}^{-1} \text{ m}^{-1} \text{ K}^{-1})(833 \text{ °C-days})/(7.52 \times 10^7 \text{ J m}^{-3})]^{1/2}$$
$$= 1.24 \text{ m} = 124 \text{ cm}$$

This is the estimated maximum depth of soil frost for a winter season for a bare mineral soil.

A method is also needed to account for effects of snow cover or surface vegetation on soil frost penetration. A trial-and-error method given by Aldrich (1956) for

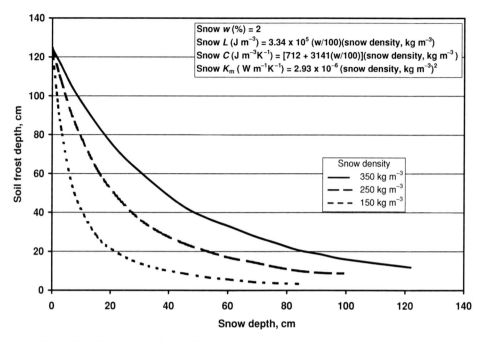

Figure 9.6 Computed relationship between seasonal soil frost depth and depth of snow cover for three snowpack densities. Calculations were based upon a trial-and-error method for the modified Berggren equation given by Aldrich (1956) for soil and climatic conditions used in the text example and snow properties given in the figure.

computing frost depth in soil with layers of varying thermal properties was adapted to show effects of an insulating snow layer over a layer of soil. In Figure 9.6 the modelled relationship between soil frost depth and snow-cover depth is given along with input data used. The same soil properties that were assumed in the above example were used to represent the soil. Computations for a snow layer were conducted for three snow densities and effective snowpack thermal conductivities to simulate a range of possible winter conditions. The effects of soil surface organic matter could also be easily added, but are not included here.

Frost depth in the soil is very sensitive to snow-cover depth, especially for shallow snowpack depths and low snowpack densities (Figure 9.6). Soil frost depth decreased rapidly with increasing snow depth from the 124-cm frost depth found in the above example for exposed mineral soil up to snow depths of about 20–50 cm depending on snowpack density. Further increases in snowpack depth produced only modest decreases in frost depth. The trial-and-error method used did not rapidly converge to zero frost depths for the range of snow depths tested which may be a limitation of the method or the input data assumptions. The rule of thumb given previously that 50 cm of snow is needed to avoid soil frost appears to be approximately correct for low-density snow. DeGaetano *et al.* (1996) presented an alternate method for computing depth of frozen ground from readily available climatic data that also includes effects of varying snow depth. DeGaetano *et al.*, found that the insulating effect of snow and surface vegetation were important parameters controlling frost depth. Whichever method is used to compute the depth of soil frost, the infiltration capacity of the frozen soil layer must also be considered.

9.5.2 *Infiltration rates and soil frost*

Seasonally frozen soils are generally permeable. The infiltration capacity of frozen soil varies inversely with the pre-melt soil moisture content that controls the ice content of soil pores upon freezing (Kane and Stein, 1983). Wet soils when frozen generally have more pore space occupied by ice and less available for infiltration of meltwater than dry soils. Like the snowpack refreezing of meltwater also occurs in the soil profile thus producing additional ice and some reduction in the infiltration rate.

Zhao and Gray (1999) proposed the following model for infiltration into frozen soil based upon numerical simulations and comparisons with field data:

$$INF = C \, S_0^{2.92}(1 - S_1)^{1.64}[(273.15 - T_1)/273.15]^{-0.45} \, t^{0.44} \qquad (9.14)$$

where:

INF = cumulative infiltration into frozen soil, mm
C = bulk coefficient
S_0 = fractional surface saturation, range 0.7–0.9
S_1 = fractional initial soil saturation, range 0.4–0.6
T_1 = initial soil temperature, K, range –4 to –8 °C
t = elapsed time, h, range 2–24 h

The bulk coefficients varied from $C = 1.3$ for boreal forest soil and $C = 2$ for prairie agricultural soils. In this relationship, cumulative infiltration increases with increasing surface saturation that controls initial capillary pressure gradients into the soil, increases with decreases in initial soil saturation that controls the amount of porosity occupied by ice, increases with increasing soil temperature that controls the heat deficit of the frozen soil layer, and increases with increasing time. For example, the cumulative 24-h amount of infiltration into a boreal forest soil ($C = 1.3$) at an initial temperature of –6 °C, a surface saturation of 0.9 and initial soil saturation of 0.6 would be:

$$INF = (1.3)(0.9)^{2.92}(1 - 0.6)^{1.64}[(273.15 - 267.15)/273.15]^{-0.45}(24)^{0.44}$$
$$= 4.8 \text{ mm} = 0.48 \text{ cm}$$

This infiltration capacity would not be adequate for daily melts that commonly exceed 2 cm. Increasing the temperature of the frozen soil layer to -4 °C and reducing the initial soil saturation to 0.4, both of which should increase infiltration capacity, increases the 24-h total to 1.1 cm which is more than adequate to infiltrate a day's melt.

A simplified approach to accounting for soil frost occurrence and permeability was used by Rekolainen and Posch (1993) in adapting the CREAMS model to conditions in Finland. A simple degree-day function was used, such that the soil was considered impermeable due to frost whenever the accumulated freezing degree-days exceeded 10 °C-days in the fall and the soil was considered thawed and permeable whenever thawing degree-days exceeded 30 °C-d in spring. The fall and spring thresholds were based upon conditions in Finland, but a similar threshold approach could be adapted to other locales.

9.6 Snow damming

In windy, low-relief basins such as found in the Arctic, snow dams can alter the timing of meltwater delivery downstream (Woo and Sauriol, 1980; Woo, 1983). In Arctic basins, streamflow generally will cease completely during winter due to freezing of the shallow soil active zone above the permafrost. During winter, redistribution of snow by the wind on these basins will fill topographic depressions with compacted wind-blown snow with very large cold contents. As melt begins in the spring these deposits of compacted snow act as dams and can slow downslope movement of meltwater as they refreeze meltwater from upslope and gradually ripen. Such deposits of snow often become saturated with meltwater causing slush flows, tunnels within the snowpack, or surface flow over the saturated snow layers. In topographic depressions and where large blocks of wind-blown snow fall or calve into channels, ponding of water by snow dams can also occur. As such dams fail, streamflow begins in steps and halts. Delays up to several days can occur due to snow dams and dam failures have been implicated in causing extreme sediment transfer events (Lewis *et al.*, 2005). Snow damming presents special problems for modelling the movement of meltwater along slopes through the snow and prediction of snowmelt runoff in such basins (Hinzman and Kane, 1991).

9.7 Snowmelt hydrographs

Streamflow hydrographs from snowmelt are strongly coupled to the energy supply for melt. Melting in steep mountain basins generally occurs first at low elevations and later at high elevations due to differences in air temperature and energy supply. Elevation may be much less important on low-relief Arctic basins; and some

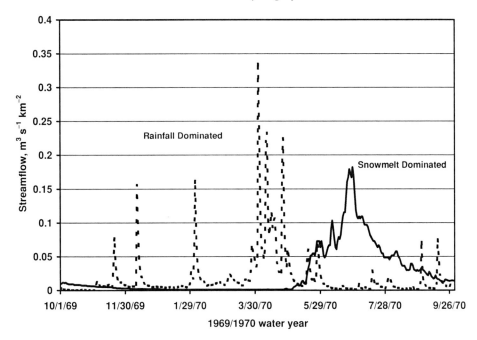

Figure 9.7 Contrasting annual streamflow hydrographs for small watersheds with hydrology dominated by rainfall (Waldy Run near Emporium PA, USA) and snowmelt (Fall Creek near Rustic, CO, USA), http://waterdata.usgs.gov.

Arctic north-flowing river basins begin melting in the headwater regions. Superimposed on variations with elevation are melt variations because of varying slope inclination and aspect. South-facing slopes melt more rapidly than north-facing slopes. Differences in vegetative cover over basins also affect snowmelt timing. Many high-elevation mountain basins have snowpacks shaded by forest cover at low elevations and relatively exposed snowpacks in the alpine at high elevations. Other basins exhibit forest cover at high elevations and open urbanized and agricultural land at lower elevations. Finally, the duration and depth of snowcover and existence of permanent snow and ice fields also influence the importance of melt runoff in overall basin hydrology.

Annual streamflow hydrographs on watersheds where hydrology is dominated by snowmelt show a distinctive pattern of flow in contrast to basins where hydrology is dominated by rainfall events. Clear examples at the large watershed scale are the north-draining Mackenzie River in Canada and the Ob, Yenisey, and Lena Rivers in Russia where the annual peak is from snowmelt. Basins dominated by snowmelt such as Fall Creek (9.3 km^2) in the Rocky Mountains of Colorado (Figure 9.7) show flow rates that reflect the seasonal availability of energy for melt. Very low rates of flow occur during the winter snowpack accumulation season, due to baseflow recession from prior periods of recharge and low rates of melt caused by ground heat conduction to the snowpack base. In the spring as energy supplies for melt increase, the flow rates gradually increase and build to a peak annual flow in late spring or early summer. Seasonal peak flows due to snowmelt occur earlier in warmer years (Westmacott and Burn, 1997) and higher and later seasonal peaks also generally occur in years with greater snowfall. Flows decline thereafter with a few minor perturbations due to occasional rain storms until the following winter at this Colorado site. Timing and magnitude of annual peaks on a particular watershed tend to vary from year to year with the sequence of climatic and snowpack conditions. Watersheds with glaciers in the extreme headwaters also display hydrographs with extended high flows (Aizen *et al.*, 1995; Moore, 1993) or even double seasonal peaks due to snowpack melting at lower elevation followed by melt of glacier ice at higher elevations (Aizen *et al.*, 1996).

On watersheds such as Waldy Run (13.6 km^2) in the Appalachian Mountains of Pennsylvania shown in Figure 9.7, hydrology is dominated by rainfall with some minor snowmelt and rain-on-snow events possible. Rainfall on this basin is evenly distributed throughout the year. Flows rise and fall more rapidly during rainfall events than snowmelt events, with about 12 major events apparent in the year shown. Peak streamflow rates still commonly occur in early spring due to rainfall on wet soils and snow, but the peak flows are still primarily due to rainfall. Energy supply limits the maximum rates of snowmelt compared to rainfall on a given day. Net energy available for melt of 100 W m^{-2}, which roughly equals average energy available for open snowpacks in mid-summer in the Sierra Nevadas (see Chapter 6), would only cause approximately 2.6 cm d^{-1} of melt. Rainfall can exceed this amount in an hour for the site in Pennsylvania. On Waldy Run low flows occur during summer due to high evapotranspiration rates. Annual stream hydrographs shown in Figure 9.7 were generated by plotting average daily flow rates for each day of the year; therefore, this approach masks the typical diurnal variation snowmelt runoff.

The timing of daily snowmelt hydrographs also reflects patterns of melt energy availability with a lag between surface melting and channel flow related to basin and snowpack conditions. The hydrograph on E. Br. Bear Brook, a small forested basin with an area about 0.11 km^2, shows that streamflow began to rise each morning before noon and peaked each day in early evening (Figure 9.8). This type of response is typical of small headwater catchments where streamflow response to melt energy inputs is relatively rapid. Streamflow response to melt on Kingsbury Stream and Piscataquis River during the same time period shows the lagged and damped response to melt due to greater basin size (longer travel distances) and possibly incomplete snow cover at lower elevations. The peak flows due to melt on one day actually occur around 0600 hours on the following day on the Piscataquis River. These curves are presented for illustration purposes only, as each

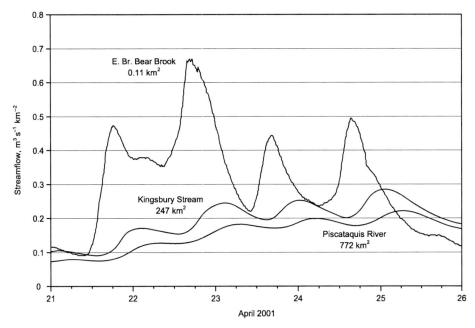

Figure 9.8 Comparison of timing of daily streamflow hydrographs during snowmelt on April 21–25, 2001 for three basins with varying area in Maine, http://water.usgs.gov/realtime.html.

watershed has a unique lag time due to varying land use and vegetation cover, topography, areal extent of snow cover, climatic conditions, etc. However, within a relatively uniform region, the lag in snowmelt runoff on a given day may be related to basin area alone. Persson (1976) found that lag times for peak runoff for basins in the Lapptrasket catchment of Sweden increased with the square root of basin area. As basin size approaches about 1000 km² in the United States, it is common for the daily flow variations due to melt to be masked by human interventions (discharges, withdrawals, and flow regulation by reservoirs). Caine (1992) found that daily amplitude of snowmelt runoff decreased and the time of daily peak flows shifted earlier in the day as the melt season advanced. These hydrograph changes were related to reduced areal snow cover and snowpack depth (Caine, 1992). Lundquist *et al.* (2005) developed a model to predict diurnal cycles in streamflow due to snowmelt. They found that time lag increased with snow depth, slope length, and channel length and decreased with channel velocity. The opportunity for more rapid translation of daily melt through permeable, near-surface soils as the watershed approaches saturation also may be a factor in advancing daily melt cycles.

Hydrograph separations to determine one day's contribution to melt have been conducted to relate melt to causal factors (Figure 9.9; Garstka *et al.*, 1958). Each

METHOD OF COMPUTATION OF SNOWMELT HYDROGRAPH

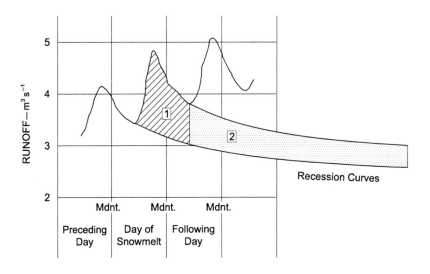

Area 1 = Volume of a day's snowmelt appearing in the first
24-hour period. (First day volume)
Area 2 = Volume of a day's snowmelt in recession flow.
Total snowmelt for the day = Area 1 + Area 2.

Figure 9.9 Graphical hydrograph separation used to determine one day's melt
contribution (adapted from Garstka *et al.*, 1958, courtesy USDA Forest Service).

day's contribution can be computed as the sum of two parts: (1) the volume of meltwater generated during the first 24 hours above the prior day's baseflow recession
curve, and (2) the volume of meltwater generated due to the difference in flows
between the current day's and prior day's baseflow recession integrated to infinity.
Baseflow recession coefficients were fit to prior streamflow data to insure accurate
prediction of recession flows. Negative daily meltwater volumes occurred occasionally. They were due to either re-freezing of meltwater in transit or lack of fit of
the chosen baseflow recession model. Regardless, meltwater volumes determined
in this manner were strongly related to meteorological factors on each day.

A ready source of streamflow data in the United States that shows variations due
to snowmelt is the US Geological Survey water website at http://water.usgs.gov/.
Daily average flow rates for the period of record are available for all gauged streams
in the United States by selecting "Historical Data" and selecting the state of interest.
Updated, "real-time" data for the past week can also be accessed at this site, which
gives flow rate variations generally at 15-minute intervals. Both historical and real-
time data can be acquired on-line in graphical or tabular form. Real-time data are
provisional data that are verified later before becoming part of the historical record.

9.8 Snowmelt and rain-on-snow floods

A combination of conditions within watersheds can lead to flooding from snowmelt or rain-on-snow events. Todhunter (2001) analyzed the 1997 Grand Forks flood resulting from snowmelt in the Red River of the North which at the time was the most costly flood on a per capita basis in US history. A combination of wet soils at the beginning of the winter, concrete frost in the soil, record snowfalls including a late-season snowfall that delayed melt, a sudden warm period, and channel clogging by river ice all occurred to produce catastrophic flooding that forced the evacuation of 75 000 people. The only factor that did not affect flooding during this event was rainfall, which was thankfully absent.

Peak flows from rain-on-snow (ROS) events can be much greater than peak flows from snowmelt or rain alone and have also been the cause of severe flooding. ROS floods are often associated with large sediment and dissolved solids yields, are sometimes associated with damage to facilities and property and on occasion cause loss of life. Anderson and Larson (1996) reported on a major ROS event in northeastern United States in January 1996 that produced the second highest flow recorded for the Susquehanna River at Wilkes-Barre PA. This ROS flood was exceeded only by the flood caused by Hurricane Agnes in 1972. ROS peak flows for the period of record on 17 watersheds in the Sierra Nevada mountains of the western United States reported by Kattleman *et al.* (1991) ranged from 0.6 to 3.9 m^3 s^{-1} km^{-2} and averaged 1.3 m^3 s^{-1} km^{-2}. Although the frequency of occurrence of severe ROS events in a region is often uncertain, MacDonald and Hoffman (1995) showed that the majority of peak annual daily flow rates in six basins in northwestern Montana and northeastern Idado were caused by ROS events. Obviously conditions leading up to the occurrence of severe ROS events deserve special consideration.

Flooding due to snowmelt and rain-on-snow events appears to be caused by a combination of circumstances that defy easy prediction. Factors contributing to these floods are many and not all contribute in all cases, but those identified include (see Kattleman, 1997):

- relatively widespread, deep snowcover extending to low elevations
- heavy rainfall extending to high elevations
- warm, windy conditions during rainfall conducive to condensation–convection melt
- rainfall occurring in mid-winter
- ripening of snowpack or loss of cold content prior to or in early stages of the rain event
- wet antecedent soil conditions
- occurrence of impermeable soil frost
- occurrence of river ice jams that can lead to overbank flooding.

Occurrence of widespread and deep snowcover can lead to prolonged melting over a large area of the watershed. Any melting that does occur is added to the rainfall and creates wet soils that lead to increased response from later rainfall inputs. Since rainy weather is accompanied by cloud cover, melt during rainfall is primarily driven by convection of latent and sensible heat to the snowpack rather than radiation. Given the typically high humidity during rainfall, wind speeds and air temperatures generally determine melting during ROS events. Although significant snowpack melting does not always occur along with rainfall, Anderson and Larson (1996) reported that melt due to warm air, high wind speeds, and high humidity associated with a near-record ROS event in the Northeast contributed as much water as the rainfall itself. On mountain basins, where rain events are often restricted to low elevations where snowcover is uncommon, ROS events that extend to very high elevations produce the largest ROS events. MacDonald and Hoffman (1995) found that mid-winter ROS peak flow events produced the greatest peak flows, but were much less frequent than ROS peak flows during the spring melt period in the mountains of northeastern Idaho and northwestern Montana. Mid-winter ROS events were associated with greater rainfall amounts than spring ROS events and possibly greater areal snow cover. Identification of unique sets of conditions that can lead to major ROS events is essential to forecasting the threat of ROS flooding.

Forest harvesting has been suggested as a possible cause of increased ROS flooding. Cutting of trees reduces canopy interception losses for both rain and snow. Cutting trees also reduces evapotranspiration rates leading to higher soil water levels and exposes the snowpack to increased wind speeds that can drive convection melt during rainfall. Available evidence from watershed studies lends support to this hypothesis (Harr, 1986; Berris and Harr, 1987; Marks *et al.*, 1998). As forests re-grow, canopy interception losses, evapotranspiration rates, and snowpack protection from wind all increase. However, at higher elevations and latitudes, re-growth rates of forests after cutting are low due to shorter, and sometimes climatically severe, growing seasons. If widespread cutting occurs on a watershed, large areas of the basin that have not sufficiently re-grown may be predisposed to yield larger peak flows during ROS events.

9.9 Applications

Hydrologic processes influencing the movement and storage of water through snowpacks and watersheds to stream channels have been described in this chapter, as well as the resulting streamflow hydrographs. These concepts and equations form the basis for modelling streamflow from snowmelt discussed in the next chapter. The current trend is to model streamflow from snowmelt on a continuous basis on large heterogeneous watersheds where spatial and temporal variations in controlling factors must be considered. Models of the processes and conditions leading up to rain-on-snow floods that cause considerable damage and loss of life are particularly important. Modelling with these objectives can only be successful with a good understanding of the behavior of hydrologic systems during melt as discussed here.

9.10 References

Aizen, V. B., Aizen, E. M., and Melack, J. M. (1995). Climate, snow cover, glaciers, and runoff in the Tien Shan, Central Asia. *Water Resour. Bull.*, **31**(6), 1113–29.

Aizen, V. B., Aizen, E. M., and Melack, J. M. (1996). Precipitation, melt and runoff in the northern Tien Shan. *J. Hydrol.*, **186**, 229–51.

Aldrich, H. P., Jr. (1956). Frost penetration below highway and airfield pavements. *Highway Res. Board*, **135**, 124–44.

Anderson, E. A. (1973). *National Weather Service River Forecast System – Snow Accumulation and Ablation Model*. NWS Hydro-17. US Dept. Commerce, NOAA, National Weather Service.

Anderson, E. and Larson, L. (1996). The role of snowmelt in the January 1996 floods in the Northeastern United States. In *Proceedings of the 53rd Annual Meeting, Eastern Snow Conf.*, pp. 141–9.

Bathurst, J. C. and Cooley, K. R. (1996). Use of SHE hydrological modelling system to investigate basin response to snowmelt at Reynolds Creek, Idaho. *J. Hydrol.*, **175**, 181–211.

Bengtsson, L. and Westerstrom, W. (1992). Urban snowmelt and runoff in northern Sweden. *Hydrol. Sci.*, **37**, 263–75.

Bengtsson, L., Seuna, P., Lepisto, A., and Saxena, R. K. (1992). Particle movement of melt water in a subdrained agricultural basin. *J. Hydrol.*, **135**, 383–98.

Berris, S. N. and Harr, R. D. (1987). Comparative snow accumulation and melt during rainfall in forested and clear-cut plots in the Western Cascades of Oregon. *Water Resour. Res.*, **23**(1), 135–42.

Buttle, J. M. (1994). Isotope hydrograph separations and rapid delivery of pre-event water from drainage basins. *Prog. Phys. Geog.*, **18**(1), 16–41.

Caine, N. (1992). Modulation of the diurnal streamflow response by the seasonal snowcover of an alpine basin. *J. Hydrol.*, **137**, 245–60.

Carey, S. K. and Woo, M. (1998). Snowmelt hydrology of two subarctic slopes, Southern Yukon, Canada. *Nordic Hydrol.*, **29**(4/5), 331–46.

Cline, D. W. (1997). Snow surface energy exchanges and snowmelt at a continental, midlatitude Alpine site. *Water Resour. Res.*, **33**(4), 689–701.

Cooley, K. R. and Palmer, P. (1997). Characteristics of snowmelt from NRCS SNOTEL (SNOwTELemetry) sites. *Proceedings of the 65th Annual Western Snow Conference*, Banff, Alberta, Canada, pp. 1–11.

DeGaetano, A. T., Wilks, D. S., and McKay, M. (1996). A physically based model of soil freezing in humid climates using air temperature and snow cover data. *J. Appl. Meteorol.*, **35**, 1009–27.

Flerchinger, G. N. and Saxton, K. E. (1989). Simultaneous heat and water model of a freezing snow-residue-soil system I. Theory and development. *Trans., ASAE*, **232**(2), 565–71.

Garstka, W. U., Love, L. D., Goodell, B. C., and Bertle, F. A. (1958). *Factors Affecting Snowmelt and Streamflow: a Report on the 1946–1953 Cooperative Snow*

Investigations at the Fraser Experimental Forest, Fraser, Colorado. US Bureau of Reclamation, US Forest Service, US Government Printing Office.

Gibson, J. J., Edwards, T. W. D., and Prowse, T. D. (1993). Runoff generation in a high boreal wetland in northern Canada. *Nordic Hydrol.*, **24**, 213–24.

Hardy, J. P. and M. R. Albert. (1995). Snow-induced thermal variations around a single conifer tree. In *Proceedings of the 52nd Eastern Snow Conference*, Toronto, Ontario, Canada: June 7–8, 1995, pp. 239–47.

Harr, R. D. (1986). Effects of clearcutting on rain-on-snow runoff in Western Oregon: a new look at old studies. *Water Resour. Res.*, **22**(7), 1095–100.

Hinzman, L. D. and D. L. Kane. (1991). Snow hydrology of a headwater Arctic basin 2. Conceptual analysis and computer modeling. *Water Resour. Res.*, **27**, 1111–21.

Kane, D. L. and Stein, J. (1983). Water movement into seasonally frozen soils. *Water Resour. Res.*, **19**(6), 1547–57.

Kattleman, R. (1997). Flooding from rain-on-snow events in the Sierra Nevada. In *Destructive Water: Water-Caused Natural Disasters, their Abatement and Control, Conference Proceedings*, IAHS Publ. No. 239, Anaheim, CA, June 1966, pp. 59–65.

Kattleman, R., Berg, N., and McGurk, B. (1991). A history of rain-on-snow floods in the Sierra Nevada., In *Proceedings of the 59th Western Snow Conference*, Juneau, Alaska April 13–15, 1991, pp. 138–41.

Kersten, M. (1949). Thermal properties of soils. University of Minnesota, *Engin. Exp. Sta. Bulletin*, **28**.

Kendall, K. A., Shanley, J. B., and McDonnell, J. J. (1999). A hydrometric and geochemical approach to test the transmissivity feedback hypothesis during snowmelt. *J. Hydrol.*, **219**, 188–205.

Kuchment, L. S. (1997). Estimating the risk of rainfall and snowmelt disastrous floods using physically-based models of river runoff generation. In *Destructive Water: Water Caused Natural Disasters, Their Abatement and Control, Conference Proceedings*, IAHS Publ. 239, Anaheim, CA, pp. 95–100.

Leydecker, A. and Melack, J. M. (1999). Evaporation from snow in the Central Sierra Nevada of California. *Nordic Hydrol.*, **30**(2), 81–108.

Lewis, T., C. Braun, D. R. Hardy, P. Francus, and R. S. Bradley. (2005). An extreme sediment transfer event in a Candadian high Arctic stream. *Arct., Antarct. Alp. Res.*, **37**, 477–82.

Lundquist, J. D., Dettinger, M. D. and Cayan, D. R. (2005). Snow-fed streamflow timing at different basin scales: Case study of the Tuolumne River above Hetch Hetchy, Yosemite, California. *Water Resour. Res.*, **41**(W07005), 1–14.

MacDonald, L. H. and Hoffman, J. A. (1995). Causes of peak flows in Northwestern Montana and Northeastern Idaho. *Water Resour. Bull.*, **31**(1), 79–95.

Marks, D., Kimball, J., Tingey, D., and Link, T. (1998). The sensitivity of snowmelt processes to climate conditions and forest cover during rain-on-snow: a case study of the 1996 Pacific Northwest flood. *Hydrol. Processes*, **12**, 1569–87.

Marsh, P. (1999). Snowcover formation and melt: recent advances and future prospects. *Hydrol. Processes*, **13**, 2117–34.

McNamara, J. P., D. L. Kane and L. D. Hinzman. (1997). Hydrograph separation in an Arctic watershed using mixing model and graphical techniques. *Water Resour. Res.*, **33**, 1707–19.

Moore, R. D. (1993). Application of a conceptual streamflow model in a glacierized drainage basin. *J. Hydrol.*, **150**, 151–68.

Persson, M. (1976). *Hydrologiska undersokningar i Lapptraskets representativa omrade* (In Swedish, English title: *Hydrological Studies in the Lapptrasket Representative*

Catchment), SMHI HB Rapport no. 13, Norrkoping, Sweden: Swedish Meteorol. and Hydrol. Instit.

Prévost, M., Barry, R., Stein, J., and Plamondon, A. P. (1990). Snowmelt runoff modelling in a balsam fir forest with a variable source area simulator (VSAS2). *Water Resour. Res.*, **26**(5), 1067–77.

Rekolainen, S. and Posch, M. (1993). Adapting the CREAMS model for Finnish conditions. *Nordic Hydrol.*, **24**, 309–22.

Rodhe, A. (1998). Chapter 12: Snowmelt dominated systems. In *Isotope Tracers in Catchment Hydrology*, ed. B. V. C. Kendall and J. J. McDonnell, Amsterdam: Elsevier, pp. 391–433.

Semadeni-Davies, A. (1998). Modelling snowmelt induced waste water inflows. *Nordic Hydrol.*, **29**(4/5), 285–302.

Schmidt, R. A., C. A. Troendle, and J. R. Meiman. (1998). Sublimation of snowpacks in subalpine conifer forests. *Can. J. For. Res.*, **28**, 501–13.

Shanley, J. B. and Chalmers, A. (1999). The effect of frozen soil on snowmelt runoff at Sleepers River, Vermont. *Hydrol. Processes*, **13**, 1843–57.

Shutov, V. A. (1993). Calculation of snowmelt. *Russian Meteorol. and Hydrol.*, **4**, 14–20.

Stadler, D., Wunderli, H., Auckenthaler, A., Flühler, H., and Bründl, M. (1996). Measurement of frost-induced snowmelt runoff in a forest soil. *Hydrol. Process.*, **10**, 1293–1304.

Todhunter, P. E. (2001). A hydroclimatological analysis of the Red River of the North snowmelt flood catastrophe of 1997. *J. Amer. Water Resour. Assoc.*, **37**, 1263–78.

US Army Corps of Engineers (1956). *Snow Hydrology: Summary Report of the Snow Investigations*. Portland, OR: North Pacific Div., US Army Corps of Engineers.

Westmacott, J. R. and Burn, D. H. (1997). Climate change effects on the hydrologic regime within the Churchill-Nelson River Basin. *J. Hydrol.*, **202**, 263–79.

Woo, M.-K. (1986). Permafrost hydrology in North America. *Atmos. Ocean*, **24**(3), 201–234.

Woo, M.-K. (1983). Hydrology of a drainage basin in the Canadian high Arctic. *Annals Assoc. Amer. Geogr.*, **73**, 577–596.

Woo, M.-K. and J. Sauriol. (1980). Channel development in snow-filled valleys, Resolute, N. W. T., Canada. *Geogr. Ann.*, **62A**, 37–56.

Zhang, Z., D. L. Kane, and L. D. Hinzman. (2000). Development and application of a spatially-distributed Arctic hydrological and thermal process model (ARHYTHM). *Hydrol. Processes*, **14**, 1017–44.

Zhao, L. and Gray, D. M. (1999). Estimating snowmelt infiltration into frozen soils. *Hydrol. Processes*, **13**, 1827–42.

Zhenniang, Y., Fungzian, L., Zhihuai, Y., and Qiang, W. (1994). The effect of water and heat on hydrological processes of a high alpine permafrost area. In *Snow and Ice Covers: Interactions with the Atmosphere and Ecosystems*, IAHS Publ. 223, ed. E. G. Jones, T. D. Davies, A. Ohmura, and E. M. Morris, pp. 259–68.

10

Modelling snowmelt runoff

10.1 Introduction

Hydrologic models used to predict streamflow can be generally classified as either deterministic or stochastic and as either lumped-parameter or distributed (Beven, 2000). Deterministic models predict a single value of streamflow from a given set of input variables, while stochastic models predict a range of possible outcomes based upon the statistical distributions of input variables. Nearly all of the snowmelt models used to continuously predict streamflow from snowmelt are deterministic, but a type of statistical model has been historically used to great advantage to predict seasonal totals of streamflow using measured snowpack and precipitation data each spring. These models are in widespread use in the western United States to forecast spring runoff.

Deterministic models can be further categorized as either lumped-parameter or distributed. Lumped-parameter models consider the entire watershed as a homogeneous unit, while distributed models break the watershed into subunits or grid points based upon a variety of criteria. Early snowmelt models were of the lumped-parameter type, but many distributed-snowmelt modelling studies have been conducted in the past decade. In terms of snowmelt prediction, the lumped-parameter type of model assumes that the snow cover and melt rates are uniform across the entire basin. These so called point-snowmelt models are still in use today and are generally found imbedded in most deterministic, distributed models. Distributed-snowmelt models generally adapt simple point-snowmelt models for conditions that vary from point to point across basins. Conditions such as varying topography (slope, aspect, and elevation), wind redistribution and sublimation of snow, vegetative cover (forest cover and type), and soil freezing conditions can all be included in distributed models.

All models also have characteristic temporal and spatial scales that can be used to classify them (Bengtsson and Singh, 2000). Statistical snowmelt models can be used to predict streamflow totals for entire runoff seasons, while deterministic melt models considering hydrologic processes may operate on time steps as short as one hour. Due to data limitations and the nature of the snowmelt-runoff process, daily time steps are often used for snowmelt models for prediction of spring snowmelt runoff. Spatial scales can range from treating the entire watershed as a single homogeneous unit down to dividing the watershed into a number of separate grid points or nodes. Vertical spatial scales also become relevant for snowmelt modelling since the snowpack can be modelled as a set of layers of varying properties that interact with themselves, the atmosphere above and soil below. Overall, each model will be developed with a spatial and temporal scale needed to achieve a specific objective, but ultimately many are constrained by availability of data and our knowledge of the processes involved.

Deterministic models used to continuously predict streamflow often explicitly consider the many hydrologic processes that control the generation of streamflow from precipitation. Although the level of detail considered in each model varies considerably, the hydrologic processes that can be considered are (DeVries and Hromadka, 1992): canopy interception, evapotranspiration, snowmelt, interflow, overland flow channel flow, unsaturated subsurface flow, and saturated subsurface flow. In this context snowmelt is seen as just one component of complex models that often involve many uncertain parameters in addition to those related to snow. Independent estimates of snowmelt parameters from those needed for simulating the rest of the hydrologic system is obviously a good idea.

This chapter begins with a review of statistical models used to forecast streamflow totals rather than prediction of continuously varying streamflow. Point-snowmelt models that predict the release of melt water from the snowpack to the soil are next described in detail. A general discussion of distributed-snowmelt-runoff modelling will follow. The chapter concludes with a brief summary of the models currently in use to predict streamflow from snowmelt. As in the prior chapters, emphasis will be on aspects of hydrologic modelling unique or important to snow hydrology. Others sources (Maidment, 1992; US Army Corps of Engineers, 1956; Beven, 2000) can be consulted for a more complete and general discussion of hydrologic modelling.

10.2 Statistical snowmelt-runoff models

Statistical methods have traditionally been used to estimate or forecast seasonal runoff volumes from snowmelt. These models rely upon statistical relationships between measured variables and can not be expected to predict runoff for unusual conditions not previously experienced. Ability to predict flows under unusual or extreme conditions is the commonly used justification for deterministic, physically based process models. As long as all the physical processes are well represented, physical models should better predict flows for extreme conditions. Regardless, given enough historical measurements very accurate statistical models for average or typical conditions can be developed using statistical methods, many of which are currently in use today. A common statistical approach has been to relate measured spring runoff to measurements of peak snowpack accumulation based upon snow-course or snow-pillow data. Snow-covered area has also been related to spring runoff. Other early statistical methods were used to simulate the entire snowmelt hydrograph (US Army Corps of Engineers, 1956; Garstka, 1958), but these methods have generally now been supplanted by the continuous-simulation process models.

One of the most useful statistical models is the relationship between maximum spring snowpack water equivalent (SWE) and the total volume of runoff during the springtime melt season (Q) as in Equation (10.1). The models generally are used to estimate total water supply available for irrigation and domestic water needs in selected western United States basins. Models generally take the form:

$$Q = a + b \, SWE \tag{10.1}$$

where empirical coefficients a and b are fit to the measured data using regression analysis. Snowpack-water-equivalent data may be derived from either manually measured snow courses or automated snow pillow (SNOTEL) sites. For example, DeWalle *et al.* (2003) used early-season SNOTEL snowpack-water-equivalent data on November 1 or December 1 to predict seasonal total snowmelt-runoff volumes for the Upper Rio Grande basin in Colorado. Other independent variables such as the amount of rainfall during the late summer and fall prior to snow accumulation and pre-melt stream baseflow rates can also be included in the model to provide some index of soil water content and groundwater storage prior to melt. Statistical methods such as principal components regression analysis are generally employed to find the "best" combination of snow course, snow pillow or precipitation sites. Since these equations are used in the forecast mode, input data must be known at the time of the forecast. However, separate equations can also be derived for forecasting flows for sequential dates on each basin, such as flow total for April–September, May–September, etc. Equations are updated each year to include the current year's data. Since peak flows are usually well correlated with total volume of spring runoff, the equations can also be easily extended to prediction of spring peak flows as well as other flow and hydrograph parameters.

An example of a statistical model derived by the USDA, NRCS, National Water and Climate Center for forecasting April–September streamflow in the Donner und Blitzen River in southeastern Oregon, USA is given in Figure 10.1 The best-fit equation was derived using principal-components regression for 18 different potential independent variables representing water-equivalent data recorded at two SNOTEL sites and snowpack depths observed at three different aerial-marker (AM)

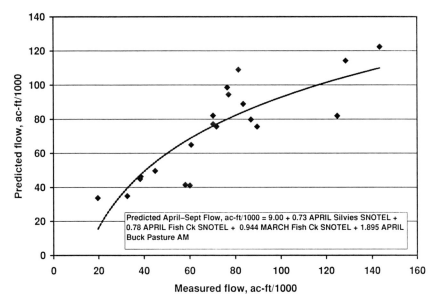

Figure 10.1 Example of a working equation used to predict total April–September streamflow for the Donner und Blitzen River in southeastern Oregon USA based upon snowpack-water-equivalent (SNOTEL) and snow-depth aerial-marker (AM) data (courtesy USDA NRCS National Water and Climate Center).

sites at different times of the year during the period 1980–1999. Snowpack water equivalent (inches) on April 1 at the Silvies SNOTEL site, water equivalent (inches) on March 1 and April 1 at the Fish Creek SNOTEL site, and the April 1 estimated water equivalent (inches) at the Buck Pasture aerial-marker site were ultimately chosen as variables. Predictions of flow are made in thousands of acre-feet (1 ac-ft. = 1234 m^3) and original units were kept in the equation. The correlation coefficient and standard error for this equation are 0.85 and 17.5 ac-ft, respectively. Jackknife or cross-validation estimates of the correlation coefficient and standard error, derived with regression coefficients fitted without the input data for the specific year being predicted, were 0.80 and 19.8 ac-ft, respectively. This equation is just one of many examples of how snow data are being used in statistical models to predict total streamflow from snowmelt.

Many other examples of the use of statistical models to predict flow totals can be found. Mashayekhi and Mahjoub (1991) present statistically derived equations developed for predicting monthly and seasonal flows into the Karadj reservoir in Iran. They found that monthly flow equations were less reliable than seasonal equations, especially flows for March, which was a transition month from accumulation to melting. Some of their equations included as variables the prior month's flow and air temperatures as well as precipitation and snowpack-water-equivalent data.

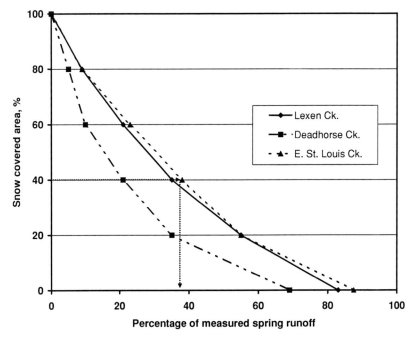

Figure 10.2 Relationship between snow-covered area and the percentage of accumulated spring runoff on three Rocky Mountain subalpine watersheds in Colorado
(see text example, modified after Leaf, 1969, courtesy of the US Western Snow
Conference).

Seasonal forecasts were made with an error of about 15%. Thapa (1993) used
a regression of March snow-covered area from satellite remote sensing against
streamflow volumes to estimate pre-monsoon streamflow amounts for several basins
in the Himalayas where other forms of data were lacking.

  Another statistical approach for snowmelt prediction is the use of standard curves
relating snow-covered area on a watershed to the fraction of total runoff occurring
during the spring melt period. Characteristic curves fit using regression analysis to
snow cover and flow data can be derived on a watershed as illustrated in Figure 10.2
for three basins in the Colorado Rocky Mountains (Leaf, 1969; Leaf and Haeffner,
1971). Such curves are quite consistent from year to year on a given watershed and
each basin has its own characteristic curve. Note the differences in curves between
Deadhorse Creek and the other two basins in Figure 10.2, East St. Louis Creek
and Lexen Creek. Whenever an estimate of snow-covered area is available during
the melt season, these curves can be used to estimate the volume of remaining
snowmelt runoff.

  An example taken from Leaf (1969) for E. St. Louis Creek helps to illustrate
the use of these curves (Figure 10.2). The snow-covered area on the watershed in

mid-June was estimated from photographs as 41% of basin area at a time when the total spring runoff based upon measured streamflow had been about 738 000 m^3 (598 acre-ft). Since this snow-covered area normally occurs when 37% of spring runoff has occurred according to the curve in Figure 10.2, the expected volume for the remainder of the melt season would be 738 000/0.37 $\cong$ 1 995 000 m^3 (1616 acre-ft). This type of relationship provides information about the timing of streamflow in addition to estimates of remaining flow volumes and provides a means for updating forecasts during the melt season. As snow-covered-area data from satellite remote sensing are now available for most basins at regular time intervals (see Chapter 5), use of this method could easily be increased in the future. When more frequent estimates of flows are needed, point- or distributed-snowmelt models of continuous streamflow are the best alternative.

10.3 Point-snowmelt models

Modelling the snowpack water equivalent and release of water from the snowpack to the soil at a point requires several modelling considerations in addition to those required for modelling response to rainfall inputs only:

- correction of snowfall amounts for wind-induced precipitation-gauge-catch errors,
- determination of precipitation type (rain, snow, or proportions of each),
- estimation of melt rates using energy-budget or temperature-index methods,
- estimation of snowpack temperature and cold content,
- estimation of snowpack liquid-water holding capacity and storage,
- routing liquid water through the snowpack to the soil.

Correction of snowfall data for measurement errors from precipitation gages with or without wind shields was previously discussed in Chapter 4. Even though corrections for gauge-catch errors are needed for both rain and snow, corrections for snowfall generally are greater. Computation of melt rates and snowpack cold content using an energy-balance approach was discussed in Chapters 6 and 7. The use of air-temperature data and the degree-day approach, though less precise, can also be used to predict melt and that method is described in detail here. Finally, modelling liquid-water behavior in snow was described in Chapters 3 and 9 and further applications are given later. All of these factors affect our ability to predict water releases from snowpacks at a point.

10.3.1 Form of precipitation

Since routine observations of the form of precipitation (rain or snow) are generally lacking, air temperature is commonly used to decide the form of precipitation in

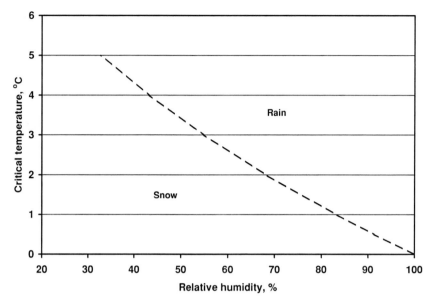

Figure 10.3 Critical temperature for separation of rain/snow events and maximum relative humidity conditions predicted with the psychrometric equation.

hydrological modelling. A critical temperature is used to determine whether precipitation fell as rain or snow. Since snow crystals falling in the atmosphere will be cooled by sublimation into relatively dry air they may survive as solid precipitation even when air temperatures are somewhat above freezing. Thus, critical temperatures based upon air temperatures are often slightly above freezing. In conditions where potential flood-causing rainstorms occur in winter with deep snow, the classification of precipitation events as either rain or snow is particularly important.

The amount of cooling of falling snow crystals due to sublimation, or the frost-bulb depression, depends in part on the relative humidity of the air near the ground. Snow-crystal temperature could be maintained at 0 °C by sublimation cooling at air temperatures of approximately 0.5, 1, 2, and 3 °C for relative humidities of 91, 83, 68 and 55%, respectively. In other words, snow falling through air at 2 °C with a 68% relative humidity could maintain a temperature of 0 °C and remain solid, while at 2 °C and 83% relative humidity the snow would melt and become rainfall. The relationship between the amount of frost-bulb cooling and relative humidity needed to maintain precipitation at 0 °C is shown in Figure 10.3. The significance of frost-bulb cooling in determining the form of precipitation will be highly variable due to humidity variations in the lower atmosphere at the beginning of precipitation events and variations in humidity during events. Regardless, the magnitude of the frost-bulb cooling is sufficient to explain observed critical air temperatures needed to separate snow and rain events.

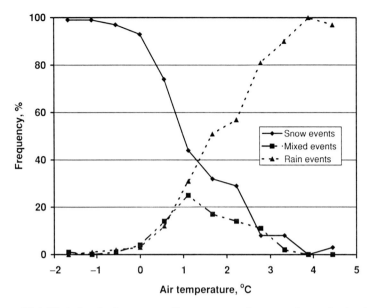

Figure 10.4 Variation in frequency of occurrence of snow, rain, and mixed precipitation events with air temperature (data from US Army Corps of Engineers, 1956).

Detailed observations by the US Army Corp of Engineers, (US Army Corps of Engineers, 1956) showed the frequency distribution of snow, rain, and mixed precipitation events as a function of air temperature (Figure 10.4). Snowfall can obviously occur over a range of air temperatures; however, a single critical temperature (T_{CRIT}) of 35 °F or 1.67 °C was recommended by the Corps of Engineers (US Army Corps of Engineers, 1956). They found that 90% of the cases were correctly categorized if precipitation occurring at air temperatures equal to or greater than T_{CRIT} was counted as rain and precipitation for temperatures below T_{CRIT} was counted as snow. Rohrer and Braun (1994) suggest that most of the error in classifying events as rain or snow is due to lack of timely air-temperature data. They reported that for hourly air-temperature data, only 2% of events were misclassified relative to using air temperature measured every 10 minutes. However, when mean daily air temperatures were used, 14% of events were misclassified. At an alpine location in their study, snowfall events were rarely observed at air temperatures above 1 °C and those that were had negligible intensity.

In general, a T_{CRIT} ranging from 0 to 2 °C is typically used. Many times the critical temperature becomes a fitting parameter in model studies. Singh and Singh (2001) described a procedure where precipitation at or below air temperatures of 0 °C was counted as snow, precipitation at or above some T_{CRIT} (e.g. 2 °C) was rain, and a linear interpolation of the fraction of rain and snow was used between 0 and 2 °C. Willen *et al.* (1971) developed a similar procedure for estimating the proportion of rain and snow in an event that has been modified and used by Leavesley *et al.* (1983) in the PRMS model. If humidity data were available, then a correction for critical temperature based upon humidity could be developed, as shown in Figure 10.3. When hourly temperature data are not available, it is customary to use mean daily air temperature as the indicator of T_{CRIT} despite the potential for misclassification of events.

10.3.2 Temperature-index methods

Air temperatures or degree-days have been widely used in hydrologic modelling to approximate snowpack energy exchange in lieu of the more data-intensive energy-budget approaches. A degree-day is defined here as one degree difference between mean daily air temperature and some base temperature. Both melt rates and changes in snowpack cold content have been estimated with the degree-day approach. Air temperatures above freezing are generally positively correlated with the meteorological conditions that control energy supply for melt. When the air is at subfreezing temperatures, the difference between air and snow temperature is also correlated with the change (positive or negative) in snowpack heat storage. The underlying assumption using the degree-day approach is that there is a linear relationship between snowmelt rates or changes in snowpack cold content and the mean daily air temperature above or below some temperature base, respectively. Although valid theoretical arguments against such linear relationships can be raised, especially at temperatures near zero (Kuz'min, 1961), the method is approximately correct and with adjustments for local and seasonal conditions the degree-day method can be of great utility.

10.3.2.1 Degree-day melt computations

Computation of melt rates using degree-days most commonly is given as Equation (10.2):

$$M = (DDF)(T_a - T_b) \qquad (10.2)$$

where:

M = melt, cm d^{-1}
DDF = degree-day factor, cm °C^{-1} d^{-1}
T_a = mean daily air temperature, °C
T_b = base temperature, °C

Mean daily air temperatures are computed as the mean of maximum and minimum daily air temperatures or hourly air temperatures for the day. Alternatively, maximum daily air temperature, mean daylight temperature, the average of hourly temperatures above freezing and other temperature parameters have also been used instead of mean daily temperatures (Martinec, 1976; Kuz'min, 1961; US Army Corps of Engineers, 1956). Base temperatures (T_b) of 0 °C are generally used for melt computation, but other base temperatures have also been employed with success for specific sites (US Army Corps of Engineers, 1956). Although Equation (10.2) is written for a daily time step, similar expressions can be developed to predict melt over longer time periods by simply accumulating degree-days, but time periods less than one day are problematical due to variations in the correlation between air temperature and snowpack energy supply between day and night periods.

Degree-day factors are affected by the air and base temperature parameters selected for use and a number of variables that are spatially or temporally related to snowmelt energy supply, as summarized in Table 10.1. Not surprisingly, a wide range of DDFs are found in the literature with most values falling between 0.1 to 0.8 cm °C^{-1} d^{-1}. In any given application it is highly desirable to obtain representative and accurate melt and air temperature data to compute local values of DDF as DDF $= M/(T_a - T_b)$. SNOTEL data in the western United States can be used to compute local DDFs for specific watersheds (DeWalle *et al.*, 2002). The degree-day approach is essentially empirical, but DDF can be related to energy exchange processes. The physical interpretation of the DDF has been derived by Brubaker *et al.* (1996) and Bengtsson (1976).

Use of Equation (10.2) is straight forward once the proper DDF is obtained. Assuming that maximum and minimum air temperature on a given day were 6 and −2 °C, respectively, the mean daily air temperature would be 2 °C. Hourly temperature data, that is often lacking, would be required to compute the mean of positive or daylight air temperatures. If the DDF for this day was 0.4 cm °C^{-1} d^{-1}, then the computed melt rate would be 0.8 cm d^{-1}. Errors in predicting daily melt can be quite large and the method works best for prediction of melt over extended time periods such as weeks to an entire snowmelt season. Application of this approach on a hourly basis using a constant DDF does not satisfactorily reproduce the diurnal changes in melt rates (Hock, 1999). Melt on days with rain can be handled using more detailed energy budget approaches as previously described in this chapter or with an adjustment to the DDF (Kuusisto, 1980).

Adjustments to DDF for seasonal changes in snowpack and environmental conditions have been developed. Martinec (1960) developed a method to adjust DDF for seasonal changes based upon snowpack density as Equation (10.3):

$$DDF = 1.1(\rho_s/\rho_w) \tag{10.3}$$

where:

DDF = degree-day factor, cm °C^{-1} d^{-1}
ρ_s = snowpack density, kg m^{-3}
ρ_w = density of liquid water $= 10^3$ kg m^{-3}

Table 10.1 *Variables affecting snowmelt Degree-Day Factors (DDF).*

Variable	Cause	DDF Response
Time of year	Decreasing snowpack cold content and albedo and increasing shortwave radiation and snow density as season advances	DDF increases during accumulation and melt season
Forest cover vs. open conditions	Shading and wind protection by forest	DDF lower and less variable in forest than in open
Topography	Variations in incoming shortwave radiation and exposure to wind	DDF higher on south-facing slopes and windy sites
Areal snow cover	Varying fraction of watershed contributing to melt	DDF for basin decreases as areal snow cover decreases
Snow surface contamination	Dust and debris on snow surface decreases the albedo	DDF increases
Rainfall	High humidity increases condensation energy supply and rain adds some sensible heat, clouds reduce solar	DDF generally lower on rainy days, as cloudiness dominates
Ice vs. snow cover	Glacier ice has lower albedo than snow	DDF higher on glacial basins when ice exposed
Other meteorological conditions for a given air temperature	Melt energy greater during unusually high wind, shortwave radiation, or humidity conditions at a given air temperature	DDF higher with greater melt energy supply for a given air temperature
Maximum daily air temperature used for T_a	Causes higher T_a and degree-day total	DDF lower, approx. half of DDF for mean daily air temperature
Mean of positive or daylight air temperatures used for T_a	Causes higher T_a and degree-day total	DDF lower than DDF for mean daily air temperature

Thus, if snowpack density gradually increased from 200 to 400 kg m^{-3} during the spring season, DDF would change from 0.22 to 0.44 cm °C^{-1} d^{-1}. It should be noted that Martinec (1960) defined degree-days based upon the average of positive air temperatures during a day which gives a lower DDF than when using mean daily air temperatures. Anderson (1973) also varied DDF seasonally between minimum and maximum DDF values during the year using a sine-wave function, while Federer

and Lash (1983) used a linear increase in DDF from January 1 to June 23. DeWalle
et al. (2002), showed that DDF derived from SNOTEL data, linearly increased with
Julian Date during spring by about 0.5% to 1% per day; a rate that exceeded the
rate of increase of potential solar irradiation.

Forest-cover effects on degree-day melt factors have also been quantified. Federer
et al. (1972) reported the following ranges of DDF based upon mean daily air
temperatures for open vs. forest conditions in northeast United States: open 0.45–
0.75, leafless deciduous 0.27–0.45, and conifer 0.14–0.27 cm $^{\circ}$C^{-1} d^{-1}. These
ranges were based upon several studies in the northeast giving the DDF ratios
for open:deciduous:conifer conditions of approximately 3:2:1. Kuusisto (1980)
showed that the DDF (cm $^{\circ}$C^{-1} d^{-1}) based upon mean daily air temperatures in
Finnish forests could be computed using Equation (10.4):

$$DDF = 0.292 - 0.164C_c \qquad (10.4)$$

where C_c = fractional canopy cover. Using this model, forest canopy cover of 0.1
and 0.7 (the limits of the forest-cover data) would represent DDFs of 0.28 and 0.18
cm $^{\circ}$C^{-1} d^{-1}, respectively. These relationships allow some corrections of DDF for
forest effects.

In order to increase the temporal and spatial resolution of DDF melt estimates,
several hybrid models using temperature and radiation parameters have also been
developed. Cazorzi and Fontana (1996) developed a multiplicative model with a
temperature term and an index to monthly average potential clear-sky shortwave
radiation for various slopes, aspects, and elevations; Hock (1999) used an additive
degree-day temperature term and a term for computed clear-sky potential shortwave
radiation for various slopes and aspects, and in studies by Martinec and deQuervain
(1975), Brubaker *et al.* (1996), and Hamlin *et al.* (1998), a component representing
net allwave radiation exchange was added to a degree-day term in melt computations. In all of these hybrid models, the DDF takes on a new physical meaning since
it is now independent or partially independent of radiation exchange. These models
enhance melt prediction over large areas by consideration of topography along with
degree-days. Models developed by Hock (1999) and Cazorzi and Fontana (1996) do
not require additional data, while models by Martinec and deQuervain (1975) and
Brubaker *et al.* (1996) require incoming shortwave radiation or cloud-cover data
at the minimum. Hock's (1999) model was applied hourly and improved ability to
simulate diurnal melt variations.

10.3.2.2 Degree-day cold-content computations

Air temperatures can also be used to track the snowpack cold content and ripeness.
Cold content is expressed as cm of liquid-water equivalent that would have to
refreeze in the snowpack to warm it to 0 $^{\circ}$C (see Chapter 3). When average daily

air temperatures are below freezing, it is generally assumed that no melt will occur and that the snowpack can lose or gain energy and cold content as a linear function of the difference in temperature between the air and snowpack. Applications of the degree-approach to cold-content computations are less common than for melt. Leavesley and Striffler (1978) estimated cold-content development for the PRMS model (see later section) using an approximation of heat conduction in a two-layer snowpack model that requires predicting snow density. According to Anderson (1976) the change in cold content of the snowpack can be computed as Equation (10.5):

$$\Delta CC = CCF(T_s - T_a)$$ (10.5)

where:

ΔCC = rate of change in cold content, cm d^{-1}
CCF = cold-content degree-day factor, cm °C^{-1} d^{-1}
T_a = mean daily air temperature, °C
T_s = snow surface temperature, °C

Optimized CCF values used in melt simulations have ranged from about 0.02 to 0.05 cm °C^{-1} d^{-1} (Anderson, 1973; Federer and Lash, 1983). The CCF values are about 1/10 of the values of DDF, indicating that snowpack warming and cooling due to heat conduction in the snow occurs at a much slower rate than melting does per degree-day. Since the snowpack gains or loses heat chiefly by conduction when there is no melt and the conductivity increases with density, a gradual seasonal increase in CCF is normally applied. Federer and Lash (1983) allowed CCF to increase linearly from 0.02 cm °C^{-1} d^{-1} on January 1 to 0.4 cm °C^{-1} d^{-1} on June 23. Anderson (1973) applied the same sine-wave function for CCF as used to seasonally adjust degree-day melt factors (DDF).

Application of Equation (10.5) to compute change in snowpack cold content is relatively direct. Assuming a CCF of 0.03 cm °C^{-1} d^{-1}, a mean daily air temperature of −6 °C and a snow surface temperature of −4 °C, the change in snowpack cold content would be an increase of 0.06 cm water equivalent for the day due to snowpack cooling. In contrast, if air temperature were −2 °C, with the same CCF and snow temperature, a −0.06 cm increment to cold content or snowpack warming would result. Operationally, changes in cold content are added to the previous day's cold content and an accounting scheme is used to determine when the snowpack is ripe (zero cold content).

An estimate of snow surface temperature is also required to use Equation (10.5). Although theoretical-based, layered-snowpack temperature models that require complete energy-budget data have been developed (Jordan, 1991; Anderson, 1976;

Horne and Kavvas, 1997), Anderson (1973) and Marks *et al.* (1992) used a simple temperature-index approach where:

$$T_{s2} = T_{s1} + \text{TSF}(T_{a2} - T_{s1}) \qquad (10.6)$$

where:

T_{s2} = snowpack surface temperature at t_2, °C
T_{s1} = snowpack surface temperature at t_1, °C
TSF = surface temperature factor, dimensionless
T_{a2} = mean daily air temperature at t_2, °C

The TSF employed by Marks *et al.* (1992) was 0.1, while Anderson (1973) initially used 0.5. The snow surface temperature is always limited to a maximum of 0 °C. In addition, surface temperature is set equal to the value of air temperature on days when snowfall exceeds about 0.5 cm water equivalent, since the new snowfall becomes the surface layer (Anderson, 1976).

An example of surface temperature computation using Equation (10.6) follows. Given a surface temperature of –4.0 °C on day 1 (T_{s1}), an air temperature of –2.0 °C on day 2 (T_{a2}), and TSF = 0.1, the surface temperature on day 2 would become:

$$T_{s2} = -4.0 + 0.1(-2.0 - (-4.0)) = -3.8\ °\text{C}$$

Larger values of TSF would allow snowpack surface temperature to track current air temperatures more closely; for example, if TSF = 0.5, then T_{s2} would increase to −3.0 °C. Following the example calculations using Equations (10.5) and (10.6), snowpack surface temperature and cold content can be approximated with only air temperature and precipitation data. Ideally, snowpack conditions such as temperatures and cold content can be periodically measured at representative locations and used to update models.

10.3.3 Point-snowmelt model algorithms

Understanding the logic behind computation of snowmelt or snowpack outflow at a point is important to snow hydrology modelling. The diagram in Figure 10.5 illustrates the components and inter-relationships for a given time interval. Four types of situations are shown depending upon air temperature (T_a) or energy-budget melt energy (Q_m), critical temperature (T_c), and precipitation amount (P):

- $P > 0$ and $T_a \geq T_{\text{CRIT}}$ (**Rainfall**) When precipitation occurs as rainfall (P_r), melting (M) during the rainfall is computed and rain plus melt become potential snowpack outflow. Melt can be computed using either the degree-day or energy-budget approach.
- $P > 0$ and $T_a < T_{\text{CRIT}}$ (**Snowfall**) If precipitation occurs as snowfall, the snowpack water equivalent (SWE) of the snowpack is incremented by the amount P, cold content is

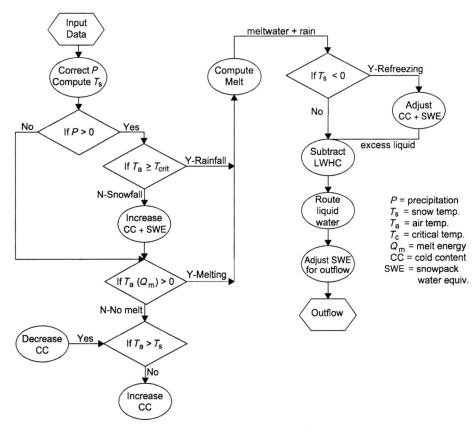

Figure 10.5 Flow diagram illustrating snowmelt-runoff computation during a time increment within a point-snowmelt model.

incremented depending upon P and temperature of snowfall relative to the snowpack temperature, and any melt during the snowfall event (M) is computed. Melt and cold-content changes are again computed using energy-budget or degree-day approaches.

• $P = 0, Q_m > 0$ or $T_a > 0\,°C$ (**Melt Event**) If no precipitation occurs and air temperatures or melt energy is greater than zero, then melting is occurring and computed melt is added to potential snowpack outflow.

• $P = 0, Q_m < 0$ or $T_a < 0\,°C$ (**Non-Event**) If no precipitation occurs and air temperatures or melt energy supply are negative, then the snowpack cold content must be either decreased or increased depending on whether the air temperature is greater or less than the snowpack temperature.

Melt and/or rain computed from any of the first three types of conditions must be compared with the snowpack cold content before it becomes outflow. If snow is at $0\,°C$ and cold content is zero, then water is directly available for routing in the pack. If snow temperature is subfreezing, then liquid water will refreeze. Any excess liquid water after the snowpack is warmed to $0\,^\circ$C then becomes available for routing. Any unfulfilled deficit in liquid-water holding capacity is subtracted from the excess liquid water. Liquid water in the ripe snowpack is routed to become outflow. Snowpack water equivalent is reduced by the amount of melt that leaves the pack in the time interval.

As implied above and in Figure 10.5, three types of storages are needed to complete the computations for each time step: snowpack-water-equivalent storage, cold-content storage, and liquid-water holding capacity storage. The snowpack water equivalent (SWE) is incremented with each new precipitation event (ΔP) and by increases in liquid water stored in the snowpack (ΔWHC, see below), and is reduced by the melt and rain water (P_r) that becomes outflow in the interval ($\Delta(M + P_r)$):

$$\text{SWE}_2 = \text{SWE}_1 + \Delta P + \Delta \text{WHC} - \Delta(M + P_r) \qquad (10.7)$$

where the minimum SWE is zero. The snowpack SWE is initially set to zero and the budget is used to determine the presence or absence of a snowpack at a point. Cold content (CC) changes with changes in snow temperature and mass changes over time (ΔCC) until the snowpack becomes isothermal as:

$$\text{CC}_2 = \text{CC}_1 + \Delta \text{CC} \qquad (10.8)$$

where the maximum CC is zero. Changes to cold content can be computed using an energy-budget or temperature-index approach. Cold-content change due to large snowfalls at temperatures different than snowpack temperature is generally negligible and is ignored here. Liquid-water holding capacity (WHC) is incremented with each new snowfall and reduced with gradual melt or rain liquid up to the maximum percentage (f) assumed as:

$$\text{WHC}_2 = \text{WHC}_1 + (f/100)\Delta P - \Delta(M + P_r) \qquad (10.9)$$

where the minimum WHC is zero. Although generally a small correction, any water stored is added to the SWE of the snowpack in the interval as indicated above.

Example calculations using the point-snowmelt model concepts just described are given in Table 10.2. The computations illustrate a brief four-day snow-accumulation period followed by a day with melt and a day with rain. No routing was employed since it was assumed that all liquid water drained from the snowpack each day. Decisions relating to rain vs. snow are straightforward with events at air temperatures greater than or equal to the $T_{\text{CRIT}} = 1\,^\circ$C counted as rain. Melt is easily computed with a DDF $= 0.4$ cm $^\circ$C^{-1} d^{-1} using Equation (10.2). However, on the last day only 0.776 cm melt occurred even though 0.8 cm was computed, because only 0.776 cm water equivalent of snow was left on that day.

Table 10.2 *Example computations illustrating use of a snowmelt budget procedure at a point.*[a]

Day	P cm	T_a °C	Snow cm	Rain cm	Melt cm	T_s °C	CC cm	WHC cm	SWE cm	Outflow cm
1	0	−2	0	0	0	0	0	0	0	0
2	0.3	−1	0.3	0	0	−0.5	−0.01	0.009	0.3	0
3	0.2	−2	0.2	0	0	−1.25	−0.02	0.015	0.5	0
4	1.2	−3	1.2	0	0	−3.0	−0.025	0.051	1.7	0
5	0	2.5	0	0	1.0	0	0	0	0.776	0.924
6	0.6	2	0	0.6	.776	0	0	0	0	1.376

[a]Equations (6.38), (6.39), (6.40), (9.6), (9.7), and (9.8) were used with DF $= 0.4$ cm °C^{-1} d^{-1}, TSF $= 0.5$, CCF $= 0.02$ cm °C^{-1} d^{-1}, $f = 3\%$, and $T_{CRIT} = 1$ °C. Assumed all liquid water drained from snowpack each day. See text for explanation of other terms.

Snow temperatures (T_s) computed with Equation (10.6) and TSF $= 0.5$ require more explanation. Snow temperature (T_s) was arbitrarily set to 0 °C on day one to allow computation between day one and two. Snow temperature is also set equal to air temperature on day four with a large snowfall, since large snowfalls dominate the snowpack temperature as previously discussed. Finally, snow temperature becomes 0 °C whenever cold content is reduced to zero, as it was on day five and six.

Cold content increased during the first three days. From day three to day four no change in CC occurred, since setting snowpack temperature equal to air temperature for big snow events reduces the temperature difference to zero. On day five all of the −0.025 cm of snowpack cold content was eliminated by melt on that day; e.g. 0.025 cm of predicted melt was refrozen in the snowpack and 0.975 cm was available for outflow. Cold content remained at zero on day six since the air temperature was above freezing.

Liquid-water holding capacity of the snow was computed using Equation (10.9) and an assumed $f = 3\%$. Each snowfall added to the water-holding capacity of the snowpack (WHC) up through day four. However, on day five melt exceeded WHC and of the 0.975 cm of melt water available after satisfying the cold content, another 0.051 cm was lost to water holding capacity, reducing liquid water available to become outflow to 0.924 cm. Water holding capacity remained at zero on day six since no new snow occurred.

The water equivalent of the snowpack (SWE) was computed using Equation (10.7). The water equivalent of the snowpack initially increased on three successive days with snowfall and no melt. On day five, 0.924 cm of meltwater was lost due to drainage from the pack and 0.071 cm was gained due to 0.025 cm of refrozen water from the cold content and 0.051 cm of liquid-water storage. Increased liquid-water

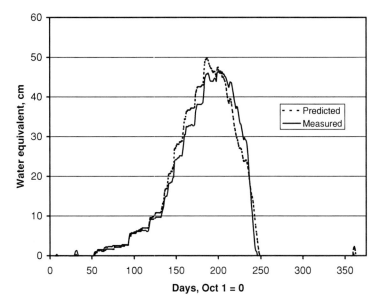

Figure 10.6 Example of prediction of daily snowpack water equivalent at a SNOTEL site using a simple degree-day approach requiring only air-temperature and precipitation data.

storage due to the refrozen water (3% of 0.025 cm) on day five was negligible and ignored. Finally, on day six when computed melt exceeded the WE, all of the remaining water stored in the snowpack was released, both stored liquid and ice. Predicted snowpack water equivalent reflects the outcome of budget computations as well as snowpack outflow.

Outflow simply reflects the sum of melt and rain available from routing in the snowpack. Clearly, from the foregoing discussion, not all melt becomes outflow because of refreezing and liquid-water storage. If the budget were being conducted on an hourly basis, rather than daily, then a routing procedure might have been needed to satisfactorily represent the outflow hydrograph.

An example of point-melt model predictions following the above discussion that has been adapted to available SNOTEL data is shown in Figure 10.6. SNOTEL data consist of daily snow pillow snowpack water equivalent, maximum and minimum daily air temperatures, and daily precipitation measured with a shielded precipitation gage at the site (see SNOTEL data at www.wcc.nrcs.usda.gov/snow/). Since outflow is not measured directly by the snow pillow, the simulation was performed to match the measured snowpack water equivalent. A degree-day approach to compute melt and cold content was employed as illustrated in Table 10.2 and given by Anderson (1973). Model parameters were adjusted by trial and error to achieve a

good fit. It was assumed that liquid water drained rapidly from the pack each day
and no routing was performed.

The overall fit for predicted versus measured snowpack water equivalent for
the year was good ($R^2 = 0.989$) and the average daily difference between predicted and measured water equivalent was 0.33 cm. The maximum and minimum
melting degree-day factors (DDF in Equation (10.2)) used were 0.4 and 0.15 cm
melt $°C^{-1}$ day^{-1}. A sine-wave variation in degree-day factors was assumed with
the minimum and maximum factors occurring on the winter and summer solstices,
respectively. The maximum and minimum cold-content degree-day factors were
0.09 and 0.034 cm $°C^{-1}$ day^{-1} (CCF in Equation (10.5)) with a similar sine-wave
variation assumed. The surface temperature factor used to predict snowpack surface temperature was 0.5 (TSF in Equation (10.6)). The critical temperature for
rain vs. snow was 1.67 $°C$ and the liquid-water holding capacity of the snowpack
was assumed to be 1%. Although the predicted melt amounts on individual days
are still subject to large errors using this approach, the example shows that with just
air-temperature and precipitation data a reasonable simulation of snowpack water
equivalent over the accumulation and melt season is possible.

10.4 Distributed-snowmelt models

Snowmelt runoff models have gradually evolved into distributed, deterministic
models. This evolution was driven in part by the need to shorten the time interval
for streamflow predictions, to accommodate the shift from simple degree-day to
energy-budget methods of melt prediction, and to enable the prediction of effects of
environmental change, such as that due to climate or land-use change, on large heterogeneous basins. The expanding availability of distributed watershed data, such
as snow-cover and digital elevation data, and increased capabilities of computers
and development of geographic information system (GIS) software and methods
have expedited this evolution.

Many approaches have been taken to model the distributed nature of snow processes across watersheds and within snowpacks. Watersheds have been represented
as a number of equally spaced grid points or nodes or the sum of processes within
discrete elevation zones, sub-watersheds, or other areas with similar distributed
properties such as vegetative cover, land use, hydrologic response, etc. Mitchell
and DeWalle (1998) used a combination of elevation and vegetation cover to define
zones within a low-relief, Pennsylvania, USA basin for melt simulation using the
SRM model (Martinec *et al.*, 1998). High and low elevation zones each with either
forest or open agricultural land-use gave better streamflow simulation results than
use of elevation zones alone. Harrington *et al.* (1995) used an iterative clustering
algorithm to cluster pixels according to elevation, snowpack water equivalent and

potential solar radiation. Snowmelt was computed separately for the resulting set
of clustered regions and summed to estimate discharge on the basin. These methods are used to account for horizontally distributed factors across watersheds, but
vertical snowpack variations due to depth-dependent snowpack properties can also
be considered.

The snowpack has generally been modelled as one homogeneous layer
(Anderson, 1973), but vertical distribution can also be modelled where the snowpack is divided into several to many layers to better estimate internal snowpack
processes. In particular, these models are quite helpful in accurately computing
snowpack surface temperature needed for surface energy-balance computations
and accounting for cold-content and liquid-water-storage changes within the snowpack. Anderson (1976) used a multiple-layer snowpack model with finite-difference
computations that considered impacts of new snowfall, compaction, metamorphic changes, and retention and transmission of liquid water. More recently, Jordan (1991) developed a numerical, one-dimensional, layered snowpack model
(SNTHERM.89) that employed improved methods for predicting liquid-water
movement in the snowpack. The snow module for the PRMS model (Leavesley
and Stannard, 1995) computes the energy balance with a two-layer snowpack.
Without such models, surface temperatures needed for energy-budget computations have to be estimated from air-temperature data, as previously described by
Equation (10.6).

Distributed-snowmelt models vary greatly in complexity. Fully distributed models could include mass and energy-budget computations for a number of relatively
homogeneous watershed subunits or regularly spaced grid points with consideration
of effects of factors such as:

- snow-covered area
- elevation
- slope and aspect
- forest cover (interception and melt effects)
- frozen ground
- wind redistribution of snow
- glaciers
- topographic shading
- avalanches

Snow-covered area needs to be considered in some manner on nearly all but essentially flat basins, while steep mountain catchments often require consideration of
the effects of elevation, slope and aspect, and topographic shading. Issues related
to frozen-ground and blowing-snow effects are more important in open alpine
and trans-arctic settings. Forest-cover effects are widespread in mid-elevation

glaciated and mountainous terrain. Distributed modelling to consider each of these factors is discussed briefly below.

10.4.1 Snow-covered area

Snow-cover depletion curves

The simplest semi-distributed models include provisions for computing or including measurements of snow-covered area. In simplest form, an equation is used to represent the changes in snow-covered area over time and melting is assigned only to the snow-covered portion. Rain inputs occur over the entire watershed, bare or snow-covered. Because snow accumulation and melt patterns are consistent from year to year across watersheds, it is possible to model changes in snow-covered area on a watershed based upon past data. The percentage of watershed area covered by snow can be related to the fraction of maximum snowpack water equivalent remaining on the basin (US Army Corp of Engineers, 1956; Anderson, 1973). A generalized curve used by Anderson (1973) with success on several basins in the National Weather Service River Forecast System model for the United States is shown in Figure 10.7 and can approximated as:

$$S = 5.9 + 194.5(SWE/SWE_i) - 139.4(SWE/SWE_i)^2 + 39.1(SWE/SWE_i)^3$$
$$(10.10)$$

where:

S = snow-covered area, %
SWE = snowpack water equivalent, cm
SWE_i = minimum snowpack water equivalent at which snow cover always exceeds 100%, cm

This equation, or one derived for local conditions, can be used to increase snow-covered area in the accumulation season and decrease snow-covered area during the melt season. Based upon experience of the US Army Corps of Engineers (1956), the degree of curvature of such a relationship should increase as a basin becomes more homogeneous. The SWE_i must be set based upon experience or maximum snowpack accumulation during the current season. Once the proper equation has been developed, use of the expression is straight forward, except for accounting for the dynamics of snow-covered-area changes at times of new snowfall during the melt season.

The effect of new snowfall (S) during the melt season was assumed by Anderson (1976) to temporarily increase snow-covered area to 100% until 25% of the new

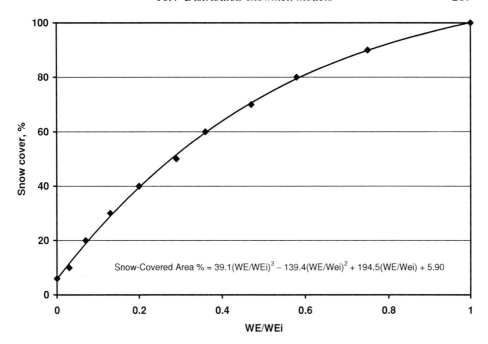

Figure 10.7 Generic snow-cover depletion curve used to relate snow-covered area on a watershed (%) to a snowpack-water-equivalent index (WE/WEi) (based upon Anderson, 1976); WE = snowpack water equivalent and WEi = minimum snowpack water equivalent where snow-cover area always equals or exceeds 100% on a basin.

snowfall water equivalent had been melted. Snow-covered area then was linearly interpolated back to the original SCA curve given by Equation (10.10) as:

$$S_2 = S_1 + [(0.75\ P - M_S)/0.75\ P](100 - S_1) \qquad (10.11)$$

where:

S_2 = interpolated new snow-covered area, $S_1 \leq S_2 \leq 100\%$
S_1 = original snow-covered area, %
P = new snowfall water equivalent, cm
M_S = subsequent melt, $\leq 0.75\ S$, cm

Once the original S_1 is reached the original non-linear function such as Equation (10.10) would be used to compute further declines in S. The decision to use 25% of S as the beginning point for linear interpolation was arbitrary and could vary among watersheds. This procedure is detailed in the following example.

If a watershed has complete snow cover whenever snowpack water equivalent exceeds 25 cm (SWE) and the current snowpack water equivalent (SWE) is 15.5 cm, then the snow-covered area by Equation (10.10) would be:

$$S = 5.90 + 194.5(15.5/25) - 53.59(15.5/25)^2 + 39.1(15.5/25)^3 = 82.2\%$$

If a major new snowfall with water equivalent of 2.2 cm occurred on the watershed, then S would increase to 100% and remain at 100% until 25% of 2.2 cm or 0.55 cm had melted. After 0.55 cm of melt occurred, the S would linearly decline as the remaining 1.65 cm of new snowfall (75% of 2.2 cm) was melted to the original water equivalent of 15.5 cm. For example, if melt of 1.80 cm were predicted, S would remain at 100% during 0.55 cm of melt. After the remaining $1.80 - 0.55 = 1.25$ cm of melt, S would decline from 100% to:

$$S = 82.2 + [(1.65 - 1.25)/1.65](100 - 82.2) = 86.5\%$$

Once S declined to 82.2% following the new snowfall, the snow-cover depletion curve (Equation (10.10)) would then be used to account for further declines.

Anderson (1973) applied the snow-cover depletion curves to an entire watershed, but it is possible to use snow-cover depletion curves to parts of a basin. Martinec and Rango (1991) used a similar procedure to estimate snow-covered area for the Upper Rio Grande Basin in Colorado, USA, except they used a family of curves, each for a different snow depth corrected for new snow rather than SWE/SWE_i. The Precipitation-Runoff Modelling System (PRMS) developed by the US Geological Survey also employs user-specified snow-cover depletion curves following Anderson's (1973) procedure within hydrologic response units on a watershed, www.brr.cr.usgs.gov/projects/SW_precip_runoff/mms/html/ snow-comp_prms.html. In this manner, distinctly different snow-cover depletion patterns can be simulated in separate regions of the same basin. Snow-covered area can also be indexed to the observed elevation of the snowline in high mountain ranges like the Himalayas (Rao *et al.*, 1996).

Snow-cover data Another approach to accounting for snow-cover depletion is to employ measured snow-covered area from satellite remote sensing, which has been done in both hydrologic and land-surface models (e.g. Rango, 1996; Turpin *et al.*, 1999; Rodell and Houser, 2004; Dressler *et al.*, 2006; Clark *et al.*, 2006). This approach obviates use of snow-cover depletion curves and substitutes measured snow-cover data for more detailed distributed computations of melt across a watershed. Further, S has been used to update the distribution of SWE by using direct-insertion approaches in which model estimates of SWE are adjusted by the satellite S (Mcguire *et al.*, 2006; Dressler *et al.*, 2006) and by using more robust approaches such as the Kalman-type assimilation strategy (Clark *et al.*, 2006) that blends satellite information and that from model simulations.

Measured snow-cover data integrates the distributed effects of varying snowpack accumulation and melt rates due to topography and can index effects of wind redistribution of snow and avalanching later in the melt season. Snow-cover data can be obtained for watershed subunits frequently enough to permit use in a forecasting mode, but some method is still needed for extending snow-covered area between data acquisitions. Although snow-cover data are not commonly used in models to date, measured snow-cover data could be used as input to verify melt computations in complex distributed models. A semi-distributed model, SRM (Snowmelt-Runoff Model) (Martinec *et al.*, 1998), uses satellite snow-cover data to directly determine snow-cover depletion for up to nine elevation zones on a watershed. In the forecasting mode, snow-cover data are updated as frequently as possible and characteristic curves based upon past remote-sensing data for each zone are used to extend the snow-cover curve until the next satellite coverage is available.

Measured snow-covered area can be compared to the computed snowpack water equivalent in that zone to adjust model parameters. Kirnbauer *et al.* (1994) found that comparisons of measured snow cover with model predictions of the presence or absence of snow cover on various slopes and aspects was much more effective in selecting model alternatives than comparisons of predicted and measured runoff. Dressler *et al.* (2006) reported that greater improvement in melt-runoff prediction from snow-cover data occurs in topographically homogeneous basins relative to more topographically complex basins in the southwestern United States while Brandt and Bergström (1994) found that greatest improvement occurred in high-elevation regions of Scandinavia with sparse climatic data resources (the cost of obtaining snow-cover data was not justified in forested lowland where climate was more homogeneous and well represented by available data).

10.4.2 Elevation

Elevation differences across watersheds affect air temperatures and other meteorological data. Many snow hydrology applications have accounted for effects of elevation by dividing watersheds into elevation zones or categorizing grid points established as equal intervals across basins by elevation (for example, see Martinec *et al.*, 1998; Bell and Moore, 1999; Williams and Tarboton, 1999). Bell and Moore (1999) found that performance in runoff prediction greatly improved when 5–25 elevation zones were used, depending upon year, for two upland British catchments. Relatively simple input such as topographic maps are required to characterize watersheds by elevation zones and digital elevation models can be used to model the watershed as a set of grid points. Ideally, climatic data are available at sites representing each elevation zone or a function can be derived to predict variables according to elevation. When representative sites are not available, then adjustments are made to available data for the effects of elevation. Typically, the elevation used to represent zones is the hypsometric mean elevation for that zone that represents the elevation where 50% of zonal area lies above and 50% lies below that elevation.

When the degree-day approach to melt computation is used, air-temperature data are generally adjusted to the hypsometric mean elevation of each zone using average lapse rates discussed in Chapter 7. This extrapolation is critical not only to computing the melt rates in each elevation zone, but in also deciding on the form of precipitation, rain or snow, in each zone. When using the energy-budget approach for melt computation, adjustment of other meteorological data for effects of elevation, such as humidity or wind speed, is also required. In all models a scheme is also needed to adjust precipitation data for elevation effects (see Chapter 2). Correcting meteorological data for elevation is complicated and local statistical methods commonly have to be relied on to relate elevation to climatic data available for a region.

Elevation adjustments can vary seasonally as well as daily and minimum temperatures can be lower at low elevation sites due to temperature inversions and cold-air drainage (Ishikawa *et al.*, 1998). Although elevation adjustment parameters for meteorological data could be fitted in the streamflow modelling process, it is suggested that adjustments be independent of overall streamflow model parameter fitting.

10.4.3 *Slope and aspect*

Slope and aspect exert a major control on the receipt of shortwave and longwave radiation as described in Chapter 7 and Appendix B. Consideration of the effects of slope and aspect is generally only needed when radiation exchange is explicitly considered, such as when the energy budget or a combined radiation–temperature index method is used to predict melt. Digital elevation models can be used to categorize grid points within a watershed by slope and aspect. Many models have now been developed that consider effects of slope and aspect on melt rates (Blöschl *et al.*, 1991a, b; Braun *et al.*, 1994, Gellens *et al.*, 2000; Harrington *et al.*, 1995; Hartman *et al.*, 1999; Hock, 1999; Link and Marks, 1999; Marks *et al.*, 1999; Ujihashi *et al.*, 1994).

Interactions between slope and aspect and other non-radiative factors could also be considered. Wind speeds that influence the rates of latent and sensible heat convection and sublimation from snowpacks vary with topography. Terrain can funnel air flow through valleys and accelerate wind over ridges. Slope and aspect also interact with the presence and depth of seasonal soil frost due to differences in depth of snow and the energy budget at the soil surface. Slope/aspect combinations facing to the north or those with shallow or no snow will have deeper soil frost. Occurrence of permafrost also varies with aspect. Many of these secondary effects of slope and aspect have yet to be considered in distributed-snowmelt modelling.

Molotch *et al.* (2005) successfully used a variable they termed "northness" to index the complex influence of slope and aspect on snow cover. Northness was derived from the product of the aspect and the sine of the slope. This variable in combination with elevation, potential solar radiation, slope, aspect and a variable describing wind redistribution was used in a binary regression tree model to predict snow depth in a Californian alpine basin.

In addition to energy-budget models that consider slope and aspect, Hottelet *et al.* (1994) developed a degree-day model that considered effects of aspect on melt rates. Degree-day melt factors within the HBV-ETH model (Bergström 1976, 1995) were adjusted for effects of aspect for several watersheds in Switzerland and Bohemia. Degree-day melt factors for south-facing slopes were multiplied by 1.5 (+50% adjustment) and those for north-facing slopes were divided by 1.5 (−33% adjustment). No corrections were used for east-facing or west-facing watershed units.

10.4.4 Forest cover

Forest cover affects both the accumulation of snow due to canopy interception losses and the melt rates due to the modification of the heat balance at the snow-pack surface. Canopy interception losses of snow were covered in Chapter 2 and forest effects on radiation exchange and melt rates were considered in Chapter 7. Simple characterizations of land-use and land-cover data, such as LULC data at http://eros.usgs.gov/products/landcover/lulc.html, or other sources, can provide information on the spatial variations of forest-cover types on large watersheds. Major differences in the distribution and density of leafless-deciduous forest, coniferous forest, and open terrain can be considered easily in distributed models. As described in Chapter 7, forest density has been considered from two points of view: boreal forests with scattered, small trees where the effects of individual trees on energy exchange are considered and more or less closed forests where the forest is treated as a relatively homogeneous medium with prescribed radiation spectral and other properties. Models for the distributed effects of forest cover have been developed by Metcalfe and Buttle (1998), Giesbrecht and Woo (2000), Link and Marks (1999). Metcalfe and Buttle (1998) were able to link melt rates to a remotely-sensed index to canopy closure obtained from Landsat TM data and ground-based measurements of canopy gap fractions in their distributed modelling of forest effects in Manitoba, Canada.

10.4.5 Frozen ground

Frozen, impermeable ground can lead to overland flow and rapid translation of melt into streamflow. Approaches to prediction of the depth of frost penetration and the infiltration capacity of frozen ground were reviewed in Chapter 9 and could be incorporated into streamflow prediction schemes. Most hydrologic models have a soil infiltration component that can also be adjusted for the effects of soil frost, but a realistic module including effects of varying air temperature, wind, snow-cover depth, and soil thermal and moisture conditions will probably be needed. Process studies of soil frost effects have been completed and distributed models of effects of soil frost on streamflow including snow-depth effects are obviously the next step in distributed modelling.

10.4.6 Wind redistribution of snow

The influence of wind on redistribution of snow across the landscape can be one of the most important distributed modelling parameters in flat to gently rolling, windy open terrain. Prevailing wind directions will often cause scour of all snow from some terrain elements facing into the wind and drift of snow into terrain elements on the lee or downwind side (Kane *et al.*, 1991). If not accounted for, such redistribution of snow often results in modelling difficulties due to occurrence of melt from deep residual drifts late in the melt season. Wind effects on accumulation of snow were discussed in Chapter 2 and several blowing-snow models (e.g. Liston and Sturm 1998 and Pomeroy *et al.* 1993) have been adapted for use in hydrologic models. Wind-redistribution effects considered here are separate from adjustments of wind speed for topographic effects that can affect convective mass and heat exchange between the snowpack and environment. Blöschl and Kirnbauer (1992) found that effects of topography like elevation and slope were more important determinants of snow cover early in the melt season and effects of wind redistribution and avalanching were more influential later in the melt season.

Several distributed-snowmelt modelling attempts have been made to include wind-redistribution effects. Blöschl *et al.* (1991a) adjusted snowpack water equivalent based upon slope steepness and slope curvature. All slopes above 60° were assumed to be snow free. All mountain tops with a radius of curvature of 50 m or less were also assumed to be snow free, while depressions with the same curvature were assigned 200% of snowpack values for open flat terrain. Snow on all other slopes was linearly interpolated between level and 60° conditions. These adjustments improved prediction of the snow-cover distribution on a steep basin in the Austrian Alps. Hartman *et al.* (1999) similarly based their correction on topographic characteristics by fitting a snow-redistribution parameter that increased snow on slopes that had a topographic similarity index (TSI) greater than the average TSI and reduced snow on slopes with a TSI less than the average. This procedure predicted greater snowpack accumulation on low slopes, valleys, and hollows. Simulations of streamflow using this approach were sensitive to the wind-redistribution parameter. Larger values caused greater redistribution and tended to predict lower early-season flows and higher late-season streamflows. Neither of these two applications considered effects of aspect explicitly. Luce *et al.* (1999) used a different approach by assigning a drift multiplier based upon field measurements to each grid cell on the watershed to adjust precipitation amounts. The drift multipliers ranged from 0.16 to 5.36, indicating quite a range in scour and drifting conditions occurred on this basin. This approach required field data not commonly available, but implicitly indexed all factors at work to influence snow-cover distribution.

Molotch *et al.* (2005) and Winstral *et al.* (2002) had success in indexing wind redistribution of snow by using the variable "maximum upwind slope" in a binary regression tree model. Maximum upwind slope was defined as the average maximum slope found using a search algorithm along 100 m long vectors, spaced at five-degree intervals, within a 90° upwind sector from each point. Using this variable, greater upwind slopes were associated with lower snow accumulations.

In the future no doubt, other creative ways of accounting for wind-redistribution effects will be employed to improve predictions of snowmelt runoff. Wind redistribution implies that some areas lose and others gain snow, not to mention sublimation losses occurring during wind redistribution. Thus the scheme used should be tied to a model that ultimately considers both redistribution and sublimation losses.

10.4.7 Glaciers

The presence of glaciers within basins adds to the complexity of streamflow modelling. Glaciers provide an additional source of meltwater for streamflow, melt at rates typically greater than snowpacks, and can route liquid water rapidly to stream channels. Moore (1993) describes a model to separately predict melt of snow on non-glaciated and glaciated portions of multiple elevation zones delineated on a British Columbia, Canada basin. Once the melting of the snowpack within the glaciated region had been completed, melt of glacier ice was modelled to occur at a rate 50% greater than for snow due to the lower albedo of glacier ice. Classification of events as rain or snow in each elevation zone was also especially critical to accurate streamflow prediction in autumn. Autumn rainfall on the glacierized portion of the basin quickly became runoff, while autumn snowfall was stored in the snowpack on the glacier until melt the following spring. Areal extent of the glacier was based upon autumn surveys of the equilibrium line altitude on nearby glaciers. Schaper *et al.* (1999) were able to distinguish snow from glacier ice using remote-sensing data to aid in the modelling of meltwater runoff by elevation zones on a basin with 67% of the area covered by glaciers.

10.4.8 Topographic shading

Topography surrounding a particular watershed grid point can significantly reduce the direct-beam solar radiation received by that point, thereby reducing the melt rate, as discussed in Chapter 7. This impact of topography is separate from that of slope and aspect considered above, that accounts for the varying angle of the direct solar beam. Basic principles involved in the corrections for slope and aspect are described in Appendix B. In steep, mountainous terrain some landscape facets can be shaded continuously by surrounding topography. In addition, at higher latitudes where solar altitude may be quite low during an entire solar day, even minor topographic relief could cause significant shading. Since these shaded-landscape components often receive snow redistributed by the wind or avalanches, the impacts of shade on maintaining snow until late in the melt season is probably significant. Several distributed-snowmelt models have included the effects of topographic shading as part of the overall radiation budget routine (Blöschl *et al.* 1991c; Hartman *et al.* 1999; Hock, 1999; Ranzi and Rosso, 1991) and other studies have used the same physiographic variables in distributing point measurements of ground-based SWE measurements (Fassnacht *et al.*, 2003), but the importance of topographic shading by itself to prediction of snow-cover distribution and snowmelt runoff has not been separately assessed.

10.4.9 Avalanches

Redistribution of snow by gravity to lower elevations, like redistribution by the wind, can significantly modify streamflow response. Avalanche snow resides in deeper packs at lower elevations to provide a source of water for streamflow later into the summer. Adjustments of snow cover for redistribution by the wind tend to mimic or include the effects of avalanches. De Scally (1992) estimated that as much as 8% of the annual snowmelt runoff in a large Himalayan basin in Pakistan came from avalanche snow. Since snow avalanches occur along well-defined pathways, it is possible that parts of the landscape of large watersheds could be routinely coded for avalanche susceptibility depending upon snowpack water equivalent.

10.5 Snowmelt-runoff models

Many models are available to continuously predict streamflow and most of these have provisions for including snowmelt-runoff. Six models in widespread use for

Table 10.3 *Comparison of snowmelt-runoff models.*

Model	Ref.[a]	Data inputs[b]	Degree-Day or Energy Budget	Lumped or Distributed	Processes Considered	Model Parameters	Min. Time Step
NWSRFS	1, 2	T_a, P, u	DD, except during rain	Lumped, but with a snow-cover depletion curve	Cold content, Snow temp., Liquid routing, Rain-on-snow	13	6 hr
SRM	3, 4	T_a, P and cloud cover for modified version	DD	Semi-distributed, snow-cover data by elevation zones	Ripening date specified, seasonal adjustments	7	daily
PRMS	5	T_a, P, incoming solar or cloud cover; or complete meteo. data	DD or EB	Distributed by hydrologic response units; two snowpack layers; elev., slope and aspect and forest effects	Snow temp., Cold content, Rain-on-snow	10	1 min in storm mode, daily otherwise
HBV-ETH	6, 7	T_a and P, monthly potential evapotransp.	DD	Semi-distributed	Parameters adjusted for slope/aspect, forest effects	11–20, varies with version	daily
SSARR	8	T_a, P, evapotransp.	DD	Lumped or semi-distributed with elevation zones	Same as NSRFS plus interception losses	15	0.1 hr
SHE	9, 10	T_a and P or complete meteo. data	DD or EB	Distributed grid network	Interception losses, liquid routing, cold content, full forest effects	>50	0.1 hr

[a] References: 1= Anderson, 1973, 2 = Burnash 1995, 3 = Martinec *et al.*, 1998, 4 = Brubaker *et al.*, 1996, 5 = Leavesley and Stannard, 1995, 6 = Bergström, 1995, 7 = Hottelet *et al.*, 1994, 8 = Speers, 1995, 9 = Bathurst *et al.*, 1995, 10 = Bathurst and Cooley, 1996.

[b] T_a = air temperature, P = precipitation, u = wind speed.

prediction of streamflow that include snowmelt routines are compared in Table 10.3. A variety of model features can be found from the simple snowmelt-runoff model (SRM) that only requires seven parameters and simple meteorological data to the complex, fully distributed SHE model that considers all physical processes, but requires many input parameters. Runoff modelling approaches used in these models in conjunction with the snow routines generally include explicit or conceptual accounting for surface, soil, and ground water storages and flow; however, SRM estimates are based upon the sum of a current day's melt or rain input plus a recession flow from the prior day. Minimum time steps also vary from 1 minute up to daily flow estimates. Many other models have been developed and tested (Singh, 1995; Singh and Singh, 2001; ASCE, 1996; Beven, 2000). Two promising point-snowmelt modules for predicting outflow from layered snowpacks that can be incorporated into distributed models for runoff prediction are SYNTHERM (Jordan 1991) and SNAP (Albert and Krajeski, 1998). Comments about each of the models compared in Table 10.3 follow.

The **NWSRFS** (National Weather Service River Forecast System) model described by Anderson (1973) is essentially a degree-day model that balances handling the complexities of snow accumulation and melt processes against the availability of data. Simple input data are required, but during rain on snow events the model shifts to an approximate energy-budget analysis that requires an additional wind-speed input. A snow-cover depletion curve is used within the model to account for varying snow-cover area. Explicit accounting for snowpack surface temperature, cold content, and liquid-water routing are also included, despite the limited data requirements. Anderson (1976) also developed a detailed, finite-difference, layered, snowpack model with a complete energy budget and compared the results to the simpler degree-day approach. Neither of these models was distributed when originally developed. The degree-day routine has been extensively validated and tested.

One of the simplest models to use in terms of parameters required is the **SRM** (Snowmelt-Runoff Model; Martinec *et al.*, 1998). This model is also a degree-day model that requires only simple temperature and precipitation data, but additionally utilizes satellite snow-covered-area data that can be input for up to eight elevation zones. Inclusion of available snow-covered-area data is an advantage because of the high correlation between snow-covered area and snowmelt runoff. Parameters can be estimated based upon experience for ungaged basins, but having some prior flow data is helpful. SRM also can be used in a forecasting mode and has a routine to allow estimates of the effects of climate change. Explicit accounting of snowpack temperature, cold content, and liquid-water routing is not included, although the date when all rain on snow is expected to percolate freely through the snowpack (ripe snow) can be specified. A modified radiation–temperature index melt routing for SRM has been developed to allow distributed modelling, but cloud-cover data are needed, at a minimum, if this option is used (Brubaker *et al.*, 1996). The radiation option has not been fully incorporated into SRM at this time. SRM has been used extensively, especially on steep mountain catchments where satellite snow-cover data can be acquired. This model will be described in detail in the next chapter.

The **PRMS** (Precipitation Runoff Modelling System) described by Leavesley and Stannard (1995) and Leavesley *et al.* (1983) includes a snow module that uses distributed energy-balance modelling for effects of slope and aspect, elevation and forest cover within hydrologic response units (HRU) on a basin. Data requirements range from minimal temperature and precipitation, to a complete set of input meteorological data. Snow-cover depletion curves are used within each HRU and more recent updates include provisions for incorporation of snow-covered-area data. PRMS is part of the US Geological Survey's Modular Monitoring System (Leavesley *et al.*, 2001) that fosters the open sharing and linking of hydrologic modules and supporting GIS, optimization, sensitivity, forecasting, and statistical analysis tools to promote distributed hydrologic modelling. PRMS has been widely used in the United States and includes provisions for modelling the effects of channel reservoirs.

The **HBV** or Swedish Meteorological and Hydrological Institute model was originally developed by Bergström (1995) to be as simple and as robust as possible. HBV has seen widespread use in Scandinavia and throughout the world over the past 20 years and many variations have been introduced to the model. The original HBV is a simple degree-day model with few parameters. It can be classified as a semi-distributed model with sub-basins included that each have a snow-cover depletion curve and a simple land-use classification (forest, open, lakes). Like SRM, snowpack temperature, cold content, and liquid-water routing are not explicitly considered in this model. Hottelet *et al.* (1994) used a HBV-ETH Version 4 model that allowed adjustments of melt parameters for aspect classes by elevation zones.

The **SSARR** (Streamflow Synthesis and Reservoir Regulation) model (Speers, 1995) developed by the US Army Corps of Engineers has essentially adapted Anderson's (1973) degree-day snow routines to a distributed format which allows up to 20 elevation zones. Alternatively, a simple lumped-parameter approach with a snow-cover depletion curve can also be used. The combination of Anderson's (1973) snow routines and SSARR's extensive river/reservoir interaction, flow routing, forecasting, and calibration capabilities make this a proven choice for large basin applications involving river and reservoir simulations.

The **SHE** (Système Hydrologique Européen) (Bathurst *et al.*, 1995) model is a grid-based, fully distributed, process-based flow model that physically models all hydrologic processes, including those pertaining to the snowpack. The snow routines (Bathurst and Cooley, 1996) can be simple degree-day or complete energy-budget routines with a layered snowpack and full consideration of snow-pack temperature, cold content and liquid-water movement using approaches given by Morris (1982). The grid-based format allows adjustment for elevation, slope and aspect, forest cover, etc. Several versions of the model exist that expand its capabilities to include erosion/sedimentation and contaminant transport (SHESED-Bathurst *et al.*, 1995, MIKE SHE-Refsgaard and Storm, 1995). However, the model requires many parameters be specified for all grid points and computer run times may be problematical (Beven, 2000). Model versions have been tested in a variety of settings throughout the world.

10.6 Snowmelt-runoff model performance and selection

Selection of a model to predict snowmelt runoff depends upon many factors includ-ing objectives of the modelling, data availability, characteristics of the watershed, and space and time scales of application, so generalizations are problematical. When project objectives are mainly focused on seasonal streamflow patterns and total volumes of water available for water supplies, then statistical models or sev-eral physical models and daily time steps could be used. However, when daily patterns of flow rates and absolute daily or seasonal peak flows are needed, say for flood forecasting or hydropower generation, then models with smaller min-imum time steps and distributed model capabilities are probably needed. Time steps of less than one day for snowmelt simulation generally mean a shift from degree-day to energy-budget methods to predict melt. A notable exception is the hybrid temperature–radiation index model developed by Hock (1999) that makes use of daily variations in potential solar irradiation for shorter-term melt estimates. Basin characteristics also influence the potential for successful snowmelt-runoff modelling. Large basins and basins with forest cover have been simulated with degree-day, lumped models with daily time steps (Bengtsson and Singh, 2000). However, choice of models to handle complexities of basin topography, land use and land cover will largely be determined by program objectives, data constraints, and space and time scales of the application.

Performance of available models in simulating seasonal or annual snowmelt runoff has been evaluated in tests conducted by the World Meterological Organiza-tion (WMO, 1986). Flow simulations with 10 models on multiple watersheds were compared to measured flows using Nash–Sutcliffe coefficients (R^2), which are sim-ilar to a statistical coefficient of determination where 1 equals perfect agreement, and percentage deviations in total runoff volumes (Dv,%). Annual prediction data abstracted from WMO (1986) Operational Hydrology Report No. 23 for all avail-able years averaged for three of the basins (Durance, W3, and Dischma) for five models that appear in Table 10.3, are given in Table 10.4. Performance of the models

Table 10.4 *Comparison of annual average prediction statistics for*
snowmelt-runoff models for the Durance, W3, and Dischma basins (WMO, 1986).

Model Criteria[a]	HBV	SRM	SSARR	PRMS	NWSRFS
R^2	0.806	0.846	0.830	0.784	0.880
Dv,%	5.6	5.0	6.3	7.6	5.8

[a] R^2 = Nash–Sutcliffe coefficient, Dv = annual flow volume error

in Table 10.4, and all 10 models tested in general, was quite comparable. The values of R^2 were based upon the mean flow rates for the calibration and verification years that tends to inflate performance in the extreme years (see Martinec and Rango, 1989), but overall values of R^2 of 0.78 to 0.88 are quite good. Volume prediction errors were a bit more variable, but the range of 5–8% again is quite acceptable for many applications. The consistency of prediction results among models, leads to the conclusion that available input data, which were common to all models, play a major role in determining errors.

Model performance in a forecasting mode is also of interest. An inter comparison of streamflow forecasting ability for daily flows up to 20 days in advance for seven snowmelt-runoff models was conducted by the World Meteorological Organization (WMO, 1992). Three of the models compared are listed in Table 10.3 (SSARR, SRM, HBV) and all were lumped-parameter or semi-distributed, degree-day models. Forecasts were based upon measured air-temperature and precipitation data for the forecast days, thus errors due to inaccurate meteorological forecasts were not included, and model updating was permitted for each new forecast over time. Comparisons were made for the Illecillewaet Basin in Canada for up to three years of data depending upon the model. Results were summarized in terms of root mean square error (RMSE) of predicted versus measured flows (Rango and Martinec, 1994). RMSE generally increased as the length of the forecast period increased for most models for the first two weeks. Average RMS errors for the models were about 30 m^3 s^{-1} which translated into errors of about 20% of average daily flow. All models outperformed simple use of the current day's flow as the forecasted flow, which would have given a RMS error of about 100 m^3 s^{-1}. Greater errors would be expected if actual forecasted air-temperature and precipitation data were used for streamflow forecasting. Updating of the models was generally done manually, but some models now incorporate automatic updating of input data, output data, model parameters and state variables (i.e. snowpack water equivalent or snow cover), during the prediction period. Overall, the WMO model intercomparisons were quite useful to show the general level of prediction possible for snowmelt runoff when operating in a forecasting or seasonal prediction mode.

Ensemble forecasting of streamflow from snowmelt has seen increasing attention in order to provide daily forecasts for the entire snowmelt season (Day, 1985). Ensemble forecasting makes use of models calibrated on specific watersheds to predict future flows based upon a set or ensemble of historical meteorological conditions measured in past years. For example, a model could be used to predict daily flow rates beginning with the flow rate on April 1 in the current year up to August 31 of the current year, using historical climatic data, such as air temperature and precipitation, measured in each of the past years. Each set of climatic data for a past year will produce a single forecast hydrograph of daily flows for April 1–August 31 and the ensemble of forecast hydrographs for all years on record can be used to estimate the probability of exceedance of any individual hydrograph in the ensemble.

10.7 Future work

Rapid advances in modelling snowmelt runoff are currently being made due to interest in simulating global climate and subsequent changes in snowmelt hydrology. A Snowmelt Modelers Internet Platform (www.geo.utexas.edu/climate/ Research/SNOWMIP/snowmip.htm) has been established to aid communications among those developing, using, and testing models of snowpack processes throughout the world. Bales *et al.* (2006) presented a detailed view of current research questions facing modellers of mountain hydrology with specific applications to snow. These questions generally fall into three topic categories: importance of and processes controlling energy and water fluxes, interactions between hydrologic fluxes and biogeochemical and ecological processes, and integrated measurement and data/information systems. Specific research questions are given in each topic category that can help guide future work.

10.8 References

Albert, M. and Krajeski, G. (1998). A fast, physically based point snowmelt model for use in distributed applications. *Hydrol. Processes*, **12**, 1809–24.

American Society of Civil Engineers (1996). *Hydrology Handbook*, 2nd edn. Manuals and reports on engineering practice 28. New York: ASCE.

Anderson, E. A. (1973). *National Weather Service River Forecast System-Snow Accumulation and Ablation Model*, NOAA Tech. Memo, NWS-HYDRO-17. US Dept. Commerce, NOAA, NWS.

Anderson, E. A. (1976). *A Point Energy and Mass Balance Model of a Snow Cover*, NOAA Tech. Rpt. NWS 19. US Dept. Commerce, NOAA, NWS.

Bales, R. C., Molotch, N. P., Painter, T. H., Dettinger, M. D., Rice, R., and Dozier, J. (2006). Mountain hydrology of the western United States. *Water Resour. Res.*, **42**, W08432.

Bathurst, J. C. and Cooley, K. R. (1996). Use of the SHE hydrological modelling system to investigate basin response to snowmelt at Reynolds Creek, Idaho. *J. Hydrol.*, **175**, 181–211.

Bathurst, J. C., Wicks, J. M., and O'Connell, P. E. (1995). Chapter 16: the SHE/SHESED basin scale water flow and sediment transport modelling system. In *Computer Models of Watershed Hydrology*, ed V. P. Singh. Highlands Ranch, CO: Water Resources Publications, pp. 563–94.

Bell, V. A. and Moore, R. J. (1999). An elevation-dependent snowmelt model for upland Britain. *Hydrol. Processes*, **13**, 1887–903.

Bengtsson, L. (1976). Snowmelt estimated from energy budget studies. *Nordic Hydrol.*, **7**, 3–18.

Bengtsson, L. and Singh, V. P. (2000). Model sophistication in relation to scales in snowmelt runoff modeling. *Nordic Hydrol.*, **31**(3/4), 267–86.

Bergström, S. (1976). Development and application of a conceptual runoff model for Scandinavian catchments. In *Swedish Meteorological and Hydrological Institute Report RHO 7*, Norrköping, Sweden.

Bergström, S. (1995). Chapter 13: the HBV model. In *Computer Models of Watershed Hydrology*, ed. V. P. Singh. Highlands Ranch, CO: Water Resources Publications, pp. 443–76.

Beven, K. J. (2000). *Rainfall-Runoff Modelling*. Chichester: John Wiley and Sons, Ltd.

Blöschl, G. and Kirnbauer, R. (1992). An analysis of snow cover patterns in a small alpine catchment. *Hydrol. Processes*, **6**, 99–109.

Blöschl, G., Gutknecht, D., and Kirnbauer, R. (1991a). Distributed snowmelt simulations in an alpine catchment: 2. parameter study and model predictions. *Water Resour. Res.*, **27**(12), 3181–8.

Blöschl, G., Kirnbauer, R., and Gutnecht, D. (1991b). Distributed snowmelt simulations in an alpine catchment: 1. Model evaluation on the basis of snow cover patterns. *Water Resour. Res.*, **27**(12), 3171–79.

Blöschl, G., Kirnbauer, R. and Gutknecht, D. (1991c). A spatially distributed snowmelt model for application in alpine terrain. In *Snow, Hydrology and Forests in High Alpine Areas*, IAHS Publ. 205, ed. H. Bergman, Lang, H., Frey, W., Issler, D., and Salm, B., pp. 51–60.

Brandt, M. and Bergström, S. (1994). Integration of field data into operational snowmelt-runoff models. *Nordic Hydrol.*, **25**, 101–12.

Braun, L. N., Brun, E., Durand, Y., Martin, E., and Tourasse, P. (1994). Simulation of discharge using different methods of meteorological data distribution, basin discretization and snow modelling. *Nordic Hydrol.*, **25**(1/2), 129–44.

Brubaker, K., Rango, A. and Kustas, W. (1996). Incorporating radiation inputs into the snowmelt runoff model. *Hydrol. Processes*, **10**, 1329–43.

Burnash, R. J. C. (1995). Chapter 10: the NWS river forecast system-catchment modeling. In *Computer Models of Watershed Hydrology*, ed. V. P. Singh. Highlands Ranch, CO: Water Resources Publications, pp. 311–66.

Cazorzi, F. and Fontana, G. D. (1996). Snowmelt modeling by combining air temperature and a distributed radiation index. *J. Hydrol.*, **181**, 169–87.

Clark, M. P., Slater, A. G., Barrett, A. P., Hay, L. E., McCabe, G. J., Rajagopalan, B., and Leavesley, G. H. (2006). Assimilation of snow covered area information into hydrologic and land-surface models. *Adv. Water Resour.*, **29**, 1209–21.

Day, G. N. (1985). Extended streamflow forecasting using NWSRFS. *J. Water Resour. Planning Manag.*, **111**(2), 157–70.

de Scally, F. A. (1992). Influence of avalanche snow transport on snowmelt runoff. *J. Hydrol.*, **137**, 73–97.

DeVries, J. J. and Hromadka, T. V. (1992). Chapter 21: Computer models for surface water. In *Handbook of Hydrology*, ed. D. R. Maidment. New York: McGraw-Hill Inc.

DeWalle, D. R., Eismeier, J. A., and Rango, A. (2003). Early forecasts of snowmelt runoff using SNOTEL data in the Upper Rio Grande basin. In *Proceedings of the 71st Annual Meeting Western Snow Conference*, Scottsdale, AZ April 22–25, 2003, pp. 17–22.

DeWalle, D. R., Henderson, Z., and Rango, A. (2002). Spatial and temporal variations in snowmelt degree-day factors computed from SNOTEL data in the Upper Rio Grande Basin. In *Proceedings of the 70th Annual Meeting of the Western Snow Conference*, Granby, CO May 20–23, 2002, pp. 73–81.

Dressler, K. A., Leavesley, G. H., Bales, R. C., and Fassnacht, S. R. (2006). Evaluation of gridded snow water equivalent and satellite snow cover products for mountain basins in a hydrologic model. *Hydrol. Processes*, **20**, 673–88.

Fassnacht, S. R., Dressler, K. A., and Bales, R. C. (2003). Snow water equivalent interpolation for the Colorado River Basin from snow telemetry (SNOTEL) data. *Water Resour. Res.*, **39**(8), 1208.

Federer, C. A. and Lash, D. (1983). *Brook: A Hydrologic Simulation Model for Eastern Forests*. Res. Rpt. 19. Water Resour. Res. Center, University of New Hampshire, Durham, NH.

Federer, C. A., Pierce, R. S., and Hornbeck, J. W. (1972). Snow management seems unlikely in the Northeast. In *Proceedings Symposium on Watersheds in Transition*, American Water Resources Association, pp. 212–19.

Garstka, W. U., Love, L. D., Goodell, B. C., and Bertle, F. A. (1958). *Factors Affecting Snowmelt and Streamflow*. Washington, DC: United States Goverment Printing Office.

Gellens, D., Barbieux, K., Schädler, B., Roulin, E., Aschwanden, H., and Gellens-Meulenberghs, F. (2000). Snow cover modelling as a tool for climate change assessment in a Swiss alpine catchment. *Nordic Hydrol.*, **31**(2), 73–88.

Giesbrecht, M. A. and Woo, M. (2000). Simulation of snowmelt in a subarctic spruce woodland: 2. open woodland model. *Water Resour. Res.*, **36**(8), 2287–95.

Hamlin, L., Pietroniro, A., Prowse, T., Soulis, R., and Kouwen, N. (1998). Application of indexed snowmelt algorithms in a northern wetland regime. *Hydrol. Processes*, **12**, 1641–57.

Harrington, R. F., Elder, K., and Bales, R. C. (1995). Distributed snowmelt modeling using a clustering algorithm. In *Biogeochemistry of Seasonally Snow-Covered Catchments*, IAHS, Publ. 228, ed. K. A. Tjonnessen, M. W. Williams and M. Tranter, pp. 167–74.

Hartman, M., J. S. Baron, R. B. Lammers, D. W. Cline, L. E. Band, G. E. Liston, and C. Tague. (1999). Simulations of snow distribution and hydrology in a mountain basin. *Water Resour. Res.*, **35**(5), 1587–603.

Hock, R. (1999). A distributed temperature-index ice- and snowmelt model including potential direct solar radiation. *J. Glaciol.*, **45**(149), 101–9.

Horne, F. and Kavvas, M. L. (1997). Physics of the spatially averaged snowmelt process. *J. Hydrol.*, **191**, 179–207.

Hottelet, Ch., Blažková, Š., and Bicik, M. (1994). Application of the ETH snow model to three basins of different character in central Europe. *Nordic Hydrol.*, **25**, 113–28.

Ishikawa, N., Nakabayashi, H., Ishii, Y., and Kodama, Y. (1998). Contribution of snow to the annual water balance in Moshiri Watershed, Northern Hokkaido, Japan. *Nordic Hydrol.*, **29**(4/5), 347–60.

Jordan, R. (1991). A one-dimensional temperature model for a snow cover, technical documentation for SNTHERM.89. In *Cold Regions Research and Engineering Laboratory Special Report 91–16*. Hanover, NH: US Army, Corps of Engineers.

Kane, D. L., Hinzman, L. D., Benson, C. S., and Liston, G. E. (1991). Snow hydrology of a headwater Arctic basin: 1. Physical measurements and process studies. *Water Resour. Res.*, **27**(6), 1099–109.

Kirnbauer, R., Bloschl, G., and Gutknecht, D. (1994). Entering the era of distributed snow models. *Nordic Hydrol.*, **25**, 1–24.

Kuusisto, E. (1980). On the values and variability of degree-day melting factor in Finland. *Nordic Hydrol.*, **11**, 235–42.

Kuz'min, P. P. (1961). *Protsess tayaniya shezhnogo pokrova (Melting of Snow Cover). Glavnoe Upravlenie Gidrometeorologischeskoi Sluzhby Pri Sovete Ministrov SSSR* Gosudarstvennyi Gidrologicheskii Institut. Main Admin. Hydrometeorol. Service, USSR Council Ministers, State Hydrol. Institute. Translated by Israel Program for Scientific Translations. Avail from US Dept. Commerce, National Tech. Inform. Service, 1971, TT 71–50095.

Leaf, C. F. (1969). Aerial photographs for operational streamflow forecasting in the Colorado Rockies. In *Proceedings of the 37th Annual Western Snow Conference*, April 1969, pp. 19–28.

Leaf, C. F. and Haeffner, A. D. (1971). A model for updating streamflow forecasts based upon areal snow cover and a precipitation index. In *Proceedings of the 39th Annual Western Snow Conference*, April 1971, pp. 9–16.

Leavesley, G. H. and Stannard, L. G. (1995). Chapter 9: The precipitation-runoff modeling system-PRMS. *Computer Models of Watershed Hydrology*, ed. V. P. Singh. Highlands Ranch, CO: Water Resources Publications, pp. 281–310.

Leavesley, G. H. and Striffler, W. E. (1978). A mountain watershed simulation model. In *Modeling of Snow Cover Runoff*, Conf. Proc., ed. S. C. Colbeck and M. Ray. Hanover, NH: US Army Corps Engineers, pp. 379–86.

Leavesley, G. H., Markstrom, S. L., Restrepo, P. J., and Viger, R. J. (2001). A modular approach to addressing model design, scale, and parameter estimation issues in distributed hydrological modeling. *Hydrol. Processes*, **16**(2), 173–87.

Leavesley, G. H., Restrepo, P., Stannard, L. G., and Dixon, M. (1983). *Precipitation-Runoff Modeling System: User's Manual*. Water Resour. Invest. Rpt. 83–4238. US Dept. Interior, Geological Survey.

Liston, G. E. and Sturm, M. (1998). A snow-transport model for complex terrain. *J. Glaciol.*, **44**(148), 498–516.

Link, T. and Marks, D. (1999). Distributed simulation of snowcover mass- and energy-balance in the boreal forest. *Hydrol. Processes*, **13**, 2439–52.

Luce, C. H., Tarboton, D. G. and Cooley, K. R. (1999). Sub-grid parameterization of snow distribution for an energy and mass balance snow cover model. *Hydrol. Processes*, **13**, 1921–33.

Maidment, D. R. (ed.) (1992). *Handbook of Hydrology*. New York: McGraw-Hill Inc.

Marks, D., Domingo, J., Susong, S., Link, T., and Garen, D. (1999). A spatially distributed energy balance snowmelt model for application in mountain basins. *Hydrol. Processes*, **13**, 1935–59.

Marks, D., Dozier, J., and Davis, R. E. (1992). Climate and energy exchange at the snow surface in the alpine region of the Sierra Nevada, 1. meteorological measurements and monitoring. *Water Resour. Res.*, **28**(11), 3029–42.

Martinec, J. (1960). The degree-day factor for snowmelt-runoff forecasting. In *Proceedings General Assembly of Helsinki*, Commission on Surface Waters, IASH Publ. 51, pp. 468–77.

Martinec, J. (1976). Chapter 4: Snow and ice. In *Facets of Hydrology*, ed. J. C. Rodda. New York: John Wiley and Sons, pp. 85–118.

Martinec, J. and deQuervain, M. R. (1975). The effect of snow displacement by avalanches on snow melt and runoff. In *Snow and Ice Symposium Proceedings*, IAHS-AISH Publ. 104, August 1971 Moscow, Russia, pp. 364–77.

Martinec, J. and Rango, A. (1989). Merits of statistical criteria for the performance of hydrological models. *Water Resour. Bull.*, **25**(2), 421–32.

Martinec, J. and Rango, A. (1991). Indirect evaluation of snow reserves in mountain basins. In *Snow, Hydrology and Forests in High Alpine Areas*, IAHS Publ. 205, ed H. Bergman, H. Lang, W. Frey, D. Issler and B. Salm, pp. 111–19.

Martinec, J., Rango, A., and Roberts, R. (1998). Snowmelt Runoff Model (SRM) User's Manual. In *Geographica Bernisia* Series P, no. 35, ed. M. F. Baumgartner and G. M. Apfl, University of Berne.

Mashayekhi, T. and Mahjoub, M. (1991). Snowmelt runoff forecasting: case study of Karadj reservoir, Iran. In *Snow, Hydrology and Forests in High Alpine Areas*, IAHS, Publ. No. 205, ed. H. Bergman, H. Lang, W. Frey, D. Issler and B. Salm, pp. 213–20.

Mcguire, M., Wood, A. W., Hamlett, A. F., and Lettenmaier, D. P. (2006). Use of satellite data for streamflow and reservoir storage forecasts in the Snake River basin, ID. *J. Water Resour. Planning Manag*, **132**, 97–110.

Metcalfe, R. A. and Buttle, J. M. (1998). A statistical model of spatially distributed snowmelt rates in a boreal forest basin. *Hydrol. Processes*, **12**, 1701–22.

Mitchell, K. M. and DeWalle, D. R. (1998). Application of the SRM using multiple parameter landscape zones on the Towanda Creek Basin, Pennsylvania. *J. Amer. Water Resour. Assoc.*, **34**(2), 335–46.

Molotch, N. P., Colee M. T., Bales, R. C., and Dozier, J. (2005). Estimating the spatial distribution of snow water equivalent in an alpine basin using binary regression tree models: the impact of digital elevation data and independent variable selection. *Hydrol. Processes*, **19**, 1459–79.

Moore, R. D. (1993). Application of a conceptual streamflow model in a glacierized drainage basin. *J. Hydrol.*, **150**, 151–68.

Morris, E. M. (1982). Sensitivity of the European hydrological system snow models. In *Proceedings Symposium on Hydrological Aspects of Alpine and High-Mountain Areas*, IAHS Publ. 138, pp. 221–31.

Pomeroy, J. W., Gray, D. M., and Landine, P. G. (1993). The prairie blowing snow model: characteristics, validation and operation. *J. Hydrol.*, **144**, 165–92.

Rango, A. and Martinec, J. (1994). Model accuracy in snowmelt-runoff forecasts extending from 1–20 days. *Water Resour. Bull.*, **30**(3), 463–70.

Rango, A. (1996). Spaceborne remote sensing for snow hydrology applications. *Hydrol. Sci. J.*, **41**, 477–93.

Ranzi, R. and Rosso, R. (1991). A physically based approach to modelling distributed snowmelt in a small alpine catchment. In *Snow, Hydrology and Forests in High Alpine Areas*, IAHS, Publ. 205, ed. H. Bergman, H. Lang, W. Frey, D. Issler, and B. Salm, pp. 141–50.

Rao, S. V. N., Ramasastri, K. S., and Singh, R. N. P. (1996). A simple monthly runoff model for snow dominated catchments in the Western Himalayas. *Nordic Hydrol.*, **27**(4), 255–74.

Refsgaard, J. C. and Storm, B. (1995). Chapter 23-MIKE SHE. In *Computer Models of Watershed Hydrology*, ed. V. P. Singh. Highlands Ranch, CO: Water Resources Publications, pp. 809–46.

Rodell, M. and Houser, P. R. (2004). Updating a land surface model with MODIS-derived snow cover. *J. Hydrometeorol.*, **5**, 1064–75.

Rohrer, M. B. and Braun, L. N. (1994). Long-term records of the snow cover water equivalent in the Swiss Alps-2. Simulation. *Nordic Hydrol.*, **25**(1/2), 65–78.

Schaper, J., Martinec, J., and Seidel, K. (1999). Distributed mapping of snow and glaciers for improved runoff modelling. *Hydrol. Processes*, **13**, (12/13), 2023–32.

Singh, V. P. (1995). *Computer Models of Watershed Hydrology*. Highlands Ranch, CO: Water Resources Publications.

Singh, P. and Singh, V. P. (2001). *Snow and Glacier Hydrology*. Dordrecht: Kluwer Academic Publishers.

Speers, D. D. (1995). Chapter 11: SSARR model. In *Computer Models of Watershed Hydrology*, ed. V. P. Singh. Highlands Ranch, CO: Water Resources Publications, pp. 367–94.

Thapa, K. B. (1993). Estimation of snowmelt runoff in Himalayan catchments incorporating remote sensing data. In *Snow and Glacier Hydrology*, IAHS Publ. 218, ed. G. J. Young, pp. 69–74.

Turpin, O., Ferguson, R., and Johansson, B. (1999). Use of remote sensing to test and update simulated snow cover in hydrologic models. *Hydrol. Processes*, **13**, 2067–77.

Ujihashi, Y., Takase, N., Ishida, H., and Hibobe, E. (1994). Distributed snow cover model for a mountainous basin. In *Snow and Ice Covers: Interactions with the Atmosphere and Ecosystems*, IAHS Publ. 223, ed. H. G. Jones, T. D. Davies, A. Ohmura, and E. M. Morris, pp. 153–62.

United States Army Corp of Engineers (1956). *Snow Hydrology, Summary Report of the Snow Investigations*. Portland, OR: N. Pacific Div., COE.

Williams, K. S. and Tarboton, D. G. (1999). The ABC's of snowmelt: a topographically factorized energy component snowmelt model. *Hydrol. Processes*, **13**, 1905–20.

Willen, D. W., Shumway, C. A., and Reid, J. E. (1971). Simulation of daily snow water equivalent and melt. *Proceedings Western Snow Conference*, Billings, MT, Vol. 39, pp. 3–8.

Winstral, A., Elder, K., and Davis, R. E. (2002). Spatial snow modeling of wind-redistributed snow using terrain-based parameters. *J. Hydrometeorol.*, **3**(5), 524–38.

World Meteorological Organization (1986). *Intercomparison of Models of Snowmelt Runoff*, Operational Hydrol. Rpt. 23, No. 646, Geneva: WMO.

World Meteorological Organization (1992). *Simulated Real-Time Intercomparison of Hydrological Models*, Operational Hydrol. Rpt. 38, WMO-No. 779, Geneva: WMO.

11

Snowmelt-Runoff Model (SRM)

11.1 Introduction

The Snowmelt-Runoff Model (SRM) is designed to simulate and forecast daily streamflow in mountain basins where snowmelt is a major runoff factor. Most recently, it has also been applied to evaluate the effect of a changed climate on the seasonal snow cover and runoff. SRM was developed by Martinec (1975) in small European basins. Thanks to the progress of satellite remote sensing of snow cover, SRM has been applied to larger and larger basins. The Ganges River Basin in the Himalayas, the largest basin where SRM has been applied so far, is about 917 444 km^2 (Seidel *et al.*, 2000). Runoff computations by SRM appear to be relatively easily understood. To date, the model has been applied by various agencies, institutes, and universities in over 100 basins situated in 29 different countries as listed in Table 11.1. About 20% of these applications have been performed by the model developers and 80% by independent users. Some of the localities are shown in Figure 11.1. SRM also successfully underwent tests by the World Meteorological Organization with regard to runoff simulations (WMO, 1986) and to partially simulated conditions of real-time runoff forecasts (WMO, 1992).

Range of conditions for model application

SRM can be applied in mountain basins of almost any size (so far from 0.76 to 917 444 km^2) and any elevation range (for example 0–8848 m a.s.l.) (Table 11.1). A model run starts with a known or estimated discharge value and can proceed for an unlimited number of days, as long as the input variables – temperature, precipitation, and snow-covered area – are provided. As a test, a ten-year period was computed without reference to measured discharges (Martinec and Rango, 1986).

In addition to the input variables, the area–elevation curve of the basin is required. If other basin characteristics are available (forested area, soil conditions, antecedent

Table 11.1 *SRM applications and results*

#	Country	Basin	Size (km²)	Elevation Range (m a.s.l.)	R^2	D_v (%)	Years Seasons	Zones	Years Applied
1	USA	EGL (Rocky Mountains)	0.29	3300–3450	N/A	N/A	N/A	N/A	1989
2	USA	WGL (Rocky Mountains)	0.6	3300–3450	N/A	N/A	N/A	N/A	1989
3	Germany	Lange Bramke (Harz)	0.76	540–700	N/A	N/A	1	1	1981
4	Germany	Wintertal (Harz)	0.76	560–754	N/A	N/A	1	1	1981
5	Czech R.	Modry Dul (Krkonose)	2.65	1000–1554	0.96	1.7	2	1	1962, 1966
6	USA	GLEES (Rocky M.)	2.87	3300–3450	N/A	N/A	N/A	N/A	1989
7	Ecuador	Antisana (Andes)	3.72	4500–5760	N/A	N/A	1	3	1996
8	Argentina	Echaurren	4.5	3000–4200	0.84	7.5	1	1	1985
9	Spain	Lago Mar (Pyrenees)	4.5	2234–3004	N/A	N/A	1	1	1965
10	Spain	Llauset dam (Pyrenees)	7.8	2100–3000	0.69	5.5	1	2	1999
11	USA	W-3 (Appalachians)	8.42	346–695	0.81	8.8	10	1	1969–1978
12	Germany	Lainbachtal (Allgauer Alps)	18.7	670–1800	N/A	N/A	5	1	1978, 1979
13	Spain	Salenca en Baserca (Pyrenees)	22.2	1460–3200	0.72	4.3	3	3	1999
14	Spain	Noguera Ribagorzana en Baserca (Pyrenees)	36.8	1480–3000	0.71	3.7	3	2	1995
15	Switzerland	Rhone-Gletsch (Alps)	38.9	1755–3630	N/A	N/A	1	4	1979
16	Switzerland	Dischma (Alps)	43.3	1668–3146	0.86	2.5	10	3	1973, 1970–1979
17	Japan	Sai (Japan Alps)	57	300–1600	0.86	N/A	3	3	1979–1981
18	Spain	Tor en Alins	60	1880–3040	0.71	7.3	1	4	1999
19	Spain	Flamisell en Capdella	84	1440–2940	0.68	8.1	1	4	1999
20	Spain	Vellós en Añisclo	85	1140–3360	0.83	1.5	1	5	1999
21	Austria	Rofenache (Alps)	98	1890–3771	0.88	2.4	1	8	1992–1993
22	United Kingdom	Feshie (Cairngorms)	106	350–1265	0.88	N/A	2	1	1979, 1980
23	Switzerland	Sedrun (Alps)	108	1840–3210	0.79	1.9	2	3	1985, 1993
24	Austria	Tuxbach	116	879–3062	0.44	12.7	3	7	1996–1998
25	Austria	Schlegeis	121	1790–3510	0.86	8.7	3	8	1996–1998

(cont.)

Table 11.1 (*cont.*)

#	Country	Basin	Size (km²)	Elevation Range (m a.s.l.)	R^2	D_v (%)	Years Seasons	Zones	Years Applied
26	Australia	Geehi River (Snowy Mtns.)	125	1032–2062	0.7	6.6	6	3	1989–1994
27	USA	American Fork (Utah)	130	1820–3580	0.90	1.7	1	4	1983
28	Austria	Venter Ache	165	1850–3771	0.82	5.4	1	8	1992
29	Switzerland	Landwasser – Frauenkirch (Alps)	183	1500–3146	N/A	N/A	1	3	1979
30	Switzerland	Massa-Blatten (Alps)[c]	196	1447–4191	0.91	−5.3	1	6	1985
31	India	Kulang (Himalayas)	205	2350–5000	N/A	N/A	N/A	N/A	–
32	Switzerland	Tavanasa (Alps)	215	1277–3210	0.82	3.1	2	4	1985, 1993
33	USA	Dinwoody (Wind River)	228	1981–4202	0.85	2.8	2	4	1974, 1976
34	Italy	Cordevole (Alps)	248	980–3250	0.89	4.6	1	3	1984
35	USA	Salt Creek (Utah)	248	1564–3620	N/A	2.6	1	5	–
36	India	Beas-Manali (Himalayas)	345	1900–6000	0.68	12	4	N/A	–
37	Argentina	El Yeso	350	2475–6550	0.91	2.6	2	3	1991, 1993
38	Norway	Laerdalselven (Lo Bru)	375	530–1720	0.86	5.2	1	5	1991
39	Norway	Viveli (Hardangervidda)	386	880–1613	0.73	11.3	2	1	1991
40	USA	Scofield Dam, Pries River (Utah)	401	2323–3109	0.8	5.0	2	1	1996, 1998
41	Spain	Cardós en Tirvia	417	1720–3240	0.8	2.6	2	4	1999
42	Japan	Okutadami (Mikuni)	422	782–2346	0.83	5.4	3	3	1984
43	USA	Joes Valley Dam, Cottonwood Creek (Utah)	435	2131–3353	0.83	18.5	2	3	1985
44	Spain	Garona en Bossost	449	1620–3080	0.75	3.0	1	4	1999
45	Spain	Noguera Pallaresa en Escaló	450	1860–2960	0.87	3.3	1	3	1999
46	USA	Bull Lake Creek (Rocky Mts.)	484	1790–4185	0.82	4.8	1	4	1976
47	Switzerland	Tiefencastel (Alps)	529	837–3418	0.55	11.3	2	5	1982, 1985
48	Spain	Valira en Seo d'Urgel (Pyrenees)	545	1740–3080	0.92	0.9	1	4	1999
49	USA	Towanda Creek[b]	550	240–733	0.78	8.3	3	6	1990, 1993, 1994
50	Spain	Garona en Pont de Rei	558	1420–3080	0.72	2.7	1	5	1999
51	USA	South Fork (Colorado)	559	2506–3914	0.89	1.8	7	3	1973–1979
52	Spain	Noguera Ribagorzana en Pont de Suert	573	920–3380	0.91	−0.6	1	6	1999

	Country	Location							
53	Argentina	Las Cuevas en Los Almendros (Andes)	600	2500–7000	N/A	N/A	N/A	N/A	1981
54	USA	Independence R. (Adirondacks)	618	261–702	0.81	5.0	1	1	1987
55	Chile	Mapocho (Andes)	630	1024–4450	0.42	29.9	1	3	1987
56	Poland	Dunajec (High Tatra)	700	577–2301	0.73	3.8	1	3	1975
57	India	Saing (Himalayas)	705	1400–5500	N/A	N/A	N/A	N/A	–
58	USA	Conejos (Rocky Mts)	730	2521–4017	0.87	1.1	7	3	1973–1979
59	Switzerland	Ilanz (Alps)	776	693–3614	0.53	8.6	2	5	1982,1985
60	Spain	Cinca en Laspuña	798	1120–3380	0.78	5.6	1	5	1999
61	China	Toutunhe	840	1430–4450	0.81	2.0	3	6	1984–1986
62	Austria	Ötztaler Ache (Alps)	893	670–3774	0.84	9.18	1	6	<1998
63	Argentina	Lago Alumin (Andes)	911	1145–2496	N/A	N/A	N/A	N/A	1985
64	China	Gongnisi (Tien Shan)	939	1776–4200	0.8	0.97	1	5	<1999
65	China	Urumqi (Tien Shan)	950	1880–4200	0.62	2.78	1	6	<1999
66	Uzbekistan	Angren	1082	1200–3800	0.63	2.3	3	6	–
67	India	Parbati (Himalayas)	1154	1500–6400	0.73	7.5	1	5	1986~1991
68	Canada	Illecillewaet (Rocky Mts)	1155	509–3150	0.86	7.0	4	4	1976, 1981, 1983, 1984
69	Spain	Segre en Seo d'Urgel (Pyrenees)	1217	360–2900	N/A	N/A	N/A	5	1996
70	India	Buntar (Himalayas)	1370	1200–5000	N/A	N/A	N/A	N/A	1985~1991
71	Chile	Tinguiririca Bajo Briones (Andes)	1460	520–4500	0.88	–0.3	1	3	1987
72	New Zealand	Hawea (S. Alps)	1500	300–2500	N/A	N/A	N/A	N/A	1989
73	Switzerland	Ticino-Bellinzona (Alps)	1515	220–3402	0.86	–0.6	1	5	1994
74	USA	Spanish Fork (Utah)	1665	1484–3277	0.85	1.0	1	4	1983
75	Switzerland	Inn at Tarasp (Alps)	1700	1235–4005	0.77	8.0	–	1	1996
76	Argentina	Tupungato (Andes)	1800	2500–6000	0.63	6.4	1	8	1981
77	Switzerland	Inn-Martina (Alps)	1943	1030–4049	0.82	4.3	1	2	1990
78	Argentina	Chico (Tierra del Fuego)	2000	N/A	N/A	N/A	N/A	N/A	N/A
79	USA	Boise (Rocky Mts.)	2150	983–3124	0.84	3.3	3	3	1976–1978
80	France	Durance (Alps)	2170	786–4105	0.85	2.6	5	5	1975–1979

(cont.)

309

Table 11.1 (cont.)

#	Country	Basin	Size (km^2)	Elevation Range (m a.s.l.)	R^2	D_v (%)	Years Seasons	Zones	Years Applied
81	USA	Madison (Montana)	2344	1965–3234	0.89	1.5	2	2	1976, 1978
82	Uzbekistan	Pskem	2448	1000–4200	0.97	−1.0	3	7	<2002
83	Morocco	Tillouguit (Atlas)	2544	1050–3411	0.84	0.5	1	3	1979
84	Austria	Salzach-St.Johann (Alps)	2600	570–3666	N/A	N/A	N/A	N/A	>1991
85	USA	Henry's Fork (Idaho)	2694	1553–3125	0.91	1.5	2	3	1976, 1979
86	USA	Cache la Poudre (Colorado)	2732	1596–4133	N/A	N/A	1	3	1983
87	Chile	Aconcagua en Chababuquito (Andes)	2900	900–6100	0.91	0.9	1	3	1987
88	USA	Sevier River (Utah)	2929	1923–3260	0.75	5.1	1	4	1983
89	Switzerland	Rhine-Felsberg (Alps)[a]	3249	562–3425	0.70	7.2	7	5	1982, 1985, 1988, 1989, 1992, 1994, 1996, 1985
90	Switzerland	Rhône-Sion (Alps)[b]	3371	491–4634	0.97	−2.1	1	7	1976–1977, 1979
91	USA	Rio Grande (Colorado)[a]	3414	2432–4215	0.84	3.8	13	3	1973–1975, 1982, 1988
92	USA	Kings River (California)[a]	4000	171–4341	0.82	3.2	5	7	
93	Chile	Maipo en el Manzano (Andes)	4960	850–5600	0.77	0.9	2	3	1982, 1988
94	India	Beas-Thalot (Himalayas)	5144	1100–6400	0.80	1.5	2	6	1986–1987
95	USA	Upper Yakima (Cascades)	5517	366–2121	0.92	2.8	1	5	1989
96	Uzbekistan	Chatkal	6591	1000–4000	0.81	1.6	3	7	–
97	Canada	Sturgeon (Ontario)	7000	N/A	N/A	N/A	N/A	N/A	1967
98	Argentina	Grande (Tierra del Fuego)	9050	N/A	N/A	N/A	N/A	N/A	N/A
99	Canada	Iskut (Coast)	9350	200–2556	N/A	N/A	N/A	N/A	N/A
100	Turkey	Karasu (Upper Euphrates)	10216	1125–3487	0.95	0.25	3	5	1997–1999
101	Uzbekistan, Kyrgystan	Karadarya	12056	1100–4568	0.87	4.0	4	8	1996, 1998, 1999
102	Tadjikistan	Zerafshan	12214	410–5500	N/A	N/A	1	8	–

No.	Country	River	Area	Elevation range	R^2	D_V			Year
103	Uzbekistan, Kazakhstan	Kafirnirgan[a]	12 369	505–3005	0.57	8.6	3	6	1988–1991
104	USA	Sevier River (Utah)	13 380	1506–3719	0.93	4.0	1	4	1983
105	USA	Snake River (Idaho)	14 897	1524–4196	0.90	0.4	11	N/A	1972–1982
106	Tadjikistan	Vakhsh[a]	37 759	1791–5291	0.63	2.8	3	8	1988–1990
107	Kyrgystan	Naryn	53 237	800–5000	0.96	1.0	2	8	–
108	Pakistan	Kabul River (Himalayas)	63 657	305–7690	0.66	6.0	1	1	1975
109	Tadjikistan, Afghanistan	Pyandzh (Pamirs and Hindu Kush)	120 534	2141–5564	0.65	5.6	3	8	1988–1991
110	China	Yellow (Anyemogen Shan)	121 972	2500–5224	N/A	N/A	N/A	3	1993
111	India, Bangladesh	Brahmaputra[a]	547 346	0–8848	0.75	–7.5	1	7	1995
112	India, Bangladesh	Ganges[a]	917 444	0–8848	0.94	8.3	1	7	1995

[a] climate change evaluated;   [b] land-use zones;   [c] separate mapping of snow cover and glaciers.
If more than one year was evaluated, averages of R^2 and averages of D_V (single values taken in absolute terms) are listed.
The accuracy criteria R^2 and D_V listed in Table 11.1 are defined as:

$$R^2 = 1 - \frac{\sum_{i=1}^{n}(Q_i - Q'_i)^2}{\sum_{i=1}^{n}(Q_i - \overline{Q})^2} \qquad D_v = \frac{V_R - V'_R}{V_R} \cdot 100$$

where:

R^2 = a measure of model efficiency
Q_i = measured daily discharge
Q'_i = simulated daily discharge
$\overline{Q}$ = average daily discharge for the simulation year or simulation season
n = number of daily discharge values
D_V = percentage difference between the total measured and simulated runoff (%)
V_R = measured runoff volume
V'_R = simulated runoff volume

Figure 11.1 Selected locations where SRM has been tested worldwide.

precipitation, and runoff data), they are of course useful for facilitating the determination of the model parameters.

SRM can be used for the following purposes:

(1) Simulation of daily flows in a snowmelt season, in a year, or in a sequence of years. The
results can be compared with the measured runoff in order to assess the performance of
the model and to verify the values of the model parameters. Simulations can also serve to
evaluate runoff patterns in ungauged basins using satellite monitoring of snow-covered
areas and extrapolation of temperatures and precipitation from nearby stations.

(2) Short-term and seasonal runoff forecasts. The microcomputer program (Micro-SRM)
includes a derivation of modified depletion curves which relate the snow-covered areas
to the cumulative snowmelt depths as computed by SRM. These curves enable the
snow coverage to be extrapolated manually by the user several days ahead employing
temperature forecasts so that this input variable is available for discharge forecasts.
The modified depletion curves can also be used to evaluate the snow reserves for
seasonal runoff forecasts. The model performance may deteriorate if the forecasted air
temperature and precipitation deviate from the observed values, but the inaccuracies
can be reduced by periodic updating.

(3) In recent years, SRM was applied to the new task of evaluating the potential effect of climate change on the seasonal snow cover and runoff. The microcomputer program has been complemented accordingly.

11.2 Model structure

Each day, the water produced from snowmelt and from rainfall is computed, superimposed on the calculated recession flow, and transformed into daily discharge from the basin according to Equation (11.1):

$$Q_n + 1 = [c_{Sn} \cdot a_n(T_n + \Delta T_n)S_n + c_{Rn}P_n]\frac{A \cdot 10\,000}{86\,400}(1 - k_{n+1}) + Q_n k_{n+1}$$

$$(11.1)$$

where:

Q = average daily discharge ($m^3 s^{-1}$)

c = runoff coefficient expressing the losses as a ratio (runoff/precipitation), with c_S referring to snowmelt and c_R to rain

a = degree-day factor ($cm\ °C^{-1}\ d^{-1}$) indicating the snowmelt depth resulting from 1 degree-day, see DDF, Section 10.3.2

T = number of degree-days (°C d)

ΔT = the adjustment by temperature lapse rate when extrapolating the temperature from the station to the average hypsometric elevation of the basin or zone (°C d)

S = ratio of the snow-covered area to the total area

P = precipitation contributing to runoff (cm)

A preselected threshold temperature, T_{CRIT}, determines whether this contribution is rainfall and immediate. If precipitation is determined by T_{CRIT} to be new snow, it is kept on storage over the hitherto snow-free area until melting conditions occur.

A = area of the basin or zone (km^2)

k = recession coefficient indicating the decline of discharge in a period without snowmelt or rainfall:

$k = k = \dfrac{Q_{m+1}}{Q_m}$ ($m, m + 1$ are the sequence of days during a true recession flow period)

n = sequence of days during the discharge computation period. Equation (11.1) is written for a time lag between the daily temperature cycle and the resulting discharge cycle of 18 hours. In this case, the number of degree-days measured on the nth day corresponds to the discharge on the $n + 1$ day. Various lag times can be introduced by a subroutine.

$$\frac{10\,000}{86\,400} = \text{conversion from } cm\ km^2\ d^{-1} \text{ to } m^3\ s^{-1}$$

T, S, and P are variables to be measured or determined each day. c_R, c_S, lapse rate to determine ΔT, T_{CRIT}, k and the lag time are parameters which are characteristic for a given basin or, more generally, for a given climate. A guidance for determining these parameters will be given in Section 11.3.3.

If the elevation range of the basin exceeds 500 m, it is recommended that the basin be subdivided into elevation zones of about 500 m each. For an elevation range of 1500 m and three elevation zones A, B, and C, the model equation becomes:

$$Q_{n+1} = \left\{ [c_{SAn} \cdot a_{An}(T_n + \Delta T_{An})S_{An} + c_{RAn} \cdot P_{An}] \frac{A_A \cdot 10\,000}{86\,400} \right.$$

$$+ [c_{SBn} \cdot a_{Bn}(T_n + \Delta T_{Bn})S_{Bn} + c_{RBn} \cdot P_{Bn}] \frac{A_B \cdot 10\,000}{86\,400}$$

$$\left. + [c_{SCn} \cdot a_{Cn}(T_n + \Delta T_{Cn})S_{Cn} + c_{RCn} \cdot P_{Cn}] \frac{A_C \cdot 10\,000}{86\,400} \right\} (1 - k_{n+1})$$

$$+ Q_n \cdot k_{n+1} \tag{11.2}$$

The indices A, B, and C refer to the respective elevation zones and a time lag of 18 hours is assumed. Other time lags can be selected and automatically taken into account as explained in Section 11.3.3.

In the simulation mode SRM can function without updating. The discharge data serve only to evaluate the accuracy of simulation. In ungauged basins the simulation is started with a discharge estimated by analogy to a nearby gauged basin. In the forecasting mode the model provides an option for updating by the actual discharge every one to nine days.

Equations (11.1) and (11.2) are written for the metric system but an option for model operation in English units is also provided in the computer program.

11.3 Data necessary for running the model

11.3.1 Basin characteristics

Basin and zone areas

The basin boundary is defined by the location of the streamgauge (or some arbitrary point on the streamcourse) and the watershed divide is identified on a topographic map. The basin boundary can be drawn at a variety of map scales. For the larger basins, a 1:250 000 scale map is adequate, whereas 1:24 000 scale maps are appropriate for small basins. All topographic maps (at 1:24 000 and 1:250 000 scales) are available by state in the United States in digital form on sets of CD-ROMs; this digital product can be used for basin and zone definition. After examining the elevation range between the streamgauge and the highest point in the basin (total basin relief), elevation zones can be delineated in intervals of about 500 m or

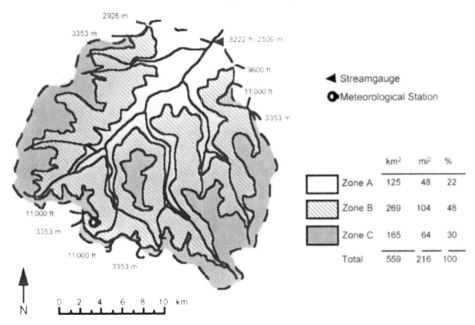

Figure 11.2 Elevation zones and areas of the South Fork of the Rio Grande Basin, Colorado, USA.

1500 ft. In addition to drawing the basin and zone boundaries, several intermediate topographic contour lines should be highlighted for later use in constructing the area–elevation curve. Once the boundaries and the contours have been determined, the areas formed by these boundaries should be planimetered manually or automatically using the digital data. Figure 11.2 shows the elevation zones and areas of the South Fork of the Rio Grande basin in Colorado, USA. The elevation range of 1408 m dictated the division of the basin into three elevation zones. Once the zones are defined, the various model variables and parameters are applied to each zone for the calculation of snowmelt runoff. To facilitate this application, the mean hypsometric elevation of the zone must be determined through use of an area–elevation curve. Many of these steps can be expedited through the use of computer analysis and a Digital Elevation Model (DEM).

Area–elevation curve

By using the zone boundaries plus other selected contour lines in the basin, the areas enclosed by various elevation contours can be determined manually by planimetering or automatically. These data can be plotted (area vs. elevation) and an area–elevation (hypsometric) curve derived as shown in Figure 11.3 for the South Fork Basin. This area–elevation curve can also be derived automatically if the user has access to digital elevation data and computer algorithms used in an image

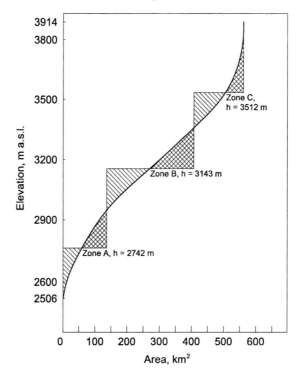

Figure 11.3 Determination of zonal mean hypsometric elevations ($\bar{h}$) using an area–elevation curve for the South Fork of the Rio Grande basin.

processing system. The zonal mean hypsometric elevation, $\bar{h}$, can then be determined from this curve by balancing the areas above and below the mean elevation as shown in Figure 11.3. The $\bar{h}$ value is used as the elevation to which base station temperatures are extrapolated for the calculation of zonal degree-days in SRM.

11.3.2 Variables

Temperature and degree-days, T

In order to compute the daily snowmelt depths, the number of degree-days must be determined from temperature measurements or, in a forecasting mode, from temperature forecasts. The program accepts either the daily mean temperature (option 0) or two temperature values on each day: T_{Max}, T_{Min} (option 1). The temperatures are extrapolated by the program from the base station elevation to the hypsometric mean elevations of the respective elevation zones.

For option 1, the average temperature is computed in each zone as

$$\overline{T} = \frac{T_{\text{Max}} + T_{\text{Min}}}{2} \tag{11.3}$$

When using daily means (option 0) or when using T_{Max}, T_{Min} (option 1), it is recommended that negative temperature values (when they occur) be used in the calculation. In line with this recommendation, the original "effective minimum temperature" alternative (automatic change of negative temperatures to 0 °C) was removed from the computer program beginning with Version 3.0. If the user still prefers this alternative, the occasional negative temperatures can be changed manually to 0 °C when inputting the data to SRM.

Because the average temperatures refer to a 24-hour period starting always at 0600 hrs, they become degree-days T (°C d). The altitude adjustment ΔT in Equation (11.1) is computed as follows:

$$\Delta T = \gamma \cdot (h_{st} - \bar{h}) \cdot \frac{1}{100} \qquad (11.4)$$

where:

γ = temperature lapse rate (°C per 100 m)
h_{st} = altitude of the temperature station (m)
$\bar{h}$ = hypsometric mean elevation of a zone (m)

Whenever the degree-day numbers ($T + \Delta T$ in Equation (11.1)) become negative, they are automatically set to zero so that no negative snowmelt is computed. The values of the temperature lapse rate are dealt with in Section 11.3.3.

The program accepts either temperature data from a single station (option 0, basin wide) or from several stations (option 1, by zone). With option 0, the altitude of the station is entered and temperature data are extrapolated to the hypsometric mean elevations of all zones using the lapse rate. If more stations are available, the user can prepare a single "synthetic station" and still use option 0 or, alternatively, use option 1. With option 1, the user may use separate stations for each elevation zone, however, the temperatures entered for each zone must have already been lapsed to the mean hypsometric elevation of the zone. Although SRM will take separate stations for each zone in this way, it is only optional. The measurement of correct air temperatures is difficult, and therefore one good temperature station (even if located outside the basin) may be preferable to several less reliable stations.

In the forecast mode of the model, it is necessary to obtain representative temperature forecasts for the given region and altitude in order to extrapolate the expected numbers of degree-days for each elevation zone.

Precipitation, P

The evaluation of representative areal precipitation is particularly difficult in mountain basins. Also, quantitative precipitation forecasts are seldom available for the forecast mode, although current efforts in this field are improving this situation. Fortunately, snowmelt generally prevails over the rainfall component in the mountain basins. However, sharp runoff peaks from occasional heavy rainfalls must be given particular attention and the program includes a special treatment of such events (see section 11.3.3).

The program accepts either a single, basin-wide precipitation input (from one station or from a "synthetic station" combined from several stations) (option 0) or different precipitation inputs zone by zone (option 1). If the program is switched to option 1 and only one station happens to be available, for example in the zone A, precipitation data entered for zone A must be copied to all other zones. Otherwise no precipitation from these zones is taken into account by the program. Further program options refer to the rainfall contributing area as explained in Section 11.3.3. In basins with a great elevation range, the precipitation input may be underestimated if only low-altitude precipitation stations are available. It is recommended to extrapolate precipitation data to the mean hypsometric altitudes of the respective zones by an altitude gradient, for example 3% or 4% per 100 m. If two stations at different altitudes are available, it is possible to assign the averaged data to the average elevation of both stations and to extrapolate by an altitude gradient from this reference level to the elevation zones. It should be noted that the increase of precipitation amounts with altitude does not continue indefinitely but stops at a certain altitude, especially in very high elevation mountain ranges.

A critical temperature (see Section 11.3.3) is used to decide whether a precipitation event will be treated as rain ($T \geq T_{CRIT}$) or as new snow ($T < T_{CRIT}$). When the precipitation event is determined to be snow, its delayed effect on runoff is treated differently depending on whether it falls over the snow-covered or snow-free portion of the basin. The new snow that falls over the previously snow-covered area is assumed to become part of the seasonal snowpack and its effect is included in the normal depletion curve of the snow coverage. The new snow falling over the snow-free area is considered as precipitation to be added to snowmelt, with this effect delayed until the next day warm enough to produce melting. This precipitation is stored by SRM and then melted as soon as a sufficient number of degree-days has occurred. The example in Table 11.2 illustrates a case where 2.20 cm water equivalent of snow fell on day n and then was melted on the three successive days.

In this example, S is decreasing on consecutive days because it is interpolated previously from the snow-cover depletion curve. In reality it should remain constant as long as the seasonal snowpack is covered with new snow, however, the model currently uses the incremental decrease of S shown in Table 11.2.

Sharp peaks of discharge are typical for rainfall runoff as opposed to the relatively regular daily fluctuations of the snowmelt runoff. SRM has been adapted to better simulate these rainfall peaks whenever the average daily rainfall calculated over the whole basin equals or exceeds 6 cm. This threshold can be changed by the user

Table 11.2 *Calculation of the melt of new snow deposited on a snow-free area ($P_n = 2.20$ cm; $T_{CRIT} = +1.0\,^\circ C$)*

Day	a (cm $^\circ C^{-1}$ d^{-1})	T ($^\circ C$ d)	S	P (cm)	Melted Depth $a \cdot T$ (cm)	P Stored (cm)	P Contributing to Runoff $a \cdot T'(1 - S)$ (cm)
n	0.45	0	0.72	2.20	0	2.20	0
$n + 1$	0.45	0.11	0.70	0	0.05	2.15	0.02
$n + 2$	0.45	2.70	0.68	0	1.22	0.93	0.39
$n + 3$	0.45	3.70	0.66	0	0.93	0	0.32

according to the characteristics of the selected basin. The procedure is described in Section 11.3.3 in connection with the recession coefficient. In spite of these precautions, rainfall runoff peaks may cause problems because local rainstorms may be missed if the network of precipitation stations is not dense enough. Also, the timing of rainfalls within the 24-hour period is frequently unknown. Actually, this period (for P_n) usually lasts from 0800 hrs on the day n to 0800 hrs on the day $n + 1$ and is published as precipitation on the day n. In some cases, however, the same precipitation amount is ascribed to the day $n + 1$ on which the measurement period ended. In such case precipitation data must be shifted backwards by one day before input to SRM.

Snow-covered area, S

It is a typical feature of mountain basins that the areal extent of the seasonal snow cover gradually decreases during the snowmelt season. Depletion curves of the snow coverage can be interpolated from periodical snow-cover mapping so that the daily values can be read off as an important input variable to SRM. The snow cover can be mapped by terrestrial observations (in very small basins), by aircraft photography (especially in a flood emergency) and, most efficiently, by satellites. The minimum area which can be mapped with an adequate accuracy depends on the spatial resolution of the remote sensor. Examples are listed in Table 11.3.

Figure 11.4 shows the snow cover in the alpine basin Felsberg mapped from Landsat 5-MSS data (Seidel and Martinec, 2004; Baumgartner, 1987). When the time interval between the subsequent satellite images becomes too long, e.g. due to visibility obscured by clouds, the depletion curves derived from the measured points may be distorted by occasional summer snowfalls. To avoid such errors, satellite images showing the short-lived snow cover from the summer snowfalls must be disregarded when deriving the depletion curves. In order to identify the new snow

Table 11.3 *Some of the possibilities of remote sensing for snow-cover mapping.*
Estimated minimum area sizes are for binary snowmapping. Sub-pixel mapping
techniques reduce the areas considerably

Platform Sensor	Spectral Bands	Spatial Resolution	Minimum Area Size	Repeat Period
Aircraft Orthophoto	Visible/NIR	2 m	1 km^2	flexible
IRS				
Pan	Green to NIR	5.8 m	2 km^2	24 days
LISS-II	1–3 Green to NIR	23 m	2.5–5 km^2	24 days
WiFS	1 Red/2 NIR	188 m	10–20 km^2	5 days
SPOT				
HRVIR	1–3 Green to NIR	2.5–20 m	1–3 km^2	26 days
Landsat				
MSS	1–4 Green to NIR	80 m	10–20 km^2	16–18 days
TM	1–4 Green to NIR	30 m	2.5–5 km^2	16–18 days
ETM-Pan[a]	Visible to NIR	15 m	2–3 km^2	16–18 days
Terra/Aqua				
ASTER	1–3 Visible to NIR	15 m	2–3 km^2	16 days[b]
	1 Red/2 NIR	250 m	20–50 km^2	1 day
MODIS	3–8 Blue to MIR	500 m	50–100 km^2	1 day
NOAA				
AVHRR	1 Red/2 NIR	1.1 km	10–500 km^2	12 hr
Meteosat				
SEVIRI	1–3 Red to NIR	3 km	500–1000 km^2	30 min
	12 Visible	1 km	10–500 km^2	30 min

Acronyms:
ASTER = Advanced Spaceborne Thermal Emission and Reflection Radiometer;
AVHRR = Advanced Very High Resolution Radiometer; HRVIR = High Resolution Visible and Near Infrared; IRS = Indian Remote Sensing; LISS = Linear Imaging Self-scanning Sensor; MIR = Middle Infrared; MODIS = Moderate Resolution Imaging Spectroradiometer; MSS = Multi-Spectral Scanner; NIR = Near Infrared; Pan = Panchromatic; SEVIRI = Spinning Enhanced Visible and Infrared Imager; SPOT = Satellite Pour l'Observation de La Terre; TM = Thematic Mapper; WiFS = Wide Field Sensor; ETM-Pan = Enhanced Thematic Mapper – Panchromatic
[a] Landsat 6 and 7 only;   [b] Depends on availability

events, coincident precipitation and temperature data should be examined. The transitory new snow is accounted for by the model as stored precipitation eventually contributing to runoff as just explained in the section on precipitation.

Figure 11.5 shows depletion curves of the snow coverage derived for five elevation zones of the alpine basin Felsberg from the Landsat imagery (Baumgartner, 1987; Seidel and Martinec, 2004; Baumgartner and Rango, 1995). Together with temperature and precipitation data, such depletion curves enable SRM to simulate

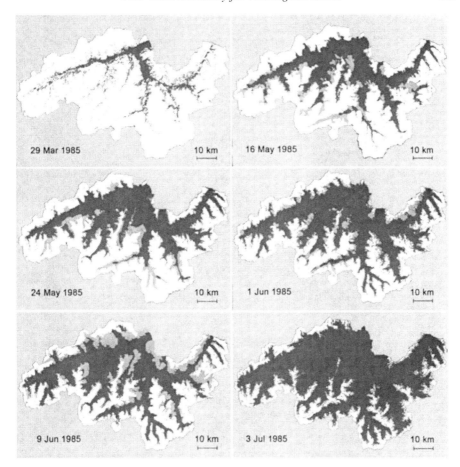

Figure 11.4 Sequence of snow-cover maps from Landsat 5-MSS, Upper Rhine River at Felsberg, 3250 km², 560–3614 m a.s.l. (courtesy Seidel and Martinec, 2004; Baumgartner, 1987).

runoff in a past year. In September, the depletion curve in zone E refers to the glacier which prevents it from decreasing any further.

For real-time runoff forecasts, however, it is necessary to know the snow-covered area within days after a satellite overflight and also to extrapolate the depletion curves of the snow coverage to the future weeks. This procedure is explained in Section 11.5.

11.3.3 Parameters

The SRM parameters are not calibrated or optimized by historical data. They can be either derived from measurement or estimated by hydrological judgment taking into account the basin characteristics, physical laws, and theoretical relations or

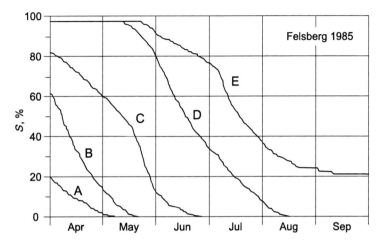

Figure 11.5 Depletion curves of the snow coverage for five elevation zones of the basin Felsberg, derived from the Landsat imagery shown in Figure 11.4. A: 560–1100 m a.s.l., B: 1100–1600 m a.s.l., C: 1600–2100 m a.s.l., D: 2100–2600 m a.s.l., E: 2600–3600 m a.s.l. (courtesy Seidel and Martinec, 2004; Baumgartner, 1987).

empirical regression relations. Occasional subsequent adjustments should never exceed the range of physically and hydrologically acceptable values.

Runoff coefficient, c

This coefficient takes care of the losses, that is to say of the difference between the available water volume (snowmelt + rainfall) and the outflow from the basin. On a long-term basis, it should correspond to the ratio of the measured precipitation to the measured runoff. In fact, comparison of historical precipitation and runoff ratios provide a starting point for the runoff coefficient values. However, these ratios are not always easily obtained in view of the precipitation catch deficit which particularly affects snowfall and of inadequate precipitation data from mountain regions. At the start of the snowmelt season, losses are usually very small because they are limited to evaporation from the snow surface, especially at high elevations. In the next stage, when some soil becomes exposed and vegetation grows, more losses must be expected due to evapotranspiration and interception. Towards the end of the snowmelt season, direct channel flow from the remaining snowfields and glaciers may prevail in some basins which leads to a decrease of losses and to an increase of the runoff coefficient. In addition, c is usually different for snowmelt and for rainfall. The computer program accepts separate values for snow, c_S, and rain, c_R, and allows for half-monthly (and, if required, daily) changes of values in each elevation zone. Examples of seasonal trends of runoff coefficients are given in Figures 11.6 and 11.7, with the half-monthly values connected by straight lines.

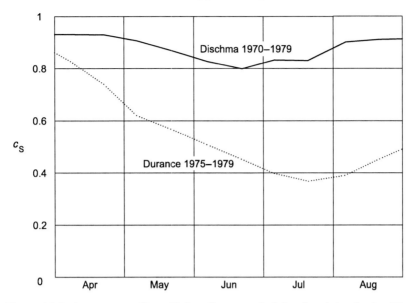

Figure 11.6 Average runoff coefficient for snow (c_S) for the alpine basins Dischma (43.3 km^2, 1668–3146 m a.s.l.) and Durance (2170 km^2, 786–4105 m a.s.l.) (Martinec and Rango, 1986).

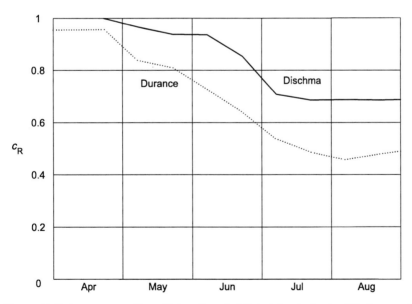

Figure 11.7 Average runoff coefficient for rainfall (c_R) for the basins Dischma and Durance (Martinec and Rango, 1986).

The runoff coefficients can even reach lower values in certain semiarid basins, particularly in the lowest elevation zone of such basins.

Of the SRM parameters, the runoff coefficient appears to be the primary candidate for adjustment if a runoff simulation is not at once successful.

Degree-day factor, a

The degree-day factor a (cm $°C^{-1}$ d^{-1}) converts the number of degree-days T ($°C$ d) into the daily snowmelt depth M (cm):

$$M = a \cdot T \tag{11.5}$$

Degree-day ratios can be evaluated by comparing degree-day values with the daily decrease of the snow water equivalent which is measured by radioactive snow gauge, snow pillow or a snow lysimeter. Such measurements (Martinec, 1960) have shown a considerable variability of degree-day ratios from day to day. This is understandable because the degree-day method does not take specifically into account other components of the energy balance, notably the solar radiation, wind speed, and the latent heat of condensation. If averaged for three to five days, however, the degree-day factor is more consistent and can represent the melting conditions. The effect of daily fluctuations of the degree-day values on the runoff from a basin as computed by SRM is greatly reduced because the daily meltwater input is superimposed on the more constant recession flow (Equation (11.1)).

The degree-day method requires several precautions:

(1) The degree-day factor is not a constant. It changes according to the changing snow properties during the snowmelt season.
(2) If point values are applied to areal computations, the degree-day values must be determined for the hypsometric mean elevation of the snow cover in question and not, for example, for the altitude of the snow line.
(3) If the snow cover is scattered, a correctly evaluated degree-day factor will produce less meltwater than if a 100 percent snow cover were assumed. A meltwater difference that arises from erroneous snow-cover information should not be compensated for by "optimizing" the degree-day factor. Instead, the correct areal extent of the snow cover should be determined and used.
(4) In large area extrapolations, point measurements should be weighted depending on how well a specific station represents the hydrological characteristics of a given zone (Shafer *et al.*, 1981).

In the absence of detailed data, the degree-day factor can be obtained from an empirical relation (Martinec, 1960):

$$a = 1.1 \cdot \frac{\rho_s}{\rho_w} \tag{11.6}$$

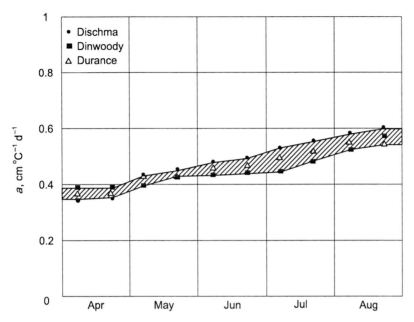

Figure 11.8 Average degree-day ratio (a) used in runoff simulations by the SRM model in the basins Dischma (10 years), Durance (five years) and Dinwoody (228 km^2; 1981–4202 m a.s.l., Wyoming, 2 years) (Martinec and Rango, 1986).

where:

a = the degree-day factor (cm $°C^{-1}d^{-1}$)
ρ_s = density of snow
ρ_w = density of water

When the snow density increases, the albedo decreases and the liquid water content in snow increases. Thus the snow density is an index of the changing properties which favor the snowmelt.

Figure 11.8 illustrates the seasonal trend of the degree-day factor in the Alps and in the Rocky Mountains. Because the geographic latitude of a basin influences the solar radiation, it may be advisable to adjust the degree-day factors accordingly. In glacierized basins, the degree-day factor usually exceeds 0.6 cm $°C^{-1}$ d^{-1} towards the end of the summer when ice becomes exposed (Kotlyakov and Krenke, 1982). The computer program accepts different degree-day factors for up to eight elevation zones which are usually changed twice a month (although daily changes are possible).

Sometimes the occurrence of a large, late-season snowfall will produce depressed a-values for several days due to the new low-density snow. The a-values in the model can be manually modified and inserted to reflect these unusual snowmelt conditions.

As is evident from Equation (11.1), the degree-day method could be readily replaced by a more refined computation of snowmelt without changing the structure of SRM. Such refinement appeared to be imperative in a study of outflow from a snow lysimeter (Martinec, 1989) but is not considered to be expedient for hydrological basins until the necessary additional variables and their forecasts become available. The degree-day method is explained in more detail in a separate publication (Rango and Martinec, 1995).

Temperature lapse rate, γ

If temperature stations at different altitudes are available, the lapse rate can be predetermined from historical data. Otherwise it must be evaluated by analogy from other basins or with regard to climatic conditions. In SRM simulations, a lapse rate of 0.65 °C per 100 m was usually employed. Slightly higher, seasonally changing values appeared to be a characteristic of tests in the Rocky Mountains. The computer program accepts either a single or a basin-wide lapse rate (option 0) or different rates for each zone (option 1). Seasonal variations can also be accommodated by inputting predetermined lapse rates every 15 days, and the lapse rate can be changed manually on selected days if a special meteorological situation (for example, a temperature inversion) requires a different value.

If the temperature station is situated near the mean elevation of the basin, possible errors in the lapse rate are to some extent canceled out because the extrapolation of temperature takes place upwards as well as downwards. If, on the other hand, the temperature station is at a low altitude, SRM becomes sensitive to the lapse rate. For some basins in Wyoming, for example, the closest temperature station was more than 100 km away and 2000 m lower than the highest snow-covered parts of the basins. In such a case, an error in the lapse rate of only 0.05 °C per 100 m causes a deviation of one degree-day from the correct degree-day value which corresponds to an error of about 0.5 cm of the computed daily meltwater depth late in the snowmelt season.

Such situations sometimes necessitate an adjustment of the originally selected lapse rate taking into account the course of the depletion curves of snow coverage. If high temperatures result from extrapolation by a certain lapse rate but no change in the snow areal extent is observed, then probably no appreciable snowmelt is taking place at this altitude. The high temperatures result from a false lapse rate which must be increased or decreased, depending on whether the temperature station is lower or higher than the mean zone elevation.

Critical temperature, T_{CRIT}

The critical temperature determines whether the measured or forecasted precipitation is rain or snow. Models which simulate the build-up of the snow cover first in order to simulate the runoff depend heavily on this parameter not only in the ablation period but particularly in the accumulation period. SRM needs the critical temperature only in order to decide whether precipitation immediately contributes to runoff (rain), or, if $T < T_{CRIT}$, whether snowfall took place. In this case, SRM automatically keeps the newly fallen snow in storage until it is melted on subsequent warm days, as explained in Section 11.3.2 under "Precipitation".

When SRM was applied in the alpine basin Dischma, T_{CRIT} started at $+3\ ^\circ$C in April at the beginning of snowmelt and diminished to $+0.75\ ^\circ$C in July. This seasonal trend with a narrower range appears to be applicable in other basins. At certain times, SRM may not take notice of a sharp rainfall runoff peak because the corresponding precipitation is determined to be snow, the extrapolated temperature being just slightly below the critical temperature. In such cases the assignment of critical-temperature and the temperature-lapse-rate values should be reviewed and logical adjustments made in order to change snow to rain. It is of course difficult to distinguish accurately between rain and snow because the temperature used is the daily mean while precipitation may occur at any time, day or night, i.e., in the warmer or colder portion of the daily temperature cycle.

As a possible refinement, formulas have been proposed (Higuchi *et al.*, 1982) to determine the proportion of rain and snow in mixed precipitation conditions.

Rainfall contributing area, RCA

When precipitation is determined to be rain, it can be treated in two ways. In the initial situation (option 0), it is assumed that rain falling on the snowpack early in the snowmelt season is retained by the snow which is usually dry and deep. Rainfall runoff is added to snowmelt runoff only from the snow-free area, that is to say the rainfall depth is reduced by the ratio snow-free area/zone area. At some later stage, the snow cover becomes ripe (the user must decide on which date) and the computer program should be switched to option 1. Now, if rain falls on this snow cover, it is assumed that the same amount of water is released from the snowpack so that rain from the entire zone area is added to snowmelt. The melting effect of rain is neglected because the additional heat supplied by the liquid precipitation is considered to be small (Wilson, 1941).

Recession coefficient, k

As is evident from Equation (11.1), the recession coefficient is an important feature of SRM since $(1 - k)$ is the proportion of the daily meltwater production which immediately appears in the runoff. Analysis of historical discharge data is usually

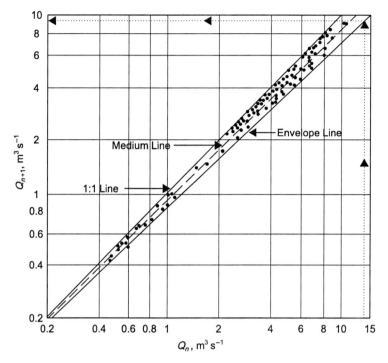

Figure 11.9 Recession flow plot Q_n vs Q_{n+1} for the Dischma basin in Switzerland. Either the solid envelope line or the dashed medium line is used to determine k-values for computing the constants x and y in Equation (11.7) (Martinec and Rango, 1986).

a good way to determine k. Figure 11.9 shows such an evaluation for the alpine basin Dischma (43.3 km², 1668–3146 m a.s.l.). Values of Q_n and Q_{n+1} are plotted against each other and the lower envelope line of all points is considered to indicate the k-values. Based on the relation $k = Q_{n+1}/Q_n$, it can be derived that for example $k_1 = 0.677$ for $Q_n = 14$ m³ s⁻¹ and $k_2 = 0.85$ for $Q_n = 1$ m³ s⁻¹. This means that k is not constant, but increases with the decreasing Q according to the equation:

$$k_{n+1} = x \cdot Q_n^{-y} \qquad (11.7)$$

where the constants x and y must be determined for a given basin by solving the equations:

$$k_1 = x \cdot Q_1^{-y}$$
$$k_2 = x \cdot Q_2^{-y}$$
$$\log k_1 = \log x - y \log Q_1 \qquad (11.8)$$
$$\log k_2 = \log x - y \log Q_2 \qquad (11.9)$$

In the given example,

$$\log 0.677 = \log x - y \log 14$$
$$\log 0.85 = \log x - y \log 1$$
$$x = 0.85$$
$$y = 0.086$$

As a formal change from the SRM User's Manual of 1983 (Martinec *et al.*, 1983), a negative sign appears in the exponent of Equation (11.7) so that the numerical values of x and y are positive.

The variability of k according to Equation (11.7) was also confirmed in other basins. This means that the recession does not exactly follow the usual equation

$$Q_n = Q_0 \cdot k^n \tag{11.10}$$

where:

$Q_0 = $ the initial discharge
$Q_n = $ the discharge after n days

but the following equation Jaccard (personal communication, 1982)

$$Q_n = x^{\frac{1}{y}} \left(\frac{Q_0}{x^{\frac{1}{y}}} \right)^{(1-y)^n} \tag{11.11}$$

where x and y are the constants of Equation (11.7).

The envelope line in Figure 11.9 and the resulting values of x and y must be determined for each basin. For ungauged basins and when historical discharge data are insufficient, x and y can be derived indirectly from the size of the basin as follows:

$$k_{Nn} = \left[x_M \left(\frac{\overline{Q}_M}{\overline{Q}_N} \cdot Q_{Nn} - 1 \right)^{-y_M} \right]^{\sqrt[4]{A_M/A_N}} \tag{11.12}$$

where: x_M, y_M are the known constants for the basin M; $\overline{Q}_M$, $\overline{Q}_N$ are average discharge values from the basin M and the new basin N; and A_M, A_N are the areas of the respective basins.

Equation (11.12) indicates that recession coefficients are generally higher in large basins than in small basins. If the increase in k with size appears to be too large, the exponent may be replaced by $\sqrt[8]{A_M/A_N}$. Even if the envelope line in Figure 11.9 can be reliably derived in a basin, it is possible that the resulting k-values may be too low to represent average conditions during the snowmelt season, especially in large basins. In such cases the SRM model will react too quickly to any change of the daily input. The simulated peaks would be too high and the simulated recession

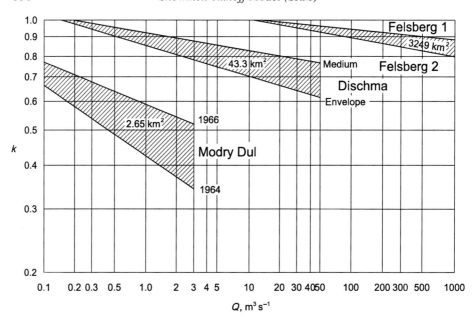

Figure 11.10 Range of recession coefficients, k, related to discharge Q resulting from various evaluations: In Dischma, the range results from using either the envelope line or the medium line in Figure 11.9. In Modry Dul, the relation varies in different years. In Felsberg, the relations (1) and (2) are derived from the size of the basin by two alternative formulas (Martinec and Rango, 1986).

too fast. A quick improvement is possible by deriving a new x and y, not from the envelope line, but from an intermediate or medium line between the envelope and the 1:1 line. This modification may especially be needed if the runoff simulation is extended to the whole year. Recession coefficients which may be right for the snowmelt period are usually too low for the winter months so that the simulated flows drop below the measured minimum values.

In very small basins, noticeable differences in the recession flow conditions and the k-values may occur from year to year. Figure 11.10 illustrates the range of k-values for varying basin size and for the mentioned alternatives of evaluation. For a very small mountain basin, the range includes the variations of k in different years (Martinec, 1970). For a larger alpine basin, the limits refer to the envelope line and to the medium line in Figure 11.9. For the largest basin, the higher limit (1) of the indicated range was derived by substituting the constants x = 0.085, y = 0.086 derived for the Dischma basin into Equation (11.12). The lower limit (2) was obtained by replacing $\sqrt[4]{A_M/A_N}$ in Equation (11.12) with $\sqrt[8]{A_M/A_N}$.

Figure 11.10 shows that k can theoretically exceed 1 for very small discharges in large basins. This does not really happen because such small discharges do not occur there. Such a situation could be produced, however, by the user inadvertently taking over the x and y values derived in a large basin and using them for a small basin without modification. In this case, if the daily snowmelt input exceeds the previous day's runoff, SRM computes a runoff decrease instead of increase. In order to avoid this error, the computer program prevents k from exceeding 0.99. However, it is advisable to avoid approaching this limiting situation by checking the x and y values with regard to the lowest flow to be expected. Recalling Equation (11.11), it follows for $n = \infty$, because $y > 0$ and $(1 - y) < 1$:

$$Q_\infty = x^{\frac{1}{y}} \left(\frac{Q_0}{x^{\frac{1}{y}}} \right)^{(<1)^\infty} = x^{\frac{1}{y}} \qquad (11.13)$$

Therefore the values x and y should fulfill the condition:

$$Q_{\text{Min}} > x^{\frac{1}{y}} \qquad (11.14)$$

where Q_{Min} = the minimum discharge in the given basin.

Adjustment of the recession coefficient for heavy rainfalls The equation (11.7) for computing the recession coefficient reflects the usual conditions characterizing the snowmelt runoff in the given basin. When a heavy rainfall occurs, the input is concentrated in a short time interval creating an abrupt rise and subsequent decline of the hydrograph. In order to simulate such events, the computer program automatically adjusts the recession coefficient whenever the daily rainfall averaged over the whole basin equals or exceeds 6 cm:

$$\text{If } P \text{ (rain)}_n \geq 6 \text{ cm} \longrightarrow \quad k_{n+1} = x(4 \, Q_n)^{-y}$$
$$k_{n+2} = x(4 \, Q_{n+1})^{-y}$$
$$k_{n+3} = x(4 \, Q_{n+2})^{-y}$$
$$k_{n+4} = x(4 \, Q_{n+3})^{-y}$$
$$k_{n+5} = x(4 \, Q_{n+4})^{-y} \qquad (11.15)$$

after which it returns to the normal equation (11.7). In this way, k gets lower so that the basin response to input becomes faster. If the precipitation is recognized by T_{CRIT} as snow and not rain, the mechanism will not be activated.

The threshold value can be changed in order to activate the rainfall peak program for smaller rainfall amounts or to delay activation until higher rainfall amounts are reached.

Time lag, L

The characteristic daily fluctuations of snowmelt runoff enable the time lag to be determined directly from the hydrographs of the past years. If, for example, the discharge starts rising each day around noon, it lags behind the rise of temperature by

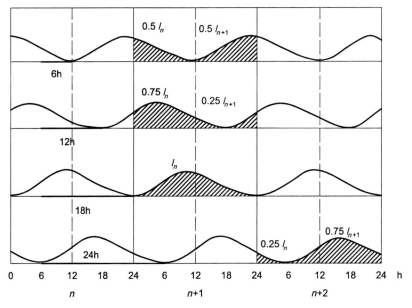

Figure 11.11 Snowmelt hydrographs illustrating the conversion of computed runoff amounts for 24-hour periods to calendar day periods. The various time lags (bold lines) are taken into account by proportions of the daily inputs, I (Martinec and Rango, 1986).

about six hours. Consequently, temperatures measured on the nth day correspond to discharge between 1200 hrs on the nth day and 1200 hrs on the $n + 1$ day. Discharge data, however, are normally published for midnight-to-midnight intervals and need adjustments in order to be compared with the simulated values. Conversely, the simulated values can be adjusted (Shafer *et al.*, 1981) to refer to the midnight-to-midnight periods. Figure 11.11 illustrates the procedure for different time lags. For $L = 6$ hour, 50% of input computed for temperature and precipitation on the nth day (I_n) plus 50% of I_{n+1} results in the $n + 1$ day's runoff after being processed by the SRM computer program:

$$L = 6\,\text{h} \quad 0.5 \cdot I_n + 0.5 \cdot I_{n+1} \rightarrow Q_{n+1} \tag{11.16}$$

Similarly:

$$L = 12\,\text{h} \quad 0.75 \cdot I_n + 0.25 \cdot I_{n+1} \rightarrow Q_{n+1} \tag{11.17}$$

$$L = 18\,\text{h} \quad I_n \qquad\qquad\qquad \rightarrow Q_{n+1} \tag{11.18}$$

$$L = 24\,\text{h} \quad 0.25 \cdot I_n + 0.75 \cdot I_{n+1} \rightarrow Q_{n+2} \tag{11.19}$$

This procedure is preferable, at least in mountain basins smaller than 5000 km^2, to evaluations of L by calculating the velocity of overland flow and channel flow. It

has been shown by environmental isotope tracer studies (Martinec, 1985) that overland flow is not a major part of the snowmelt runoff as previously believed. There
is increasing evidence that a major part of meltwater infiltrates and quickly stim-
ulates a corresponding outflow from the groundwater reservoir. With this runoff
concept in mind, the seemingly oversimplified treatment of the time lag in the
SRM model is better understood. If the hydrographs are not available or if their
shape is distorted by reservoir operations, the time lag can be estimated according to the basin size and by analogy with other comparable basins. Generally,
the time lag in a basin increases as the snow line retreats. In the WMO intercomparison test (WMO, 1986), most models calibrated the time lag. However,
these results appear to be of little help to determine the proper values. Contradictory time lags have been calibrated by different models. However, if the time
lags for all models participating in the WMO intercomparison test are averaged for
each basin, the resulting values support the expected relation between L and basin
size:

$$
\begin{aligned}
&\text{Basin W-3 } (8.42 \text{ km}^2) : 3.0 \text{ h} \\
&\text{Dischma } (43.3 \text{ km}^2) \quad : 7.2 \text{ h} \\
&\text{Dunajec } (680 \text{ km}^2) \quad\ : 10.5 \text{ h} \\
&\text{Durance } (2170 \text{ km}^2) \ : 12.4 \text{ h}
\end{aligned}
$$

If there is some uncertainty, L (percentages in Equations (11.16)–(11.19)) can be
adjusted in order to improve the synchronization of the simulated and measured
peaks of average daily flows. It should be noted that a similar effect results from
an adjustment of the recession coefficient.

11.4 Assessment of the model accuracy

Accuracy criteria

The SRM computer program includes a graphical display of the computed hydrograph and of the measured runoff. A visual inspection shows at the first glance
whether the simulation is successful or not. SRM additionally uses two well-
established accuracy criteria, namely, the coefficient of determination, R^2, and the
volume difference, D_v.

The coefficient of determination is computed as follows:

$$
R^2 = 1 - \frac{\sum\limits_{i=1}^{n}(Q_i - Q_i')^2}{\sum\limits_{i=1}^{n}(Q_i - \overline{Q})^2} \tag{11.20}
$$

where:

Q_i = the measured daily discharge
Q'_i = the computed daily discharge
$\overline{Q}$ = the average measured discharge of the given year or snowmelt season
n = the number of daily discharge values

Equation (11.20) also corresponds to the Nash–Sutcliffe coefficient in which case $\overline{Q}$ is a long-term average measured discharge applied to the respective years or seasons.

The deviation of the runoff volumes, D_v, is computed as follows:

$$D_v[\%] = \frac{V_R - V'_R}{V_R} \cdot 100 \qquad (11.21)$$

where:

V_R is the measured yearly or seasonal runoff volume
V'_R is the computed yearly or seasonal runoff volume

Numerical accuracy criteria are never perfect, as illustrated by Figure 11.12. From the visual judgment both simulations look good because the fundamental

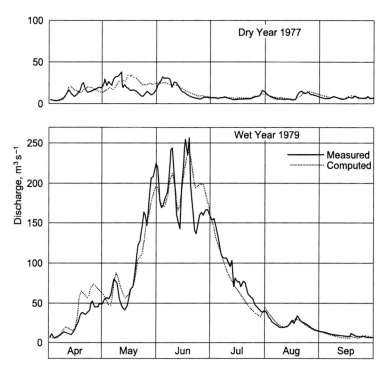

Figure 11.12 Runoff simulations in the basin of the Rio Grande near Del Norte, Colorado (3419 km^2, 2432–4215 m a.s.l.) (Martinec and Rango, 1989).

difference between two extreme years is well reproduced. However, $R^2 = 0.95$ in 1979 while it amounts only to 0.48 in 1977. In spite of this unfavorable value, the simulation (or forecast) in 1977 would certainly be useful for water management because it correctly reveals an extremely low runoff.

With $\overline{Q}$ as a long-term average substituted into Equation (11.20) (Nash–Sutcliffe) instead of the average for the specific year, a much more favorable but deceptive value for R^2 of 0.97 results for the year 1977. This value is high because the long-term $\overline{Q}$ is much higher than any Q in 1977. Consequently, the daily deviations of the simulated runoff become relatively insignificant although in absolute terms they are not negligible. In addition to these criteria which are automatically computed and displayed after each model run, the coefficient of gain from daily means, DG, could be computed by the user as follows:

$$
DG = 1 - \frac{\sum\limits_{i=1}^{n}(Q_i - Q_i')^2}{\sum\limits_{i=1}^{n}(Q_i - \overline{Q}_i)^2}
\tag{11.22}
$$

where:

$Q_i = $ the measured daily discharge
$Q_i' = $ the computed daily discharge
$\overline{Q}_i = $ the average measured discharge from the past years for each day of the period
$n = $ the number of days

Thus, R^2 compares the performance of a model with "no model" (average discharge) and DG with a "seasonal model" (long-term average runoff pattern). Negative values signal that the model performed worse than "no model" or worse than the "seasonal model."

Model accuracy outside the snowmelt season

SRM has been designed to compute runoff during the snowmelt season, but it can be run for the whole year if required. According to the mentioned WMO intercomparison test (WMO, 1986), about the same accuracy as for the snowmelt season can be achieved for the entire year in mountain basins with a low winter runoff. An example in Figure 11.13 shows that SRM can even be run without updating for ten years.

Elimination of possible errors

It is not possible to give threshold values of accuracy criteria which would determine whether a model run is successful or whether something must be changed. With good data, a value like $R^2 = 0.80$ might still be improved. In unfavorable conditions, with incomplete data, a user may be satisfied even with lower R^2 values.

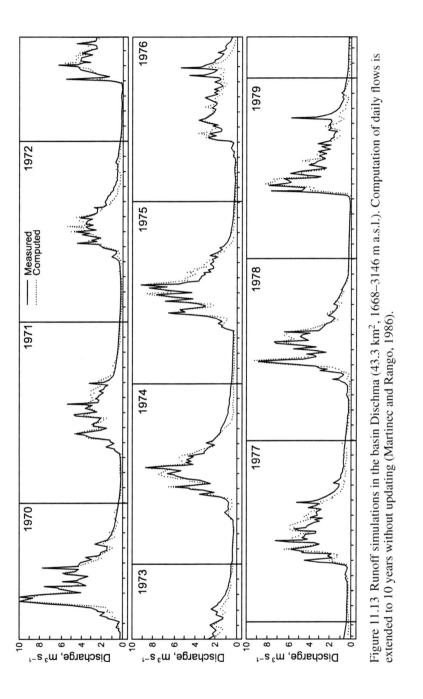

Figure 11.13 Runoff simulations in the basin Dischma (43.3 km², 1668–3146 m a.s.l.). Computation of daily flows is extended to 10 years without updating (Martinec and Rango, 1986).

Sometimes, however, the graphical display as well as the numerical criteria indicate that something went definitely wrong. Before adjusting one or the other parameter, it is recommended to check several probable sources of error first. Examples in Table 11.4 are based on the actual experience of various users.

11.5 Operation of the model for real-time forecasts

In order to be applied for real-time discharge forecasts, a model should be able to simulate the runoff not only in selected test basins with good data or in specially equipped experimental basins where a particular calibration model was developed, but also in basins where such forecasts are required by the user. SRM has relatively modest requirements for input variables (temperature, precipitation, and snow-covered area) and therefore it was easily possible to shift the runoff simulations to the basins delivering water for hydroelectric schemes, as required by an electric company. An example of such simulations is shown in Figure 11.14.

As mentioned in Section 11.1, SRM can be used for short-term (for example weekly) forecasts of daily flows as well as for longer time period forecasts such as monthly runoff volumes or seasonal runoff volumes. For short-term forecasts, temperature, precipitation, and snow-covered area must be forecasted or predetermined for the coming days and substituted into the model. Temperature and even precipitation forecasts are becoming increasingly available from meteorological services, but the snow-covered areas must be extrapolated by the model user.

Extrapolation of the snow coverage

The future course of the depletion curves of the snow coverage can be evaluated from the so-called modified depletion curves (MDC). These curves are automatically derived by SRM from the conventional curves (CDC) by replacing the time scale with cumulative daily snowmelt depths as computed by the model. Consequently, if SRM is run in a whole hydrological year, the derivation of MDC from CDC starts with the summer half year and not earlier. The decline of the modified depletion curves depends on the initial accumulation of snow and not on the climatic conditions, as is the case with the conventional depletion curve.

Procedure for weekly forecasts of daily flows

Assumption: A family of modified depletion curves has been derived from the past snow-cover monitoring and temperature measurements in the given basin. Two of these curves representing the initial water equivalents $H_W = 20$ cm and $H_W = 60$ cm are plotted in Figure 11.15.

Table 11.4 *Errors experienced by SRM users and their correction*

	Error	Cause
1	Runoff simulation went too high (Dinwoody, 1976; Dischma, 1977).	Snow-cover depletion curves distorted by summer snowfall: Eliminate satellite observation after a snowfall event and redraw the depletion curves.
2	Runoff simulation suddenly deteriorated (Dischma).	Input of snow-coverage data S shifted by 1 month: Correct the S-input.
3	The simulated runoff hydrograph declined uncontrollably in September (Illecillewaet basin).	Input of snow-coverage data broken off by the end of August: Complete the S-input.
4	Runoff simulation far below the measured runoff (Illecillewaet).	Decimal point in the measured precipitation and discharge data shifted and thereby increasing the measured runoff 10×, but the simulated runoff only slightly: Correct the decimal point.
5	After the start of snowmelt, the simulated runoff kept decreasing in spite of snowmelt input (small tributary basins of upper Rhine, 1985).	Values x, y for the recession coefficient formula (Equation (11.7)) were taken over from a much larger basin so that k exceeded 1.0 for low Q: Correct x, y with regard to the basin size. Possibility of this error was eliminated in the computer program Version 2.01 and later versions (3.0, 3.1, 3.11, and 4.0).
6	Frequent deviations from the measured runoff, periodical lows of the hydrograph not simulated (Felsberg, 1982).	SRM simulates natural runoff while the measured runoff was influenced by storage (on weekends) and release from artificial reservoirs for hydropower. These interventions must be corrected in order to compare simulated and measured runoff.
7	Rainfall peaks inadequately simulated (Illecillewaet).	Rainfall input is concentrated to shorter periods than snowmelt input which accelerates the basin response. Program Versions 2.01 and later versions take this feature automatically into account whenever rainfall exceeds a preselected threshold.
8	Rainfall peak inadequately simulated even with the special rainfall peak program (small tributary basins of upper Rhine August 25, 1985).	Temperature extrapolated from station 800 m a.s.l. to zone D 2380 m a.s.l. by 0.65 °C/100 m → T = + 0.43 °C while T_{CRIT} is + 0.75 °C. By decreasing T_{CRIT}, snowfall was converted to rainfall and the runoff peak was better simulated.

9	Rainfall peak inadequately simulated because the daily amount was below the special program automatic threshold. Threshold was lowered but the peak simulation deteriorated further (small tributary basins of upper Rhine, 1985).	Only one precipitation station was available but the "by zone" option was switched on. Consequently precipitation input took place only in one zone. With one station, select option "basin wide."
10	Runoff simulation went very high in a basin with a large elevation range of 7400 m (Kabul River, 1976).	Compact snow cover was assumed above the snow line so that the snow-covered area was exaggerated. Temperature was extrapolated by too low a lapse rate. Snowmelt was not computed separately for elevation zones but for the entire basin. This usually leads to overestimation of meltwater input. Simulation was improved by re-evaluating the snow coverage and selecting a higher temperature lapse rate corresponding to the climate. With regard to the large elevation range, the basin should have been divided in several elevation zones.
11	Distorted runoff simulation (outflow from a snow lysimeter).	Values of recession coefficient false: Negative y values used with the equation $k = x \cdot Q^{-y}$ so that the exponent became positive. Consequently, k increased with the increasing Q while it should have decreased. Use equation $k = x \cdot Q^{-y}$ and positive values of y, so that the exponent is always negative.
12	Difficulties with the timing of rainfall-runoff peaks (WMO test for simulated operational forecasts, Illecillewaet basin).	Precipitation data from one of the stations were ascribed to the date on which it was measured at 0800. Data had to be shifted one day backwards. See also Section 5.2.2.
13	Forecasted runoff for hydroelectric stations Sedrun and Tavanasa too low in April–June 1994. In one run, snow cover was not completely melted by the end of September.	Snowmelt was computed by smooth long-term average temperature thus decreasing the number of degree days. Reintroduce the daily fluctuations of temperatures.
14	Discrepancies in runoff simulations for Rio Grande at Del Norte.	Erratic precipitation data. Automatic extrapolation by altitude gradients from two stations resulted in negative precipitation amounts in the highest zone when gradient was negative. Extrapolate averaged precipitation data by a uniform gradient as recommended in Section 5.2. Always visually inspect input precipitation data.
15	Discrepancies in runoff simulations for Rio Grande at Del Norte	Occasionally missing temperature data were interpreted by the computer program as $0 = 0\,°C$. Inspect temperature data and complete missing values by interpolation.

(*cont.*)

Table 11.4 (cont.)

	Error	Cause
16	Effect of climate change (see Section 8.3): Decrease of runoff computed by an increased temperature (Rio Grande at Del Norte)	The "winter deficit" was exceeded on a warm day with a high snowmelt depth, so that too much was cut off from MDC_{EXCL}. Consequently CDC_{CLIM} was shifted too much and less runoff resulted. Cut off MDC_{EXCL} by the value of the previous day ("nearly equalled").
17	Overestimation of runoff in the Rhine-Felsberg basin 1–15 May 1993 with usual temperature lapse rate of 0.65 °C/100 m.	Frequent föhn-wind in this period, temperature differences between Weissfluhjoch (2693 m a.s.l.) and Davos (1560 m a.s.l.) correspond to a lapse rate as high as 0.95 °C/100 m. The value of 0.8 °C/100 m was used.
18	Evaluation of the climate effect in the Kings River basin 1973: For a temperature increase +4 °C, snow cover on April 1 disappears in the C zone (1700–2300 m.a.s.l.) but still exists in the lower B zone (1100–1700 m a.s.l.)	The decrease of the snow water equivalent on April 1 due to warming is evaluated from uniform amounts of precipitation and snowmelt depths in winter in each zone (Figures 11.20, 11.21). The snow in the B zone was completely melted for T °C on April 1 so that no winter deficit resulted for $T + 4$°. However, in reality, some snow survived for T °C due to variable snow depths as seen by satellites ($S = 27\%$ on April 1). With no computed snow for T °C as well as for $T + 4$ °C, no winter deficit resulted and therefore the existing snow cover was falsely preserved for $T + 4$ °C. A manual correction (elimination of CDC_{CLIM}) was necessary. In the C zone there was enough snow to avoid the error from uniform, instead of variable, snow depths in winter, so that the snow cover duly disappeared as a result of the evaluated winter deficit. This error can occur only with very little snow in a zone on April 1 and can be readily corrected. Whenever satellites see some snow on April 1 and the winter accumulation of snow from precipitation and snowmelt is zero, eliminate CDC_{CLIM}.
19	Failure of runoff simulations in the basin Pskem: the derived constant y in the recession formula (Equation 11.7) is negative, so that the recession is steeper for low flows than for high flows. The envelope line in Figure 11.10 was derived taking into account evidently false points in the low range of flows. Such points may occur, for example, when the river flow is artificially stopped by freezing.	The recession of the natural runoff slows down with a decreasing discharge. To this effect, y must be > 0. In a log–log representation (Figure 11.10), the points outside the range between the 1:1 line and a parallel line for points in the range of high flows (below the dashed line in Figure 11.10) must be disregarded. With this measure, the problem of the runoff simulation was eliminated.

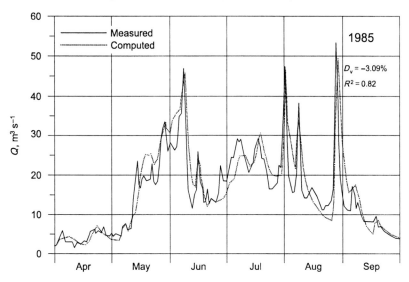

Figure 11.14 Runoff simulation in the catchment area of the hydroelectric station Tavanasa (Swiss Alps, 215 km², 1277–3210 m a.s.l.) (Baumann *et al.*, 1990, courtesy K. Seidel).

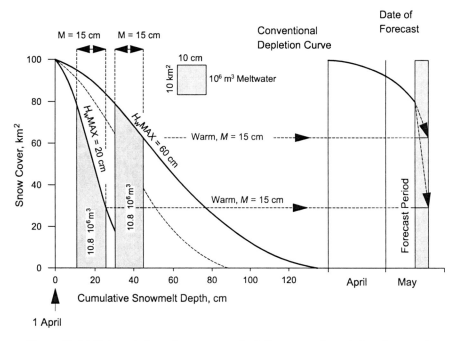

Figure 11.15 Extrapolation of snow-cover depletion curves in real-time from modified depletion curves with the use of temperature forecasts. $H_{W\ MAX}$ is the water equivalent of the snow cover at the beginning of the snowmelt season (courtesy D. Hall and J. Martinec).

Example 1 – Snow accumulation in the basin unknown, snow coverage measured by Landsat on May 15, $S = 80\%$, cumulative snowmelt depth (from degree-days and degree-day ratios) to date: 30 cm. Temperature forecast: 30 degree-days for the next week, converted to meltwater depth $M = 15$ cm by a degree-day ratio $a = 0.5$ cm $°C^{-1}d^{-1}$, $S = 80\%$ and $\Sigma M = 30$ cm indicate that the curve for $H_W = 60$ cm is applicable. The snow coverage will drop to 64% in seven days. Extrapolated conventional depletion curve indicates values for day-to-day discharge computations.

Example 2 – As above, but the cumulative snowmelt depth to date is only 10 cm. Consequently, the curve for $H_W = 20$ cm is applicable and the snow coverage will drop to 33% in seven days, which leads to a different extrapolation of the conventional depletion curve and to a different weekly total of forecasted daily runoff volumes. If the initial water equivalent is known, for example from SNOTEL (a system of data transmission using meteor paths for reflecting the signals and operated in the USA) or from manual snow surveys taken near the beginning of the forecast period, the appropriate modified depletion curve can be selected at the start of the snowmelt season. Otherwise, the average curve (dotted line in Figure 11.15) is used until the correct curve can be identified by satellite data.

Figure 11.16 shows that it is possible to derive a nomograph of modified depletion curves (Rango and van Katwijk, 1990) for a given basin from the past years. As

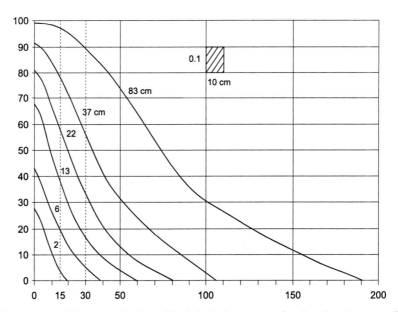

Figure 11.16 Nomograph of modified depletion curves for the elevation zone B (1284 km², 2926–3353 m a.s.l.) of the Rio Grande basin. The curves are labelled with the areal average water equivalent (April 1) of the snow cover which they represent (Rango and van Katwijk, 1990).

noticed in Figure 11.15, the area below a modified depletion curve indicates the water volume stored in the snow cover if the *y*-axis scale is in km^2. If the *y*-axis scale is in percent snow coverage, it indicates the water equivalent of the snow cover as an areal average. Therefore each curve in Figure 11.16 can be labelled by the water equivalent which it indicates. The rectangular shaded area means 0.1×10 cm $=$ 1 cm. Because the area below the highest curve is 83 times larger, this curve indicates that at the beginning of computations of the cumulative snowmelt depth (usually on April 1), the snow accumulation corresponded to the water equivalent of 83 cm. The values for each curve are automatically determined by the computer program.

The nomograph is used for real-time forecasts as follows: In a current year, the snow-covered area is monitored from the start of the snowmelt season and simultaneously the cumulative snowmelt depth is computed. The snow-covered area must be evaluated as quickly as possible after each satellite overflight. The degree-days necessary for melting the temporary snow cover from intermittent snowfalls are disregarded. If, after some time, for example on May 15, the snow coverage amounts to 80% and the cumulative snowmelt depth amounts to 15 cm, the modified depletion curve labeled by 37 cm is identified to be valid for that year. This curve can be used for extrapolating the snow-covered area. For example, if another 15 cm will be melted in the next week according to temperature forecasts (total 30 cm), the snow coverage will drop to 55%. The snow-covered areas thus extrapolated are used for real-time forecasts of daily flows. The modified curve also indicates the water equivalent of snow (37 cm) at the start of the snowmelt season for seasonal runoff forecasts. If, in another future year, the cumulative computed snowmelt depth coincides for example with the snow coverage of only 36%, the curve labeled 13 cm is valid. The appropriate curve can thus be identified but with a certain time delay. If the initial water equivalent of the snow cover can be evaluated from point measurements, the proper curve can be selected at the start of the snowmelt season with no time delay.

The computer program also provides an option for plotting a modified depletion curve in which the totalized melt depth includes new snow that falls occasionally during the snowmelt period. While the new snow-excluded MDC is used to evaluate the water equivalent of the seasonal snow cover at the start of the snowmelt period, the new snow-included MDC can be used to evaluate the shifting of the conventional depletion curves by changed temperatures, as will be explained in Section 11.6. The depletion curves of snow-covered areas are dealt with in more detail in an earlier publication (Hall and Martinec, 1985).

Figure 11.17 shows a simulated runoff forecast for the Rio Grande basin in which the forecasted temperatures were replaced by seasonal average temperatures, the precipitation was 110% of the average precipitation randomly distributed over each month, and the snow-covered area was forecasted by using temperatures and the appropriate modified depletion curve. Evidently the seasonal average temperatures

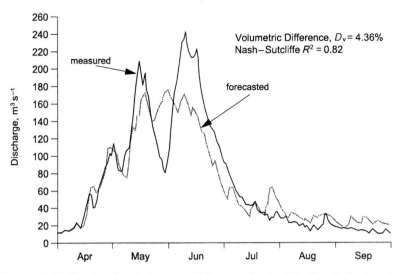

Figure 11.17 Simulated real-time runoff forecast for the Rio Grande basin using long-term average temperature instead of actual temperatures for the year 1983 (Rango and van Katwijk, 1990).

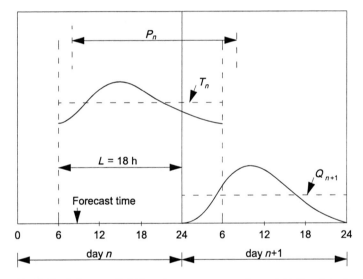

Figure 11.18 Real-time availability of temperature and precipitation data for short-term runoff forecasts in contrast to runoff simulation.

do not show the cold spell in the second half of May 1987 and therefore the runoff decline is not simulated.

The difference between the runoff simulation and short-term forecast is illustrated by Figure 11.18. The temperature T_n, precipitation P_n, and snow-covered area S_n are used to compute Q_{n+1} with $L = 18$ h. At the time of simulation, these values are known. When Q_{n+1} is forecasted in the morning of the day n, T_n and P_n are not yet known and forecasted values must be used. In order to forecast further ahead (Q_{n+2}, Q_{n+3}), the forecasted values T_{n+1} and P_{n+1}, T_{n+2} and P_{n+2} are used. The snow-covered areas S_n, S_{n+1}, S_{n+2} are extrapolated by using temperature forecasts and the modified depletion curve MDC.

For other lag times, SRM automatically combines Q in the appropriate proportions of two subsequent inputs, as explained in Section 11.3.3. For example, if $L = 24$ h, the input from T_{n-1} and P_{n-1} (which might be already known at the time of the forecast) is represented by 25% and the input from T_n and P_n (forecasted values) by 75%.

In the absence of temperature and precipitation forecasts, runoff forecasts can be issued on condition, for example, that long-term average values or extreme values (maxima, minima) will occur. It is also possible to use fictitious values as will be shown in the section dealing with climate change.

The feasibility of real-time forecasts was demonstrated for two hydroelectric stations in the Swiss Alps (Brüsch, 1996). With the use of snow-cover monitoring by Landsat as well as of temperature and precipitation forecasts from the Swiss Meteorological Office, the daily runoff was forecasted always for four subsequent days. The runoff volume from April through September was forecasted and updated with the use of modified depletion curves (Martinec and Rango, 1995).

Updating

The model performance in the forecasting mode is naturally affected by the reduced accuracy and reliability of temperature and precipitation forecasts. The propagation of errors can be avoided by periodical updating. In earlier versions of the computer program (starting with Versions 3.11 and 3.2 and continuing to the present), the computed discharge can be replaced every one to nine days by the measured discharge which becomes known for the corresponding day so that each subsequent forecast period is computed by using a correct discharge value.

Even without this updating, SRM prevents persistent large errors by a built-in self-adjusting feature which is efficient if Equation (11.7) is carefully assessed: Figure 11.19(a) shows a model runoff simulation starting with computed discharge of only one half of the correct value. Updating by actual discharge improves the simulation as shown in Figure 11.19(b). Even without updating, however, the initial discrepancy is soon eliminated automatically.

Further possibilities of updating will be made available to users when more experience in real-time situations is accumulated. For example, it should be possible to adjust some parameters (e.g., the runoff coefficient) in the progress of the forecast, but only within hydrologically and physically acceptable limits. In any case, false

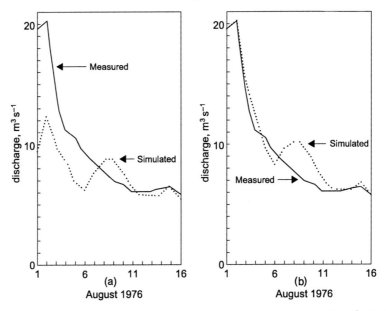

Figure 11.19 Discharge simulation in the Dinwoody Creek basin (228 km², 1981–
4202 m a.s.l.) in Wyoming, (a) without updating, and (b) with updating by actual
discharge on August 1.

forecasts of temperature and precipitation should be updated whenever a correction
by new data is indicated.

What is generally called "updating" can be thus divided into three categories:

(1) Updating the computed discharge by the measured discharge when it becomes known,
   i.e., checking with the measured discharge to avoid carry-over of errors when the next
   forecast is issued.
(2) Adjustment of model parameters in the process of forecast.
(3) Correction of temperature, precipitation, and snow-cover forecasts according to actual
   observations.

Short-term discharge forecasts can be updated as frequently as each day
(Baumann *et al.*, 1990).

11.6 Year-round runoff simulation for a changed climate

SRM uses a real snow cover from satellite monitoring in the present climate in
order to produce a snow cover and runoff in a changed climate. This requires
a rather sophisticated procedure, but uncertainties arising from a fictitious snow
cover simulated from precipitation and arbitrary threshold temperatures are avoided.
In any event, the SRM program finishes this task, including printout of figures,

tables, numerical results and hydrographs, within minutes. Another advantage is the independence from calibration, enabling the model parameters to be meaningfully shifted in time or adjusted if so indicated by climate scenarios.

If SRM is used only to predict snow-covered areas and regional snow accumulation in a changed climate, for example, in mountain areas which are not hydrological basins, the computer program requires the following data:

- number of elevation zones, their areas, and hypsometric mean elevations
- current snow-covered areas, daily values (S)
- daily average or Max/Min temperatures (T)
- precipitation, daily (P)
- degree-day factor (a)
- temperature lapse rate (γ)
- critical temperature (T_{CRIT})

For runoff computation, the remaining SRM parameters are required: c_S, c_R, RCA, k, L and the initial Q.

So far, SRM has usually been applied to simulate or forecast runoff during the snowmelt season. It was only run year-round for international tests of model performance (WMO, 1986). However, climate change, as a new field of application, requires SRM to be run during the whole hydrological year. Therefore, the problem of winter runs are dealt with in the next section.

11.6.1 Snowmelt-runoff computation in the winter half year

In the winter half year (usually October–March in the Northern hemisphere), the evaluation of the snow coverage is more difficult than during the snowmelt season. Satellite data, if available, are not frequent enough to distinguish the stable snow cover from frequent transitory snowfalls which are subsequently melted. An assumption of a stable snow cover between two available satellite measurements may lead to an overestimation of the snow coverage.

Option 1: Put $S = 0$ so that each precipitation event recognized by T_{CRIT} as snow is automatically stored and subsequently melted over the whole area $(1 - S = 1)$. This method is applicable only if precipitation data are good enough to represent the input or if they can be adequately adjusted. This is in many cases not possible due to the well-known catch deficit of precipitation gauges and due to the lack of measurements in the high elevations of mountain regions. It is for this reason that SRM uses the snow-covered area whenever possible for computing the runoff input.

Option 2: Assume a stable snow cover $(S = 1$ or a little less in a rugged terrain) in January and February, for example, and $S = 0$ in October–December. If $S = 1$

on April 1 from satellite data, assume $S = 1$ in March as well. If it is less than 1 on April 1, the snow coverage in March is put to 0 or, which may be more accurate, it is interpolated from 1 on March 1 to the S-value on April 1. Naturally, $S = 1$ can be assumed for a longer part of the winter in higher elevation zones while in the lowest zone option 1 may be preferable. The present program keeps the snow coverage estimated for the present climate unchanged for climate runs. It is therefore recommended to assume a complete snow coverage only in months in which it is expected to last in a warmer climate as well.

Whenever $S = 1$ is introduced, the SRM program cancels any existing storage of preceding temporary snowfalls because such snow becomes a part of the seasonal snow cover. Snow storage is also automatically canceled on April 1 because all existing snow is then accounted for by the depletion curve of the snow coverage. With the variables thus accounted for, SRM is run as usual. Because the estimated snow coverage is less accurate than CDC in the summer, the water equivalent of new snow is reduced at once by the simultaneous $(1 - S)$ to prevent an inadequate S from influencing the computation on a later melt day. This deviates from the summer procedure explained in Section 11.3.3, Table 11.2.

The model parameters should be adapted to winter conditions. In particular, the constants x, y for the recession formula (Equation (11.7)) derived usually for summer conditions sometimes allow the discharge to decrease too low so that a slower recession formula is indicated. By using the equation

$$Q_{Min} > x^{\frac{1}{y}} \tag{11.14}$$

x, y can be adjusted in order to prevent the discharge from sinking below a selected level after a long recession period.

In view of frequent snowfalls, values of the degree-day factor lower than those used in the summer are recommended. Values of c_S and c_R higher than in summer can be expected. The main purpose of the winter runoff simulation is the evaluation of the runoff redistribution in the winter and summer half years. Consequently it is more important to compute the winter runoff volume as accurately as possible than to try to improve the daily accuracy (R^2).

Change of snow accumulation in the new climate

As the first step, the effect of a climate change on snow-covered areas and runoff was evaluated in the summer half year only, assuming an unchanged initial snow cover on April 1 (Martinec *et al.*, 1994; 1998). For a year-round temperature increase, the seasonal snow cover on April 1 is deprived of a certain snow water equivalent by additional snowmelt and by a conversion of some precipitation events from snow to rain in October through March. This decrease of the snow water equivalent is computed by rewriting the input part of the SRM formula as follows:

$$\Delta H_W = \sum_{n=1}^{182} [a_n \cdot T_n \cdot S_n + a_n \cdot T_n(1 - S_n) + P_{Rn}]$$

$$- \sum_{n=1}^{182} [a_n \cdot T_n' \cdot S_n' + a_n \cdot T_n'(1 - S_n') + P_{Rn}'] \qquad (11.23)$$

where:

ΔH_W = difference between the present and future areal water equivalent of the snow cover on April 1 (cm).

a = degree-day factor (cm $°C^{-1}$ d^{-1})

T = temperature in the present climate at mean hypsometric elevation, as degree-days (°C d)

T' = temperature in a warmer climate as degree-days (°C d)

S = ratio of snow-covered area to total area, present climate

S' = ratio of snow-covered area to total area, warmer climate

P_R = rain according to T_{CRIT}, present climate

P'_R = rain according to T_{CRIT}, warmer climate

182 = number of days October through March

Equation (11.23) thus summarizes the SRM input to runoff which consists of snowmelt from the stable snow cover (S), melting of snow which temporarily covers the snow-free area ($1 - S$) and rain. The distinction between a stable and temporary snow cover during the snow accumulation in winter is rather arbitrary due to insufficient satellite monitoring, but the total of both snowmelt inputs always equals 100% of the occurring snowmelt M:

$$M \cdot S + M(1 - S) = M(S + 1 - S) = M \qquad (11.24)$$

In March, however, if S is put to 0 while there is a stable snow cover, and there happens to be little snowfall, the snowmelt input may be underestimated.

Figure 11.20 illustrates the areal water equivalent of the snow cover on April 1 as a difference between the winter precipitation and the winter input to runoff. This water equivalent can be also computed by accumulating the daily zonal melt depths. In this hypothetical example there is an agreement of the water equivalent determined either way. In natural conditions, discrepancies are to be expected mainly due to difficulties in evaluating the areal precipitation value for mountainous regions. In such cases, it is recommended that the value from the accumulated zonal melt be considered as more reliable. The accumulated zonal melt value might even be used to correct the winter precipitation data and to estimate the altitude precipitation

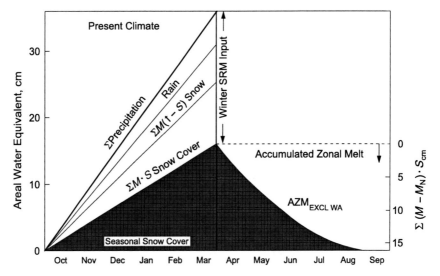

Figure 11.20 Illustration of the snow accumulation in the winter and snowmelt in the summer in the present climate (hypothetical example).

gradient. Another advantage of this method is that it takes into account a possible redeposition of snow by wind during the accumulation season.

On the other hand, no losses are normally considered (losses indicated by the runoff coefficient are assumed to take place after meltwater has left the snow cover). This may lead to underestimation of the retrospectively computed water equivalents if significant evaporation from the snow surface takes place. However, if degree-day ratios are used which have been derived from lysimeter measurements under similar evaporation conditions, this error is eliminated.

In a warmer climate, the winter input to runoff increases, as shown in Figure 11.21, so that there is less snow on April 1 if winter precipitation remains unchanged. A combined effect of a warmer climate and increased winter precipitation can result in rare cases (low temperatures in high mountains with little or no effect of the temperature increase) in an increased snow accumulation.

Runoff simulation for scenarios of the future climate

In order to evaluate the effect of a warmer climate on runoff in mountain basins, the SRM program uses the real seasonal snow cover of the present as monitored by satellites and models a climate-affected seasonal snow cover.

The future snow conditions in terms of snow-covered areas and areal water equivalents in different elevation zones constitute useful information for planning water management and winter tourism. Using snow-coverage data as model input, the climate-affected runoff in the whole hydrological year is computed. It allows

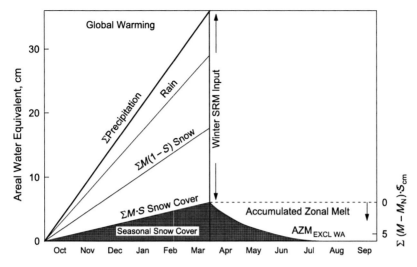

Figure 11.21 Illustration of the snow accumulation in the winter and snowmelt in the summer in a warmer climate (hypothetical example).

examination of the changes of daily runoff peaks and the redistribution of runoff volumes in the winter and summer half years.

The procedure is illustrated by evaluating the effect of a temperature increase of +4 °C on runoff in the Rio Grande basin at Del Norte (3419 km², 2432–4215 m a.s.l., elevation zones A, B and C) in the hydrological year 1979 (Figure 11.22). The following symbols are used:

CDC = conventional depletion curve of snow-covered area interpolated from periodical snow-cover mapping.

MDC$_{INCL}$ = modified depletion curve of snow covered area with new snow included. This curve is derived from CDC by relating the snow coverage to the accumulated computed snowmelt depth. It indicates how much snow, including new snow falling during the snowmelt period, must be melted (in terms of computed snowmelt depth) in order to decrease the snow-covered area to a certain proportion of the total area and ultimately to zero. The shape of this curve depends on the initial water equivalent of the snow and on the amount of new snow.

MDC$_{EXCL}$ = modified depletion curve of snow-covered area with new snow excluded. This curve is derived from MDC$_{INCL}$ by deducting the melt depths of new snow from the accumulated snowmelt depth. The shape of this curve depends on the initial water equivalent of the snow cover and is independent of subsequent snowfalls. The area below this curve indicates the areal water equivalent of the initial snow cover.

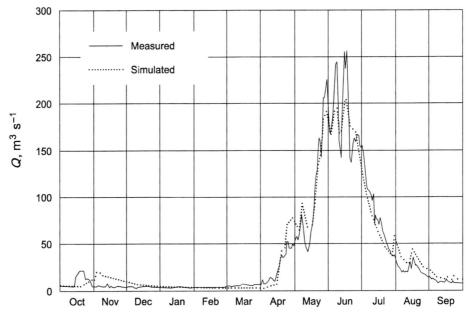

Figure 11.22 Measured and simulated runoff in the Rio Grande basin near Del Norte in the hydrological year 1979.

AZM_{INCL} = accumulated zonal melt with new snow included. This curve accumulates daily computed snowmelt depths multiplied by the respective snow coverage (as decimal number) and shows the totals on a time scale.

AZM_{EXCL} = accumulated zonal melt with new snow excluded. This curve is derived from AZM_{INCL} by deducting the zonal melt of new snow from the accumulated zonal melt. Again it relates the successive totals to time. The final total is the areal water equivalent of the initial snow cover (as also indicated by MDC_{EXCL}).

MDC_{CLIM} = modified depletion curve of snow covered area for a changed climate. This curve takes into account the amount of snowfalls changed by the new climate. If there is no change, it is identical with MDC_{INCL}.

CDC_{CLIM} = conventional depletion curve of snow-covered area in a changed climate.

$CDC_{CLIM\ MA}$ = conventional depletion curve of snow-covered area in a changed climate (derived from MDC_{INCL}) adjusted for the input to SRM runoff computation (model adjusted). It appears in publications disregarding the winter effect of a changed climate.

$MDC_{EXCL\ WA}$ = winter adjusted curve. The effect of a warmer winter is taken into account by decreasing the curve according to the "winter deficit." With a simultaneous increase of winter precipitation a positive balance of the winter snow accumulation may result in which case the curve is increased. The area below this curve indicates the areal water equivalent of the initial snow cover in a changed climate.

$AZM_{EXCL\ WA}$ = winter adjusted curve. The effect of a warmer winter and, if necessary, of a changed precipitation is taken into account. The final total is the water equivalent of the initial snow cover in a changed climate.

$MDC_{CLIM\ WA}$ = winter adjusted curve. It is derived from $MDC_{EXCL\ WA}$ by taking into account snowfalls "surviving" in the new climate.

$CDC_{CLIM\ WA}$ = conventional depletion curve of snow-covered area in a changed climate. The curve is derived from $MDC_{CLIM\ WA}$.

$CDC_{CLIM\ WA,\ MA}$ = curve adjusted for the model input (model adjusted). If the derivation from $MDC_{CLIM\ WA}$ results in more than one S-value on a date, the first (highest) value is used. If there is no new S-value on a date, the previous day's S-value is repeated until there is a new S-value. With this adjustment $CDC_{CLIM\ WA,\ MA}$ can be used as input to SRM which requires one S-value on each day, like provided by the original CDC. This curve is used to compute the year-round climate-affected runoff.

Due to the time shift by the changed climate, the derivation of $S_{CLIM\ WA}$ values from $MDC_{CLIM\ WA}$ may stop before the end of the computation period. If $S_{CLIM\ WA} < S$, the program decreases the depletion curve $CDC_{CLIM\ WA,\ MA}$ to the last S-value of the original CDC, which is typically zero, and repeats it for the missing days. If an elevation zone contains glaciers or permanent snow cover, $CDC_{CLIM\ WA,\ MA}$ drops to the residual snow or ice coverage, again taken over from the original CDC. In less frequent cases (no effect of a temperature increase in a high elevation zone, increased precipitation), $S_{CLIM\ WA} > S$. The program then determines $\Delta S = S_{CLIM\ WA} - S$ and extrapolates $CDC_{CLIM\ WA}$ as $CDC + \Delta S$ on the missing days.

In order to evaluate the effect of a temperature increase on the Rio Grande basin, the following steps are to be taken:

(1) Runoff in the whole hydrological year is simulated (Figure 11.22) in order to verify the preselected parameters and the estimated snow coverage in winter.
(2) Conventional depletion curves of the snow coverage (CDC) used as input variable in the summer are plotted (Figure 11.23).
(3) Winter runoff is simulated separately for T and $T + 4\,°C$ in order to obtain the respective runoff volumes (hydrographs are printed).
(4) The decrease of the snow water equivalent on April 1 due to the increased snowmelt in winter ("winter deficit" or "negative winter adjustment") is computed as explained earlier in Section 11.6.1 under "change of snow accumulation in the new climate".
(5) Summer runoff is simulated separately for T in order to obtain the runoff volume (hydrographs are printed).

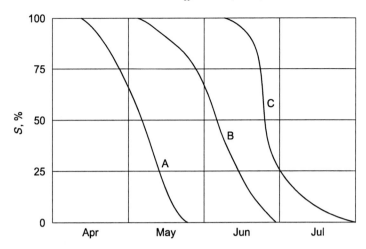

Figure 11.23 Conventional depletion curves of the snow coverage from Landsat data in the elevation zones A, B, and C of the Rio Grande basin near Del Norte in 1979.

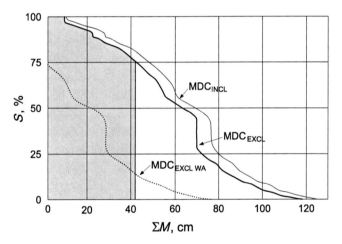

Figure 11.24 Modified depletion curves for zone A: MDC_{INCL} derived from CDC, therefore including new snow, MDC_{EXCL} with new snowmelt excluded, $MDC_{EXCL\ WA}$ with "winter deficit" (shaded area) cut off.

At this point, the climate-affected conventional depletion curves, CDC_{CLIM}, which are needed as input variables for computing the summer part (since they do not exist in the accumulation period in the winter) of the climate-affected runoff, are derived as follows:

(6) The modified depletion curve MDC_{INCL} is derived from the CDC. This curve relates the snow coverage with cumulative snowmelt depths including new snow in the summer (Figure 11.24).

(7) MDC$_{EXCL}$ is derived by eliminating melt depths referring to new snow from cumulative snowmelt depth (Figure 11.24). The area below this curve indicates the initial areal water equivalent of the snow cover, as shown in Section 11.5.

(8) The climate effect is taken into account by depriving MDC$_{EXCL}$ of the "winter deficit" computed in step (4), and MDC$_{EXCL\ WA}$ (winter adjusted) is derived. The program prints both MDC$_{EXCL}$ and MDC$_{EXCL\ WA}$. In zone A, $\Delta H_W = -36.94$ cm was computed so that MDC$_{EXCL}$ is cut off on the day when this value was equalled or exceeded. This happened on April 27, when ΣM for MDC$_{EXCL}$ was 42.04 cm (not reduced by S, therefore higher). MDC$_{EXCL\ WA}$ thus derived is shifted to start on April 1.

(9) The cumulative snowmelt is printed in relation to time as zonal snowmelt, that is to say reduced each day by the respective percentage of the snow coverage. The accumulated zonal melt curve AZM$_{EXCL}$ indicates graphically (Figure 11.25) that the "winter deficit," in this example $\Delta H_W = -36.94$ cm, was exceeded on April 27, when the accumulated zonal melt amounted to 37.69 cm. The previous day's total 35.9 cm is also printed so that the user can cut off the MDC$_{EXCL}$ at the previous day's value if the next day's value is much higher than the computed ΔH_W.

(10) After melt depths of new snow of the present climate and the winter deficit had been taken out of MDC$_{INCL}$ to derive MDC$_{EXCL\ WA}$, melt depths of new snow "surviving" in the warmer climate are put back to derive MDC$_{CLIM\ WA}$ as illustrated for zone A in Figure 11.26.

(11) The climate-affected conventional depletion curves adjusted for the "winter deficit", CDC$_{CLIM\ WA}$, are derived as follows: MDC$_{CLIM\ WA}$ indicates, for example, that a snowmelt depth of 22 cm is needed to decrease the snow coverage to 50%

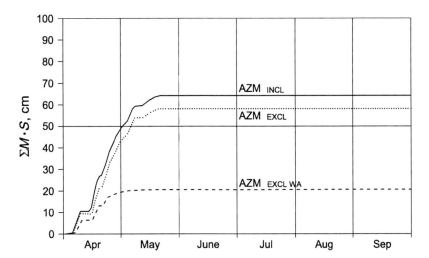

Figure 11.25 Accumulated zonal melt curves for zone A: AZM$_{INCL}$: computed daily melt depth multiplied by S from CDC (= zonal melt). AZM$_{EXCL}$ with new snow zonal melt excluded and AZM$_{EXCL\ WA}$ derived from AZM$_{EXCL}$ by cutting it off on April 27 and transferring it to April 1.

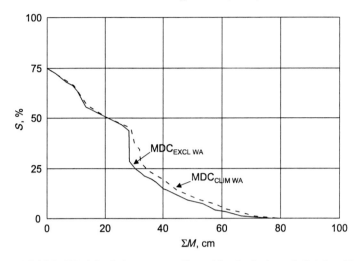

Figure 11.26 Modified depletion curve, adjusted for the "winter deficit" and including new snow of the changed climate ($MDC_{CLIM\ WA}$) derived from $MDC_{EXCL\ WA}$ for zone A.

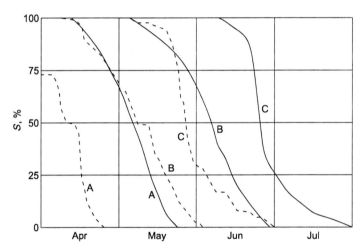

Figure 11.27 Effect of a changed climate ($T + 4\ °C$) on snow-covered areas of 1979 in elevation zones A, B, and C of the Rio Grande basin near Del Norte. $CDC_{CLIM\ WA}$ is shifted from the original CDC due to a reduced snow cover on April 1 and due to increased temperatures in the snowmelt season.

(Figure 11.26). This occurs in the present climate, according to CDC in Figure 11.23 on May 5. In a warmer climate ($T + 4\ °C$ in this example) a cumulative snowmelt depth of 22 cm and a corresponding decrease of the snow coverage to 50% are reached already on April 9, so that the 50% point is shifted to that date (Figure 11.27). The program takes the cumulative snowmelt depth computed by present temperatures on each day and searches for the date on which this snowmelt depth was equalled or exceeded when the higher temperatures are used for computation. If the new climate implies changes of the degree-day factor (see discussion under "Model parameters in a changed climate" later in this section), the cumulative snowmelt depth must be computed not only by higher temperatures, but also by changed (usually higher) a-values. The computer program takes care of this matter. Comparable snowmelt depths are reached about one month earlier so that $CDC_{CLIM\ WA}$ is shifted in time against the original CDC as illustrated in Figure 11.27 for all elevation zones. The method of CDC shifting in the summer is also explained by a numerical example elsewhere (Rango and Martinec, 1994). In view of the stepwise character of the cumulative snowmelt depths, a slightly higher snow water equivalent than the "winter deficit" may be cut off to derive $MDC_{EXCL\ WA}$ (see step (9) above) which would ultimately accelerate the decline of $CDC_{CLIM\ WA}$. On the other hand, searching for cumulative snowmelt depths equalled or exceeded in deriving $CDC_{CLIM\ WA}$ may result in a very slight delay of the decline.

(12) $CDC_{CLIM\ WA}$ is used to compute the climate-affected runoff in the summer half year. It should be noted that in contrast to the model runs in the simulation mode, R^2 in the climate-change runs does not indicate the model accuracy, but results from the difference between the hydrographs computed for the present and changed climate.

Figure 11.28 shows the climate-affected runoff computed by original precipitation, temperature $T + 4\ °C$ and snow-covered areas according to $CDC_{CLIM\ WA}$ compared with the original runoff simulation in Figure 11.22. The run is started by discharge computed on March 31 with $T + 4\ °C$ (see step (3)). The year-round climate-affected hydrograph thus consists of a winter simulation with $T + 4\ °C$ and

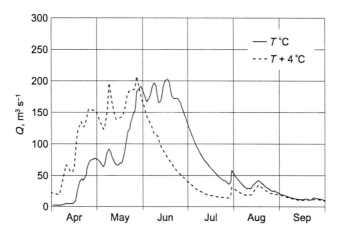

Figure 11.28 Climate-affected runoff ($T + 4\ °C$) in the Rio Grande basin near Del Norte, compared with the runoff simulated by data of 1979 (as shown in Figure 11.22) for April–September.

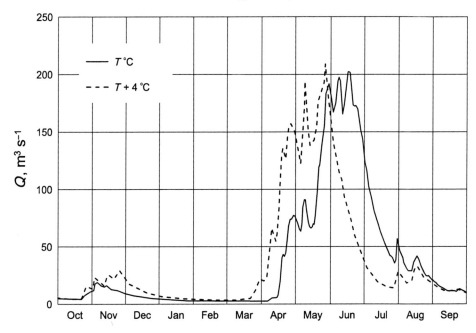

Figure 11.29 Simulated runoff in the Rio Grande basin near Del Norte in the hydrological year 1979 and climate-affected runoff computed by increased temperatures
$(T + 4 \,°C)$ and correspondingly changed snow conditions.

estimated S and of a summer computation by the climate program with $T + 4\,°C$
and S from $CDC_{CLIM\ WA}$.

Figure 11.29 shows year-round hydrographs computed with temperatures of
1979 and with temperatures increased by $+4\,°C$. As listed in Table 11.5, the future
winter runoff would be increased at the expense of the summer runoff. The actual
effect is greater because the increased climate-affected runoff in late March is carried over to April as recession flow as explained elsewhere (Rango and Martinec,
1997). Proportionally the redistribution of runoff is relatively small because the
cold winter of 1979 did not allow much snowmelt even with the increased temperatures (see Section 11.6.1, Normalization of data to represent the present
climate).

In the given example, precipitation was not changed in the new climate but the
climate program can also handle the combined effect of changed temperatures and
changed precipitation. Usually the temperature effect prevails but, as mentioned
earlier in Section 11.6.1 under "change of snow accumulation in the new climate",
increased precipitation and absence of melting conditions in high altitudes in winter
may convert the "winter deficit" to "winter gain" or "positive winter adjustment."
In this case, the computer program derives $MDC_{EXCL\ WA}$ by stretching MDC_{EXCL}

Table 11.5 *Seasonal redistribution of runoff of 1979 in the Rio Grande basin near Del Norte, CO (3414 km^2) due to climate change*

1979	Winter		Summer		Hydrological Year	
	10^6 m^3	%	10^6 m^3	%	10^6 m^3	%
Measured	86.53	7.2	1122.43	92.8	1208.96	100
Computed, T	91.87	7.6	1120.15	92.4	1212.02	100
Computed, $T + 4\,°C$	146.76	12.3	1046.16	87.7	1192.92	100

proportionally to this gain:

$$g = \frac{P' - I'}{P - I} \tag{11.25}$$

where:

P, P' = precipitation in the present and changed climate
I, I' = winter input (see Equation (11.23)) in the present and changed climate

If there is no winter input in either case due to low temperatures, a precipitation increase by 20% results in $g = 1.2$. The x-coordinates of MDC$_{EXCL}$ (cumulative snowmelt depth) are multiplied by g to derive MDC$_{EXCL\ WA}$ which conforms to the increased water equivalent of the snow cover on April 1. In the summer half year, the position of CDC$_{CLIM\ WA}$ as opposed to CDC results from the balance of the contradictory effects of an increased initial snow cover and of increased melting.

Model parameters in a changed climate

SRM parameters are predetermined which requires more hydrological judgment than mechanical calibration or optimizing. However, as has been pointed out elsewhere (Klemes, 1985; Becker and Serban, 1990; Nash and Gleick, 1991), calibration models are not suitable for climate-effect studies because the parameters cannot be meaningfully adapted to the conditions of a changed climate. In the given example from the Rio Grande basin near Del Norte, the seasonal change of the degree-day factor a and of the runoff coefficient for snow c_S was taken into account. The degree-day factor gradually increases in line with snow density while c_S reflects the decline of the snow coverage and the stage of vegetation growth. Since the original CDCs are moved by about one month earlier (see Figure 11.27), the values of both parameters were shifted accordingly by 31 days in the climate run. For example, $a = 0.45$ cm $°C^{-1}$ d^{-1} selected for May in the present climate was used in April in the warmer climate. The climate program provides for automatic shifting by any number of days. When September values are shifted to August, the value of September 30 is repeated in September. The shifting is stopped in January in order to prevent winter conditions being transferred to the autumn. There is no consensus yet whether a warmer climate will increase losses in which case the values of c_S and c_R would be generally decreased, because a decreased evapotranspiration due to the CO_2 increase might offset the temperature effect (Carlson and Bunce, 1991; Gifford, 1988).

Selected parameters can be shifted or changed in accordance with the expected conditions of a future climate. Future versions will also enable the constants x, y for the recession formula to be adjusted if, for example, a steeper recession would be indicated by drier soil conditions.

Normalization of data to represent the present climate

The climate effect is evaluated by comparing present snow and runoff conditions with conditions modelled for a climate scenario. The current climate can be represented by precipitation, temperatures, snow-covered areas, and simulated runoff of a single year, especially if this year appears to be "normal". Average data from a number of years would seem to be more representative. However, the use of long-term average daily temperatures, with day-to-day fluctuations smoothed out, may considerably underestimate snowmelt depths and the runoff as mentioned in Table 11.4.

A normalized data set can be prepared by adjusting daily temperatures of a selected year by monthly deviations from long-term averages and by multiplying daily precipitation amounts by ratios of long-term and actual monthly totals. The normalized depletion curves of the snow coverage can be derived from the curves of the selected year by considering the normalized temperatures and precipitation as a "new climate" for that year and running the SRM climate program. The normalized temperatures, precipitation, and snow-covered areas are then considered as a standard year and the climate program is run again with specifications of a changed climate. The present climate program Version 4.0, however, allows only uniform changes of temperature and precipitation for the winter and summer half year. The next version will be adjusted in order to carry out the outlined normalization of data as well as to take into account more detailed climate scenarios.

Outlook

The SRM program Version 4.0 evaluates the snow coverage, the areal water equivalent, and computes runoff for any increase of temperature and any change of precipitation in the winter and summer half year. Options are open concerning model parameters to be shifted in time or adjusted in magnitude in response to a changed climate. Currently, data from actual years are used to represent today's climate. The next program version will enable temperature and precipitation to be changed monthly. With this refinement, it will be possible to derive daily temperatures, precipitation and snow-covered area for a "normalized" year which will represent the current climate better than a selected actual year. A refined computation of snowmelt and glacier melt with a radiation component is under preparation in order to take into account climate scenarios with changed cloud conditions and to predict the behavior of glaciers in the next century.

Information about new developments will be available on the Internet and in SRM-Workshops.

11.7 Micro-SRM computer program

Background

The Martinec/Rango Snowmelt-Runoff Model (SRM) was originally a FORTRAN model designed to operate on an IBM 370-series mainframe computer. The first computerized version of the model was developed by Martinec *et al.* (1983). In 1986 the model's FORTRAN code was downloaded to an IBM PC and modified to operate in the PC environment. That same year a decision was made to develop a unique PC-oriented version of the model, taking full advantage of the PC's inherent capabilities. The results of that decision was Micro-SRM, Version 1.0.

Additional refinements have been incorporated in several subsequent Micro-SRM Versions up to and including 4.0. However, SRM itself remains unchanged and relatively simple, so that the computations by Equation (11.1) can still be performed by any pocket calculator which has a function x^y. Of course the PC program automatically handles the multiple input of temperature and precipitation for up to eight elevations zones of a basin, any desired lag time, and complicated snow/rain situations. A model run for up to 365 days is finished within several seconds, the computed hydrograph is immediately displayed in comparison with the measured discharge and, if desired, quickly printed. Also, the achieved accuracy is automatically computed and displayed. A summary of parameter values can be displayed after each run so that adjustments can be made and their effect assessed.

SRM does not require numerous runs because calibration is not necessary. The ease with which the results are obtained should not lead to a replacement of the deterministic approach of SRM by a "try and see" philosophy. SRM is designed to operate with physically based estimates of parameters which should not require much change after the initial selection. As mentioned earlier, seemingly unsatisfactory results have been frequently improved not by adjusting the parameters but by correcting errors in data sets and in the input of variables.

A prime consideration in the design of Micro-SRM was to develop a snowmelt modelling "environment" such that the model user was provided not just model algorithms, but a complete set of tools for managing the associated model processes: data entry, storage and retrieval, display of data, and results. Traditionally, the most time-consuming and error-prone activity involved in using any physically based model has been the accumulation of large amounts of input data in the form and format needed to drive the simulation, with actual execution of the model a trivial task by comparison. Recognizing this, the developers chose to pattern the design of Micro-SRM after that originally developed during the automation of the Soil Conservation Service's (SCS) Technical Release Number 55 (TR-55), Urban Hydrology for Small Watersheds (Soil Conservation Service, 1986). This joint Agricultural Research Service (ARS)/SCS effort provided valuable experience in developing highly interactive "front-ends" for interfacing complex models with model users. The approach used by the ARS/SCS programming team that automated TR-55 was to develop an efficient, easy-to-use, highly interactive data entry/manipulation environment, and include model algorithms as just one of many functions that support and use that environment (Cronshey *et al.*, 1985). Micro-SRM consists of an integration of the mainframe SRM FORTRAN algorithms converted to Basic and a variation of TR-55's data entry/data management algorithms. The current version of the model, Micro-SRM, Version 4.0, was developed using Microsoft QuickBA-SIC 4.5 and contains several subroutines from QuickPak Professional, a BASIC toolbox developed by Crescent Software, Inc.[1]

Micro-SRM and WinSRM availability

The Hydrology and Remote Sensing Laboratory supports and distributes the Snowmelt-Runoff Model free of charge to any interested party. The program and related files are available via the Internet by accessing the SRM home page at, http://ars.usda.gov/services/docs.htm?docid=8872. This site also includes an electronic version of the SRM User's Manual. A distribution diskette, available on either 5″ or 3½″ media, containing the executable code, supporting files, and example data files can be provided upon request. To obtain the latest version of Micro-SRM, contact:

USDA-ARS, Hydrology and Remote Sensing Laboratory
Bldg. 007, Room 104, BARC-W
10300 Baltimore Avenue
Beltsville, MD 20705–2350 USA
(301) 504–7490

or via electronic mail to:

ralph.roberts@ars.usda.gov
alrango@nmsu.edu

---

[1] Trademarks used in this document (e.g., IBM PC, IBM CORP.; MS-DOS, QuickBASIC, Microsoft Corp.; QuickPak Professional, Crescent Software) are used solely for the purpose of providing specific information.

11.8 References

Baumann, R., Burkart, U. and Seidel, K. (1990). Runoff forecasts in an Alpine catchment of satellite snow cover monitoring. In *Proceedings International Symposium on Remote Sensing and Water Resources*. Enschede, The Netherlands: International Association of Hydrogeologists, pp. 181–90.

Baumgartner, M. F. (1987). Schneeschmelz-Abflusssimulationen basierend auf Schneeflächenbestimmungen mit digitalen Landsat-MSS and NOAA-AVHRR Daten. (Snowmelt runoff simulations based on snow cover mapping using digital Landsat-MSS and NOAA-AVHRR data). In *Remote Sensing Series 11*, German version, Tech. Report HL-16, Zurich, Switzerland: Department of Geography, University of Zurich. English summary: Beltsville, MD: USDA, Agricultural Research Service, Hydrology Laboratory.

Baumgartner, M. F. and Rango, A. (1995). A microcomputer-based alpine snow cover analysis system. *Photogramm. Eng.*, **61**(12), 1475–86.

Becker, A. and Serban, P. (1990). Hydrological models for water resources system design and operation. *Operational Hydrology Report 34*. Geneva: WMO.

Brüsch, W. (1996). Das Snowmelt Runoff Model ETH (SRM-ETH) als universelles Simulations- und Prognosesystem von Schneeschmelz-Abflussmengen. (The Snowmelt Runoff Model ETH as a universal simulation and forecast system for snowmelt runoff). In *Remote Sensing Series 27*, German version, Zurich: University of Zurich Remote Sensing Laboratories.

Carlson, T. N. and Bunce, J. A. (1991). The effect of atmospheric carbon dioxide doubling on transpiration. In *Proceedings American Meteorological Society, Special Session on Hydrometeorology*, Salt Lake City, Utah, pp. 196–9.

Cronshey, R. G., Roberts, R. T., and Miller, N. (1985). Urban Hydrology for Small Watersheds (TR-55 Revised). In *Proceedings ASCE Hydraulic Division Specialty Conference*, Orlando, Florida.

Gifford, R. M. (1988). Direct effects of CO_2 concentrations on vegetation. In *Greenhouse Planning for Climate Change*, ed. G. L. Pearlman. Melbourne: CSIRO, pp. 506–19.

Hall, D. K. and Martinec, J. (1985). *Remote Sensing of Ice and Snow*. New York: Chapman and Hall Ltd.

Higuchi, K., Ageta, Y., Yasunari, T., and Inoue, J. (1982). Characteristics of precipitation during the monsoon season in high-mountain areas of the Nepal Himalaya. In *Hydrological Aspects of Alpine and High Mountain Areas*, Proc. Exeter Symposium, IAHS Publ. 138, pp. 21–30.

Klemes, V. (1985). *Sensitivity of Water Resources Systems to Climate Variations, WCP Report 98*. Geneva: WMO.

Kotlyakov, V. M. and Krenke, A. N. (1982). Investigations of the hydrological conditions of alpine regions by glaciological methods. In *Hydrological Aspects of Alpine and High-Mountain Areas*, Proc. Exeter Symposium, IAHS Publ. 138, pp. 31–42.

Martinec, J. (1960). The degree-day factor for snowmelt runoff forecasting. In *IUGG General Assembly of Helsinki*, IAHS Commission of Surface Waters, IAHS Publ. 51, pp. 468–77.

Martinec, J. (1970). Study of snowmelt runoff process in two representative watersheds with different elevation range. In *IAHS-UNESCO Symposium*, IAHS Publ. 96, Wellington, New Zealand, pp. 29–39.

Martinec, J. (1975). Snowmelt-Runoff Model for stream flow forecasts. *Nordic Hydrol.* **6**(3), 145–54.

Martinec, J. (1985). Time in hydrology. In *Facets of Hydrology*, vol. II, ed. J. C. Rodda. London: John Wiley and Sons, pp. 249–90.

Martinec, J. (1989). Hour-to-hour snowmelt rates and lysimeter outflow during an entire ablation period. In *Snow Cover and Glacier Variations*, Proc. Baltimore Symp. IAHS Publ. No. 183, pp. 19–28.

Martinec, J. and Rango, A. (1986). Parameter values for snowmelt runoff modelling. *J. Hydrol.* **84**, 197–219.

Martinec, J. and Rango, A. (1989). Merits of statistical criteria for the performance of hydrological models. *Water Resour. Bull.* **25**(20), 421–32.

Martinec, J. and Rango, A. (1995). Seasonal runoff forecasts for hydropower based on remote sensing. In *Proceedings of the 63rd Annual Western Snow Conference*, Sparks, NV, pp. 10–20.

Martinec, J., Rango, A., and Major, E. (1983). *The Snowmelt-Runoff Model (SRM) User's Manual, NASA Reference Publ. 1100*. Washington, DC.

Martinec, J., Rango, A., and Roberts, R. (1994). Snowmelt Runoff Model (SRM) user's manual. In *Geographica Bernensia*, ed. M. F. Baumgartner. Berne: Department of Geography, University of Berne.

Martinec, J., Rango, A., and Roberts, R. (1998). Snowmelt Runoff Model (SRM) user's manual. In *Geographica Bernisia*, ed. M. F. Baumgartner. Berne: Department of Geography, University of Berne.

Nash, L. L and Gleick, J. A. (1991). Sensitivity of streamflow in the Colorado Basin to climatic changes. *J. Hydrol.*, **125**, 221–41.

Rango, A. and Martinec, J. (1994). Areal extent of seasonal snow cover in a changed climate. *Nordic Hydrol.*, **25**, 233–46.

Rango, A. and Martinec, J. (1995). Revisiting the degree-day method for snowmelt computation. *Water Resour. Bull.* **31**(4), 657–69.

Rango, A. and Martinec, J. (1997). Water storage in mountain basins from satellite snow cover monitoring. In *Remote Sensing and Geographic Information Systems for Design and Operation of Water Resources Systems*, Proc. 5th Scientific Assembly of IAHS, Rabat, Morocco, IAHS Publication 242, ed. M. F. Baumgartner, G. A. Schultz and A. L. Johnson, pp. 83–91.

Rango, A. and van Katwijk, V. (1990). Development and testing of a snowmelt runoff forecasting technique. *Water Resour. Bull.*, **26**(1), 135–44.

Seidel, K. and Martinec, J. (2004). *Remote Sensing on Snow Hydrology*. New York: Springer-Verlag.

Seidel, K., Martinec, J. and Baumgartner, M. F. (2000). Modelling runoff and impact of climate change in large Himalayan basins. In *Proceedings of the International Conference on Integrated Water Resources Management for Sustainable Development*, New Dehli, India, pp. 1020–28.

Shafer, B. A., Jones, E. B., and Frick, D. M. (1981). *Snowmelt Runoff Simulations Using the Martinec-Rango Model on the South Fork Rio Grande and Conejos River in Colorado*, AgRISTARS Report CP-G1–04072. Greenbelt, MD: Goddard Space Flight Center.

Soil Conservation Service. (1986). *Urban Hydrology for Small Watersheds*, Tech. Release 55, 2nd edn, Washington, DC: Department of Agriculture.

Wilson, W. T. (1941). An outline of the thermodynamics of snow-melt. *Trans. Am. Geophys. Union*, **1**, 182–95.

World Meteorological Organization (1986). *Intercomparison of Models of Snowmelt Runoff*. Operational Hydrol. Report 23, No. 646, Geneva: WMO.

World Meteorological Organization (1992). *Simulated real-time intercomparison of hydrological models*. Operational Hydrol. Report 38, WMO – No. 779, Geneva, Switzerland: WMO.

12

Snowpack management and modifications

Snow in the global environment is altered by human activity in many ways, some intentional and others inadvert. Among the deliberate changes made to snowpacks are blowing-snow management to benefit agriculture and transportation, forest management to protect and enhance water yields from snowpacks, urban snow management to improve transportation and habitation in developed areas, albedo modification to enhance snowmelt rates, and ski-area management for enhanced and safe recreational opportunities. The major inadvert change that may impact snowpacks is the possibility of climate change that is being forecast for the twenty-first century. This final chapter generally discusses these human–snow interactions from the viewpoint of the snow hydrologist.

All management of snow should be based upon a thorough understanding of the processes involved at a given site and how they may be varied to accomplish a given goal. Furthermore, the management recommendations must consider acceptability, cost, ease of use, risks, maintenance requirements, and long-term environmental impacts. Many processes affecting snow that can be altered to meet management goals have been described in earlier chapters.

12.1 Blowing-snow management

Blowing-snow management generally has the objective of creating snow drifts in useful or non-problematic locations by using some type of barrier to the wind. Barriers take the form of constructed snow fences, rows of living shrubs or trees, and crop stubble. Farmers have realized for centuries that seeds and seedling plants survive and grow faster when located in furrows that accumulated small drifts of snow that supplemented soil moisture. More recently, agriculturalists have used interplantings of tall and short crops (Bilbro and Fryrear, 1988; Scholten, 1988), planted rows of trees or shrubs (Scholten, 1988; Shaw, 1988), or constructed snow fences to accumulate and store more snow on the landscape. Constructed snow

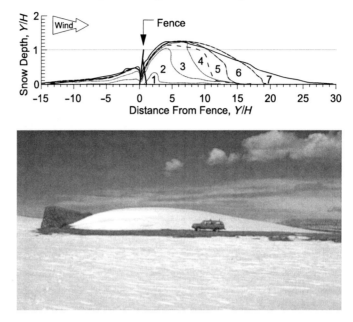

Figure 12.1 Stages of snow-drift growth upwind and downwind from a 50% poros-ity Wyoming snow fence. Drift depth and length are scaled to fence height (H). (Tabler, 2003, © 2003 by Ronald D. Tabler, all rights reserved. Reprinted with permission).

fencing has been used on rangeland to create drifts in natural drainages that supply stock ponds with water and to reduce water losses from sublimation of blowing snow (Sturges and Tabler, 1981). Finally, use of snow fences to improve visibility and prevent drifting of blowing snow on roads or others areas with public access has become standard practice in regions with blowing-snow problems. Factors control-ling the transport of snow by the wind were described in Chapter 2, basic principles governing functioning and use of constructed snow fences, living windbreaks, and crop stubble to drift snow are reviewed below.

12.1.1 Growth of snow drifts

Impacts of snow fences or windbreaks on snow accumulation depends upon the alteration of air flow through, around and over the barrier (Heisler and DeWalle, 1988). Tabler (1991, 1994) described stages of growth of snow drifts around a con-structed snow fence (Figure 12.1) that can illustrate how drifts form in general. Stage One represents initial growth of a drift due to slowing of the wind in the lee of the snow fence which allows accumulation of creeping or saltating ice particles extend-ing for distances up to $5H$ upwind and $15H$ downwind, where H is vertical fence height. After sufficient initial drift growth, Stage Two begins with the formation of an eddy or recirculation zone downwind from the fence. Recirculating air creates a steep, concave slip face on the downwind drift and causes further accumulation of snow transported by turbulent transport that primarily deepens the downwind drift. Stages Two to Three continue the accumulation until the downwind drift reaches a maximum depth of about $1H$ or $1.2H$ depth for a 50% fence porosity. Stages Four to Six are characterized by a gradual reduction in size of the recirculation zone as the downwind drift lengthens to a position of $20H$ downwind. Once recirculating air is eliminated due to growth of the drift, further drifting of snow marks the beginning of Stage Seven that smoothes and elongates the downwind drift to lengths up to about $30H$. During the entire process a smaller drift gradually forms upwind of the snow fence that can reach a distance up of 10–$15H$ upwind. Upwind drifts generally only store about 15% of total drift water. The process is completed with the formation of equilibrium snow drifts upwind and downwind of the fence, which represent the maximum storage of snow for a specific type of fence and location. Formation of equilibrium drifts is dependent upon having a sufficient supply of blowing snow.

12.1.2 Constructed snow fences

Several design features of constructed snow fences such as height, porosity, and bottom gap are important to maximize snow storage and control drift dimensions. Drift size generally increases with fence height (H) up to the maximum height of blowing snow above the ground. Constructed fences up to about 4 m height have been used. Snow fence porosity combined with height generally controls the downwind extent of snow-drift formation. Solid barriers produce relatively short, deep drifts that can bury the fence, while fences with around 40–50% porosity produce drifts that extend to $30H$ and are considered to be most efficient. Bottom gaps in the fence of about $0.1H$ are also considered necessary to allow wind to sweep snow away from the base of the fence.

One design that has been extensively studied and used is the Wyoming snow fence (Figure 12.2), which is constructed with horizontal wooden slats arranged to give 40–50% porosity and 3.78 m vertical height (Tabler, 1980). The Wyoming fence is inclined about $15°$ in the downwind direction and has a $0.1H$ bottom gap. Total storage of snow water equivalent in drifts on level terrain around the Wyoming fence can be computed as (Tabler *et al.*, 1990):

$$Q = 8500 \, H^{2.2} \tag{12.1}$$

where Q is the maximum volume of stored water in upwind and downwind drifts at equilibrium, expressed as kg of water per m length of fence; and H is fence height in m. These types of relationships in combination with estimates of the total

Figure 12.2 Wyoming snow fence that has been extensively tested and used to control blowing snow in Western United States (Tabler 2003, © 2003 by Ronald D. Tabler, all rights reserved. Reprinted with permission).

amount of blowing snow at a site (see Chapter 2) are extremely useful in planning installation of snow fences. Fence height needed to store water as well as volumes of water stored for a fence of finite length can be computed.

Given a seasonal volume of blowing snow of 150 000 kg of water per meter perpendicular to wind direction (see Equation (2.4) in Chapter 2), determine the height of snow fence needed to store this snow. By setting Equation (12.1) equal to the seasonal blowing-snow estimate, the height of the fence needed can be found as:

$$Q = 8500\,H^{2.2} = 150\,000 \text{ kg m}^{-1}$$
$$H = 3.68 \text{ m}$$

Thus, a single Wyoming snow fence of 3.68 m height, could conceivably store all of the blowing snow during a winter season for this site. Total volumes of water stored for a 100 m long fence segment would be:

$$(150\,000 \text{ kg m}^{-1})(100 \text{ m}) = 1.5 \times 10^7 \text{ kg}$$

Where calculated fence heights exceed the practical limits for construction, parallel fences of lesser height could be used with suitable spacing of 20–30H to avoid overlap of upwind and downwind drifts between fences. Within limits, a single tall fence is often more cost effective and improves downwind visibility more than several short fences.

The fraction of water stored in drifts around snow fences that eventually becomes streamflow or groundwater recharge depends upon sublimation and evaporation losses during the melt season and watershed soils and geology. Sturges (1992) showed that an 800 m long Wyoming snow fence deployed upwind and parallel to a major channel on a 306 ha rangeland catchment increased streamflow from snowmelt by 129%. Fence placement caused accumulation of snow directly into the channel area and, combined with soil and geologic conditions that favored shallow or surface flow of meltwater, total volumes of streamflow from melt and the duration of melt runoff were significantly increased. Under conditions where meltwater from drifts around snow fences becomes groundwater recharge, direct streamflow increases from snow fencing may be much less. Regardless, snow fences increase water from blowing snow that becomes groundwater or streamflow at some point downstream.

Deployment of snow fences to prevent drifting and blowing snow on roadways, around buildings or in other public access areas must be conducted with prevailing wind directions in mind. Wind directions during blowing-snow events differ from general wind directions during winter and observations of the orientation of natural snow drifts in an area are needed prior to construction or planting of snow fences and windbreaks. When prevailing winds during blowing snow are perpendicular to the long axis of any area to be protected, fences can be simply placed parallel to the long axis with suitable space for the downwind drift to form. When prevailing winds during blowing snow are parallel to the long axis of the area or structure to be protected, then multiple parallel fences located at about 45° to the long axis are needed. Many more practical considerations for deployment of snow fences for roads and buildings, beyond the scope of this text, are given in Tabler (1991, 1994, 2003), Verge and Williams (1981), and Tabler *et al.* (1990).

12.1.3 Living windbreaks

Living windbreaks are more difficult to manage to achieve design criteria, but offer a natural and often lower-cost alternative for control of blowing snow. Living windbreaks are used for two general purposes: (1) field windbreaks to spread snow across downwind fields to augment soil moisture and (2) living snow fences to accumulate snow in deep, dense drifts upwind from areas used for public access. Effective field windbreaks to drift snow across downwind fields can be obtained with one row of deciduous hardwood trees with a parallel row of low shrub planted about 3 m upwind. The deciduous hardwood trees offer low wind resistance and allow for long downwind drifts to form over the downwind field, while the upwind row of shrubs is helpful in drifting snow as the trees grow tall and lose their lower branches. For living snow fences a single dense row of conifer trees with persistent foliage is often sufficient. A living snow fence of trees or shrubs must of course be planted a sufficient distance upwind so that large drifts do not form in the area to be protected. Trees and shrubs selected for planting must be those adapted to the region and local plant nurseries or government agencies can often give good advice on species to plant for each type of windbreak. Deployment of these windbreaks follows general criteria given earlier for constructed fences. Living snow-fence programs have been established in several states in the western United States that provide more specific advice, especially on plant species selections, planting and maintenance.

12.1.4 Crop stubble as a windbreak

Crop stubble can also be an effective windbreak to accumulate snow on fields in winter in dryland farming regions. Stubble height, just like snow fence height, is an important determinant of the amount of snow accumulated. Grain stubble 20–30 cm high can increase soil water content at spring planting time by 2–4 cm in the Great Plains of North America which can significantly increase crop yields. Alternating swaths of tall and short stubble oriented perpendicular to the wind is even more effective in accumulating snow than uniform stubble height (Bilbro and Fryrear,1988; Scholten, 1988). Stubble that protrudes through the drifted snow helps to reduce melt rates and evaporation/sublimation losses from accumulated snow by reducing wind speeds and shading the snow. Increased snow within stubble can also insulate the soil and reduce frost heaving of root systems for crops like alfalfa.

12.2 Forest management

Forest management activities, chiefly related to timber cutting, can have profound effects on the snow hydrology of small watersheds. Timber harvesting that reduces or temporarily eliminates forest canopy cover can influence both snow interception losses and melt rates of snowpacks. Changes in snow interception loss and melt rates in turn will influence the timing and yield of water as streamflow. Early studies clearly demonstrated the influence of forest management on streamflow and principles derived therefrom have been codified into best management practices for foresters to follow. Forest management in cold regions is generally conducted using methods that seek to minimize impacts of cutting on peak flows and increase water yields from snowpacks by altering the distribution and amount of snow that accumulates. Concern with forest-cutting effects on snow hydrology is generally confined to regions where snowfall represents over 20% of annual precipitation and natural snowpacks accumulate in forests through much of the winter season.

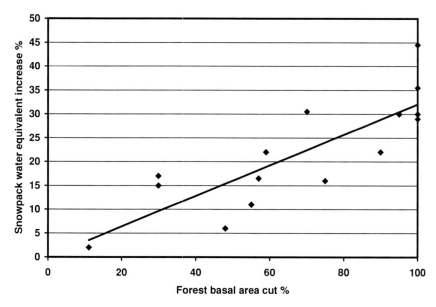

Figure 12.3 Effects of conifer forest cutting on peak snow accumulation at Fraser Experimental Forest, Colorado, USA (modified from Troendle, 1987 courtesy USDA Forest Service).

12.2.1 Forest cutting and snow accumulation

Impacts of forestry on snow hydrology largely vary with the amount and pattern of forest cutting. Individual tree removal for selective regeneration cutting, thinnings, timber stand improvement and salvage cuttings, will create many small openings that reduce the biomass available for snow interception and allow somewhat greater transmission of solar radiation and increased melt and evaporation/sublimation. Larger clear-cut openings within watersheds also reduce canopy interception losses, but expose surface snowpacks to increased wind speeds which affects scouring by the wind and evaporation/sublimation losses. Figure 12.3 summarizes results of experiments at the Fraser Experimental Forest in Colorado, USA and shows effects of a variety of partial forest cuttings on peak snowpack accumulation. A roughly linear percentage increase in snowpack accumulation is indicated with increasing percentage of basal area removal, although there is much scatter in the data. Anderson *et al.* (1976) summarized data from the western USA that showed similar results; heavy forest cutting created increases in maximum snowpack accumulation of 22–45%, while light cutting increased snowpack accumulation by 10–17%.

Increased accumulations in partially cut forests should be viewed as the <u>net effect</u> of gains from reduced canopy interception losses and losses due to increased

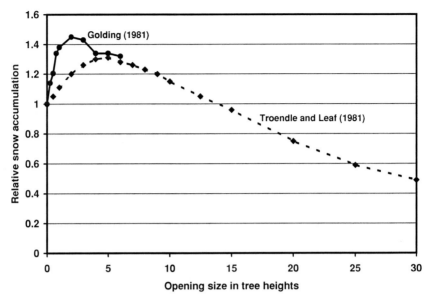

Figure 12.4 Maximum snow accumulation in conifer forest openings as a function of opening size, redrawn from Troendle and Leaf (1981) and Golding (1981). Opening size is measured in the downwind direction in multiples of tree height (*H*).

evaporation/sublimation. Variations in snowpack sublimation with slope and aspect may help explain variations in maximum forest snowpack accumulations due to partial cutting. Schmidt *et al.* (1998) showed 20% greater sublimation on south-facing than north-facing slopes during the winter season at Fraser Experimental Forest in Colorado. Such differences could help reduce peak snowpack accumulations due to cutting on south slopes.

Snow accumulation in clear-cut openings varies with opening size due to the effects of blowing snow and sublimation. Figure 12.4 shows effects of forest opening size on maximum snowpack accumulation based upon studies by Troendle and Leaf (1981) in Colorado and Wyoming, USA and Golding (1981) in Alberta, Canada. In both studies, snow accumulation increases with opening size to a maximum at a small opening size, 2*H* in Canada and 5*H* in Colorado and Wyoming, and then decreases with further increases in opening size. In small openings, most of the increases in snow accumulation are due to reduced canopy interception losses. In very large openings, snowpack accumulations are lower due to wind scour and sublimation losses. Maximum snowpack accumulation in spring is generally thought to occur at an opening size where the net effects of reduced canopy interception of trees removed in the opening, wind redistribution of snow out of the opening, increased sublimation from snow in the opening due to wind, and possibly wind transport into the opening from the upwind forest produce maximum accumulation.

The contributions of wind transport into the opening from the upwind forest remain unclear. Schmidt and Troendle (1989) showed, by measuring the flux of blowing-snow particles, that a plume of blowing snow from the upwind forest appeared to add to the accumulation of snow in a downwind clearing in Colorado subalpine forest. Existence of blowing-snow conditions may add to variability of snowpack accumulation experienced when openings are created, but other direct evidence of wind transport of intercepted snow from the upwind forest into openings is not available.

Some studies show patterns of reduced snowpack accumulations immediately downwind from openings (Golding and Swanson, 1986; Gary, 1980; Gary and Troendle, 1982; Golding, 1981). An example from Gary and Troendle (1982) in Figure 12.5 based upon experiments in Wyoming shows gains in the opening that were essentially completely offset by losses in the downwind forest. One interpretation of such patterns is that a clear-cut opening causes a back eddy that scours snow from the forest snowpack immediately downwind from the cut and adds it to the accumulation in the opening or transports snow farther downwind (Gary, 1980). Another interpretation of reduced snow accumulation downwind from openings is that snowpacks in the downwind forest are subjected to increased early-season melting and evaporation/sublimation that reduces peak springtime accumulations due to increased snowpack exposure to wind and radiation (Meiman, 1987; Golding and Swanson, 1986).

Whether forest cutting in a patchwork of openings and uncut areas actually increases snowpack available for streamflow across large watersheds has been debated. Hoover and Leaf (1967) presented arguments for Fool Creek watershed in Colorado that increased snow accumulations in openings were offset by reduced accumulations in the downwind forest such that the net precipitation to the partially cut watershed was unaffected. However, analysis of more data from the same watershed by Troendle and King (1985) did show that maximum snow accumulation across the Fool Creek watershed (cut and uncut areas) was increased 9% due to cutting. Thus, it appears that increased snowpack accumulation does occur over large partially cut watersheds, but relative roles of processes leading to increased accumulations are difficult to sort out.

Overall, it is clear that either clear-cut openings or selective removal of individual trees, can lead to increased snowpack accumulations. Increases in clear-cut openings are best achieved with $2H$ to $5H$ opening sizes. See Chapter 2 for a discussion of snow interception and blowing-snow processes.

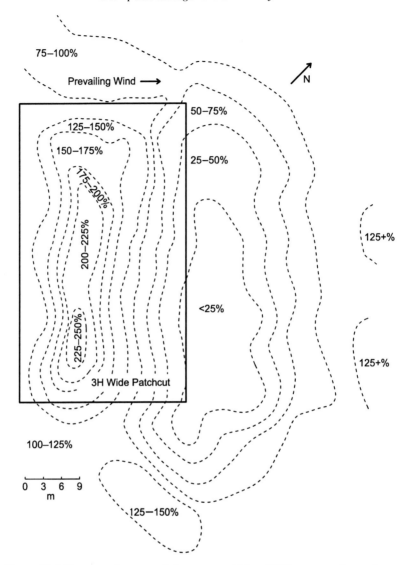

Figure 12.5 Patterns of snowpack increases within a *3H* forest opening and snowpack depletion in the downwind forest (Gary and Troendle, 1982, courtesy USDA
Forest Service).

12.2.2 Forest cutting and snowpack ablation

Ablation of snowpacks is the sum of losses due to melting and the net effect of vapor
exchange. Ablation rates are largely controlled by the energy budget at the snowpack surface in forest or in clear-cut openings. Basic principles controlling energy

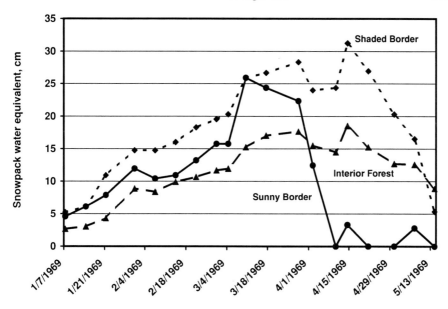

Figure 12.6 Snow accumulation and ablation patterns in interior forest and along the shaded and sunny border of a 1*H* wide, E–W oriented clear-cut strip in Englemann spruce (*Picea engelmanni* Parry) forest in New Mexico (redrawn from Gary, 1974 courtesy USDA Forest Service).

exchange were considered in Chapter 6 and forest effects on energy exchange were covered in Chapter 7. Ablation rates are increased as the forest canopy is selectively thinned due to increased solar radiation transmission and increased wind speeds. Increased accumulations due to reduced interception are often offset by greater rates of snowpack ablation in thinned stands, so that snow disappearance occurs at approximately the same time in thinned and uncut stands. Effects of changing forest density can be indexed by changes in leaf-area index (LAI), radiation extinction coefficients or canopy density. Based upon periodic measurements of snowpack water equivalent in eastern US forests, ablation rates in open/deciduous hardwood forest/conifer forest cover types can be approximated using the ratios 3:2:1 (Federer *et al.*, 1972).

Shadow patterns cast by forest edges play a major role in controlling ablation rates in clear-cut openings, even though some solar radiation is transmitted through the forest. Snowpacks along forest edges that are shaded, melt and sublimate more slowly. Those in more exposed locations ablate rapidly due to more frequent and intense direct beam solar radiation received during mid-day (Figure 12.6).

Figure 12.6 shows typical variations in accumulation and ablation within a clear-cut strip in conifer forest that was 1*H* wide and oriented E–W in New Mexico,

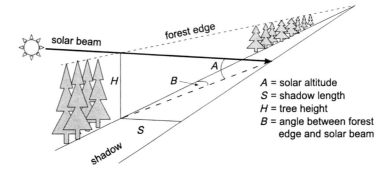

Figure 12.7 Shadow length–solar geometry relationships along infinite forest edge.

USA. Accumulation totals, until major melting began in March, were lowest in the interior forest, intermediate along the sunny border, and greatest along the shaded border due to interception, evaporation/sublimation and blowing-snow effects. Gary (1974) found that daily snowpack evaporation/sublimation rates were 87% higher along the sunny than shaded border during a 12-week winter period in this opening. After ablation began, the greatest ablation rates occurred along the sunny border, intermediate rates occurred along the shaded border, and lowest rates were found in the interior forest. Snow disappeared several weeks earlier along the sunny border of the cut strip, but both interior forest and shaded border retained snow until mid-May.

Widths of forest openings can be varied to control shadow length and average melt rates across an opening. For level terrain, the shadow length cast by an infinite forest edge at any time of day or year can be computed as (Figure 12.7):

$$S = H \sin B \cot A \qquad (12.2)$$

where:

$S =$ shadow length measured perpendicular to the forest edge
$H =$ tree height
$A =$ solar altitude
$B =$ acute angle between the solar beam direction and the forest edge direction

Equation (12.2) applies when a shadow is cast by the forest edge. Shadow lengths change with solar altitude and solar azimuth during the day and during the winter and spring melt seasons at a given latitude. Shadow lengths can be quite long at high latitudes where solar altitudes are low during the accumulation and melt seasons.

If trees were 18 m tall along a forest edge oriented SW to NE (edge azimuth = 45° east of north) where the Sun's altitude above the horizon was 60° at an azimuth of 180° (= solar noon, Sun from due south) on May 1, what length shadow would be cast by the edge? The acute angle between the solar beam and edge would be 45° and thus:

$$S = 18 \sin 45° \cot 60° = (18)(0.7071)(0.5774) = 7.35 \text{ m}$$

Thus, at noon on this day only a relatively narrow band of snowpack surface, about 7.35/18 or 0.4H wide, would be shaded.

Orientation of long rectangular cut openings and tree height obviously are also important controls on shadow lengths. Slope angle will also affect shadow length but is not considered here.

12.2.3 Forest cutting and snowmelt runoff

Forest cutting can affect both the volume and timing of streamflow and ground-water recharge from snowmelt. Changes in the natural timing of snowmelt runoff due to cutting will depend upon the pattern of cutting and synchronization of melt runoff from cut and uncut portions of a watershed. Volumes of snowmelt water yielded as streamflow also can be increased due to reduced canopy interception losses and wind redistribution of snow into openings in combination with reductions in growing-season evapotranspiration rates in the openings due to absence of trees.

General recommendations for forest harvesting using clear-cut openings in subalpine conifer forest are available. Opening size and orientation should vary somewhat with objectives of the harvesting: increasing the volume of streamflow or preventing changes in flow regimen and peak flows. In all cases, intervening uncut forest areas at least as wide as clear-cuts should be used to provide shade and a source of blowing snow. When cutting to primarily enhance water yields, foresters should use: wider openings about 5H wide (H = average tree height) to optimize snow accumulation, forest edges oriented perpendicular to prevailing winds to take advantage of blowing-snow effects, and relatively large portions of watershed area cut to effect greater evapotranspiration losses, but never more than 50% area cut at any time.

When cutting to primarily protect the natural flow regimen and minimize changes in peak flows, foresters should use: smaller openings, about 2H, to keep average melt rates in openings low and wherever possible use strip cuts orientated to provide shade, e.g. E–W orientations. Smaller total fractions of watershed area cut ($<1/4$ to $1/3$ of basin area in a cycle) should also be used. Successive cuttings generally are made in the direction of the Sun to provide shade from original or regrowing forest to produce what is known as a wall-and-step cutting pattern (Figure 12.8). To further control peak flows, advantage can be taken of the natural melt desynchronization by cutting wider strips at lower elevations and on south-facing slopes that naturally

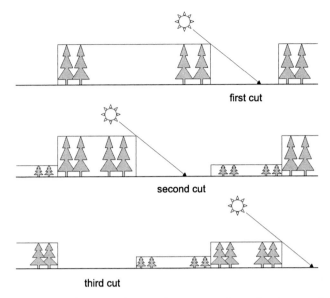

Figure 12.8 Wall-and-step cutting used to remove forest in three cutting cycles
while providing shade to control melt rates and minimize impacts on peak flows.

melt earlier. Cutting narrower strips at high elevations and on north-facing slopes
that naturally melt later also can help to prevent changes to flow regimen.

Fool Creek watershed in the Rocky Mountains of Colorado, USA (Troendle
and King, 1985) is a classic paired-watershed experiment on impacts of forestry
practices on snow hydrology in subalpine conifer forest (Chapter 1; Figure 1.8).
The 289 ha basin had 50% of timbered area cut in forest strips ranging in width
from $1H$ to $6H$. Over a 28-year period since cutting, annual water yields had been
increased an average of 8.2 cm per year or an increase in flow of about 36%.
Annual peak flows in spring on the basin were increased about 20% and advanced
7.5 days due to cutting. Flow increases were greater in years with greater winter
and spring precipitation and appear to be very long lasting. Estimates indicate that
nearly 70+ years will be required for flows to return to pre-cut levels due to slow
growth rates in the subalpine forest. Interestingly too, flow increases generally
occurred during the early spring melt season due to reduced evapotranspiration
during the previous growing season in cut areas. In the Colorado subalpine, summer
rainfalls are generally insufficient to recharge the soil and thus increases in soil water
content in cut areas due to reduced evapotranspiration have a delayed impact on
streamflow until melt begins and recharges soil the following spring. Troendle and
Kaufmann (1987) indicate that selective forest cutting, rather than using clear-cut
strips, can also increase annual flows by similar amounts in Colorado subalpine
forest.

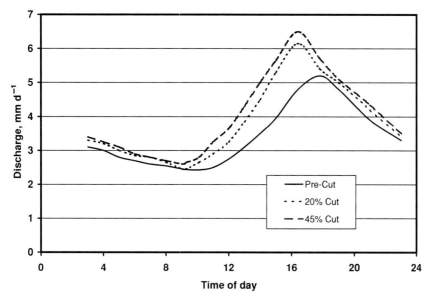

Figure 12.9 Impact of harvesting of deciduous forest in Pennsylvania on daily snowmelt hydrographs (based upon data from DeWalle and Lynch, 1975).

In regions where peak runoff and total water yields are not greatly affected by snowmelt, clear-cutting in large blocks can still produce significant changes in streamflow for selected melt events. For example, DeWalle and Lynch (1975) found for deciduous hardwood forest in Pennsylvania, that daily peak streamflows for melt events during complete snow cover were increased by 18% and 29% following successive large block clear-cuts covering 20% and 45% of basin area, respectively (Figure 12.9). Snowpack melt rates and net allwave radiation were more than doubled in the clear-cut area compared to adjacent uncut forest (DeWalle *et al.*, 1977). Reductions in peak daily flows on watersheds with large block cuts would be expected later in the spring season, due to desynchronization of melt runoff, after snow was depleted in the openings (Hornbeck and Pierce, 1970).

Inadvertent forest disturbances such as mortality or defoliation due to fire, insects and diseases, and windthrow can in some cases cause changes in snow hydrology similar to that due to cutting. Loss of foliage obviously affects shading and the snowpack energy balance. Mortality followed by windthrow reduces shading by tree boles even further. Any aerodynamic changes to the forest canopy can affect blowing-snow transport and have implications for snowmelt due to convection.

12.3 Urban snow management

The snow hydrology of urban and suburban areas is modified significantly from that of undeveloped watersheds. Interconnected impervious areas in developed areas

Figure 12.10 Urban snow requires dedicated areas for snow storage (left) and may lead to accumulation of surface contaminants later in the spring (right) (photos by D. R. DeWalle). See also color plate.

can lead to more direct delivery of meltwater into receiving streams, lakes, and wetlands. Snow interference with transportation and access to structures also leads to snow removal from roads, parking areas, sidewalks, etc. in urban areas (Figure 12.10). Urban snow can become contaminated with urban waste, deicing salts and anti-skid abrasives. Direct dumping of plowed urban snow into water bodies is not recommended due to pollution threats.

Melting in urban areas is affected by modifications to the radiation balance (Semadeni-Davies *et al.*, 2001). Pollution of urban snow by particulates lowers the snowpack albedo which increases absorbed shortwave energy available for melt. Particulates contained in snow gradually accumulate on the snowpack surface as melting occurs and continue to reduce albedo and increase melt. Urban snow around structures undergoes significant shading and piles of urban snow also reduce surface area available for melt and cause variations in angles for receipt of shortwave radiation at the snowpack surface. Buildings alter the longwave radiation receipt at the snowpack surface. Melt prediction in urban areas is thus quite complicated; being dependent upon the specific snowpack management practices used, the physical nature and layout of the urban area, and the ultimate radiation balance and distribution of urban snowpacks. Ho and Valeo (2004) categorized urban snow into four types: snow piles, snow on road shoulders, snow on sidewalk edges, and snow in open. For each type they developed different albedo functions to help predict snowmelt runoff.

The quality of meltwater from urban snow is a significant environmental issue. Deicing chemicals and abrasives become incorporated in urban snowpacks. Sodium and calcium chloride are primarily used for deicing, but calcium magnesium acetate made from dolomitic limestone and acetic acid has shown promise as a less toxic substitute. High chloride concentrations in urban meltwater can be toxic for plants and aquatic organisms. Anti-caking compounds, such as cadmium, added to deicing salts to aid in salt spreading can exacerbate pollution problems. In addition to deicing chemicals, urban snow receives significant particulate inputs from human activity and anti-skid sand or cinder applications, especially snow that accumulates curbside. Dry deposition of trace metals from human activity in urban areas also contributes to contamination. Presence of snow in winter reduces the opportunity for street sweeping, which can increase the particulates, solid organic debris (leaves and trash) and trace-metal pollutant load in urban snow. See guidelines established by transportation departments in many regions, such as the *Road Salt and Snow and Ice Control Primer* among others published by the Transportation Association of Canada.

As in natural snowpacks, the first flush of meltwater from urban snow will generally yield much higher concentrations of dissolved solids than later melt (see Chapter 8). Particulates, trash, and trace metals adsorbed on particulates often remain on the soil at the snow disposal site. High chloride content of urban meltwater can increase mobility of trace metals in melt water, which enhances the threat of downstream contamination.

Snow removed from streets, roads, and parking areas in central business districts where large snowfalls are commonly encountered, should be transported and disposed of in dedicated storage areas. Large snow piles can result that often persist late into the melt season. Since snow storage areas can be the depository for pollutants in urban snow, it is recommended that a small retention basin be employed downslope from snow piles to help retain meltwater pollutants (see Figure 12.10). Snow storage sites should also be fenced and located to avoid wellhead protection and playground areas.

12.4 Albedo modification

Albedo modification to influence shortwave radiation exchange at the snowpack surface has been used for centuries to increase melt rates (Slaughter, 1969). Increases in melt can augment soil water content for early-season crop production, clear sidewalks and streets to improve accessibility, reduce ice jams in rivers, and control snowmelt runoff in streams. Air pollutants that settle onto the snowpack surface can also contribute to acceleration of melt rates. Dusting of river ice and snow to speed melt and prevent ice jams at critical locations in channels is still commonly practiced.

Various types of blackening agents such as wood ash, cinders, fly ash, coal dust, etc. applied to the snowpack surface can greatly reduce the surface albedo, thereby increasing the amount of absorbed shortwave radiation and melt. Small reductions in albedo can have a disproportionately large effect on absorbed shortwave radiation and melt. A 25% reduction in albedo, e.g. a decrease from 80% to 60% reflection, would cause a 100% increase in absorbed shortwave from 20% to 40% absorption. Since melt rates are generally closely linked to absorbed shortwave radiation, melt rates generally also increase by large amounts. For example, Beyerl (1999) showed that 0.03 kg m^{-2} surface application of wood ash reduced snowpack albedo by 35% and increased melt rates by 70%. Warren and Wiscombe (1981) found that fine particulates such as soot and soil dust could reduce snow albedo by 5–15% in the wavelength bands up to about 0.9 μm without affecting radiation exchange at larger wavelengths where snow is a good absorber.

Practical application of albedo modification to enhance melt is not as straightforward as might be imagined (Slaughter, 1969). Properties of ideal surface additives are: high shortwave absorptivity, high thermal conductivity, low solubility, low toxicity, and high particle density (nonfloatable). In addition, the material should be low cost, readily available, easily applied, and otherwise environmentally benign. Materials that meet all or most of these criteria are not easy to find. Timing of surface applications also must be considered so that fresh snowfall does not cover treated surfaces, although dark layers rapidly reappear unless the fresh snowfall is deep. Finally, surface accumulations of debris can insulate the surface and decelerate melt, as evidenced by large piles of urban snow covered by thick layers of leaves, landscape mulch, and other particulates that can endure into late spring (Figure 12.10). Kongoli and Bland (2002) found that manure applications to snow actually decreased melt rates compared to untreated snow. Prediction of the albedo modification effects of various surface additives on melt rates is difficult, but the method can be quite effective in altering melt rates.

12.5 Snow management in ski areas

Snow is modified or managed in many ways in ski-area developments to enhance recreational opportunities and provide a safe environment for visitors. A major activity at ski areas is snow-making to augment snow accumulation. In addition, skiers in steep terrain with large natural snow accumulations also require protection from avalanches which is achieved by managing snow in avalanche startup, track, and runout zones.

Snow making is accomplished using specialized machines to spray a fine mist of water droplets into cold air to create snowfall where desirable for skiing. Various types of snow-making equipment are available, but two basic types are compressed-air snow guns and fan-type sprayers (Figure 12.11). Snow making requires fairly significant water supplies and energy for compressors, pumps, and fans. Generally a wet-bulb temperature in the air of less than −2 °C is required for proper freezing. Nucleating agents may be added to the sprayed water to enhance freezing of

Figure 12.11 Ratnik low-energy modified snow gun (left) and modified fan-type Lenko snow gun (right) being used to augment snowpack for skiing (courtesy of Ratnik Industries, Inc. and Lenko Quality Snow, Inc., respectively). See also color plate.

droplets. The man-made snow generally must be groomed or smoothed to create a useful surface for skiing. Ideally, return flows from melting on ski slopes run back to ponds or infiltrate to recharge aquifers used to provide snow-making water.

Snow making can significantly affect the water balance of ski areas. Some additional water is lost or consumed due to evaporation/sublimation losses during snow making and due to increased durations of snow cover on ski slopes. Eisel *et al.* (1988) estimated that about 6% evaporation/sublimation loss occurred during actual snow making in Colorado, USA. In a later study, Eisel *et al.* (1990) estimated total losses from snow making were 13–37% of water used at several ski areas in Colorado, USA; including losses due to snow making, evaporation/sublimation loss from ski slopes, and transpiration loss from return flows. Since natural evaporation/ sublimation losses will occur from snow and exposed soil on ski runs, the added effect of snow-making activities on water losses should be the net loss, e.g. total evaporation minus natural evaporation without snow making. Regardless, it is clear that snow making will increase consumptive use of water to some extent.

In ski areas with steeper terrain and large snowpack accumulations, reduction in the risk caused by snow avalanches is needed. Two types of snow avalanches occur: (1) loose-snow avalanches that move on the surface and cause relatively low risk and (2) the larger slab avalanches where large blocks of snow slide downslope with sometimes devastating effects (Figure 12.12). Areas where slab avalanches

Figure 12.12 Large slab avalanche in Colorado, USA that began in the alpine startup zone (photo courtesy of Richard Armstrong, US National Snow and Ice Data Center, Boulder, CO). See also color plate.

typically occur are easily distinguished with well-defined startup zones at high elevation (often above timber line), avalanche tracks that follow confined drainage-ways or sometimes slide down broad slopes, and runout zones in valleys where the avalanches generally come to rest. Avalanche tracks in subalpine regions are often recognizable by the absence of older trees.

Snow management practices employed to reduce risk from avalanches in ski areas or to protect structures and transportation corridors vary for the startup, track, and runout zones (Table 12.1). Risk at ski areas can be reduced by simply closing facilities when avalanche hazards are deemed highest; however, artificial triggering of avalanches by using explosives and artillery under very carefully controlled

Table 12.1 *Summary of snow management practices for reduction of avalanche hazards (Perla and Martinelli, 1978)*

AVALANCHE ZONE	OBJECTIVE	PRACTICE
Startup Zone	Prevent snow cornice formation	Wind accelerators and deflectors, explosives, upwind snow fences
	Retard avalanche initiation	Walls, fencing, netting, reforestation
	Trigger avalanches at safe times	Explosives, artillery
Track Zone	Prevent damage to new structures	Land-use zoning by avalanche hazard classes
	Protect roads and rail lines	Tunnels, shed roofs
	Prevent damage to existing structures	Deflecting walls and foundations at base of structures
Runout Zone	Stop snow movement	Earthen mounds and constructed walls
	Protect new structures	Land-use zoning by avalanche hazard classes

conditions can be quite effective. In addition, snow accumulations in high-elevation avalanche startup zones can be stabilized by using snow fences, nets, and walls. Formation of cornices of wind-blown snow and ice above avalanche startup zones, that may collapse and trigger avalanches, can also be minimized using explosives or wind deflectors that accelerate blowing snow past the cornice-formation regions. Reforestation of startup zones to stabilize snowpacks can also be attempted, but is extremely difficult given the typically severe growing conditions in subalpine regions. Within avalanche track zones, various types of deflecting walls can be used to divert avalanching snow away from the base or foundation of structures. Avalanching snow in track zones can also be deflected over roads and rail lines by using tunnels or shed roofs. Finally, within avalanche runout zones, deflecting walls and sets of rounded earthen mounds can be used to slow and confine avalanches to safe areas. The most effective method for protecting new human development from snow avalanches is land-use zoning based upon avalanche hazard classes. Readers are referred to the *Avalanche Handbook* (1978) by Perla and Martinelli for more details related to snow-avalanche forecasting and protection practices. Also see McClung and Schaerer (2006) for further details on mountain weather and formation and occurrence of avalanches. Finally, a handbook by Weir (2002) that deals with avalanche hazards and forest practices is another excellent source of information.

When large volumes of snow are displaced by inducement of snow avalanches, the timing of snowmelt runoff may be altered (Martinec, 1989). Avalanched snow deposited in flat, exposed runout zones at low elevations, generally is exposed to earlier and higher surface melt rates, which could accelerate snowmelt runoff in spring. However, avalanched snow that is concentrated in large deposits with reduced surface area, especially if it accumulates in narrow, steep-sided shaded runout zones, may extend the snowmelt-runoff season.

12.6 Climate change and snow hydrology

Climate change due to increases in atmospheric greenhouse gases could cause increases in air temperatures and changes in precipitation regimes that would have important impacts on snow hydrology. Increases in air temperatures can cause greater occurrence of rainfall than snowfall events as well as increases in snowpack melt rates. Any accompanying alterations in precipitation regimes due to climate change can also influence total snowfall and meltwater runoff amounts, the seasonal timing of that runoff, and/or the occurrence of extreme melt and rain-on-snow events. Loss of snow and ice cover over large land areas due to warming and atmospheric pollution also can greatly reduce the global albedo and enhance absorption of solar radiation at the Earth's surface; thereby producing positive feedback to the climate-change processes (see Figure 12.13). The likelihood of occurrence of climate change is currently very uncertain and implications of climate change for many aspects of present-day society are being actively researched (IPCC, 2001).

The impact of climate change on streamflow from mountainous areas and high latitudes where snowfall is significant will depend upon the combination of air temperature and precipitation changes that occur. Estimates of air-temperature changes using various global climate models are reasonably consistent, at least in sign, but precipitation change forecasts are highly variable. Increases in air temperature alone would causes decreases in annual flows due to increases in evaporation. Changes in mean air temperature of $+2$ to $+4\,^{\circ}\text{C}$, which are commonly predicted using General Circulation Models for the year 2100 (Wigley, 1999), would result in annual flow declines up to about 10%. However, such warming trends in combination with annual precipitation changes ranging from -25% to $+25\%$, could cause annual streamflow changes to range from -51% to $+67\%$, respectively, for mountainous basins in western United States (Frederick and Gleick, 1999). Increases in annual mean streamflows are more likely at higher latitudes because warming is expected to be accompanied by increases in precipitation at these latitudes. Seasonal estimates of climate change for specific regions across the world

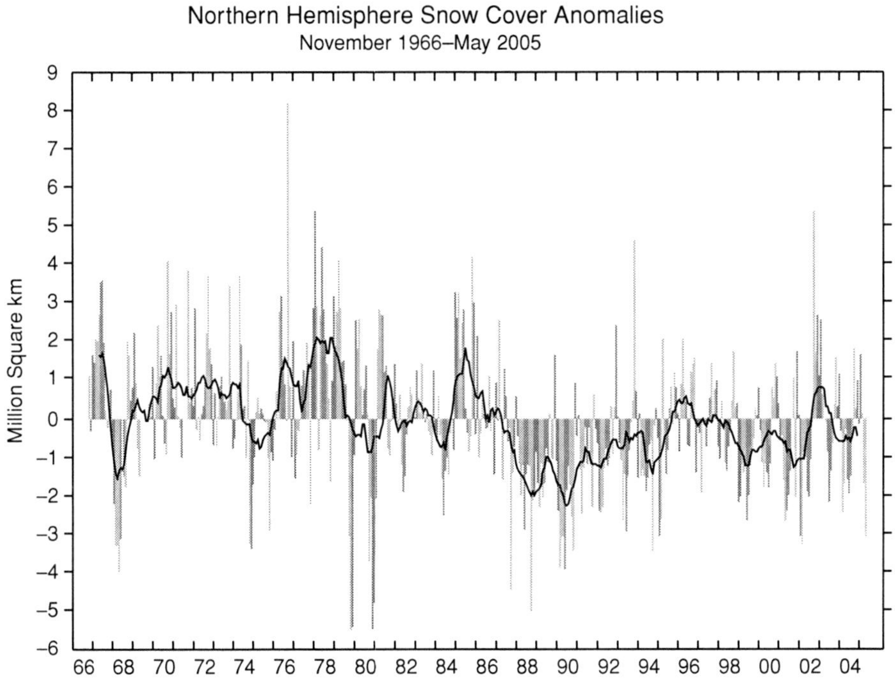

Figure 12.13 Variations in monthly snow cover over the Northern Hemisphere including Greenland for Nov 1966–May 2005 showing a decline in snow-cover possibly due to global warming about a mean snow-covered area of 25.6 million km^2 (courtesy of Dr. David A. Robinson, Rutgers University, Snow Lab.).

are needed in the future to improve estimates of climate-change effects on snow hydrology.

The timing as well as volume of streamflow from snowmelt would be affected by the climate-change scenario assumed. For example, Rango *et al.* (2003) estimated the impacts of climate change on seasonal snowmelt runoff in the Rio Grande at Del Norte, Colorado for 2050 and 2100 using the Snowmelt-Runoff Model (Figure 12.14). Based upon global climate model forecasts for the area, they assumed a 2 °C increase in air temperature and a 5% increase in precipitation by year 2050 and shifts of 4 °C and 10%, respectively, by year 2100. Increased air temperatures caused earlier and more concentrated snowmelt runoff due to higher melt rates and more rainfall events earlier in the winter. Increased peakflows also would be expected by year 2100 for the assumed climate-change scenarios. More frequent occurrence of large rain-on-snow events, often associated with serious flooding, could be a further consequence, but the forecasting of large individual storm events due to climate change is currently not possible.

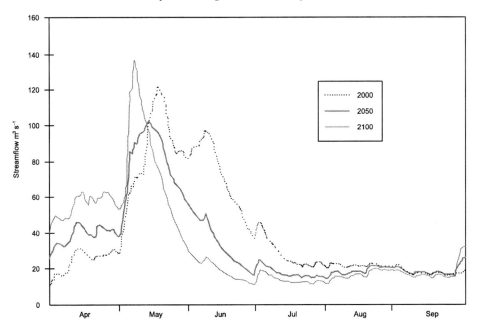

Figure 12.14 Simulation of climate-change impacts on streamflow from snowmelt for the Rio Grande at Del Norte, Colorado for years 2000, 2050 and 2100 (Rango *et al.*, 2003, courtesy USDA, ARS).

12.7 References

Anderson, H. W., Hoover, M. D., and Reinhart, K. G. (1976). *Forests and Water: Effects of Forest Management on Floods, Sedimentation, and Water Supply*, Gen. Tech. Rpt. PSW-18. Pacific Southwest For. Range Exp. Sta.: US Department of Agriculture, Forest Service.

Beyerl, H. (1999). Increased melt rate due to wood ash on snow. Unpublished M. S. Thesis, Lule Univ. Technology, Sweden.

Bilbro, J. D. and Fryrear, D. W. (1988). Chapter 8: Annual herbaceous windbarriers for protecting crops and soils and managing snowfall. In *Agriculture, Ecosystems and Environ.*, **vol. 22/23**, 149–61.

DeWalle, D. R. and Lynch, J. A. (1975). Partial forest clearing effects on snowmelt runoff. In *Proceedings Symposium Watershed Management*, Irrig. Drainage Div., American Society of Civil Engineers, pp. 337–46.

DeWalle, D. R., Parrott, H. A., and Peters, J. G. (1977). Effect of clearcutting deciduous forest on radiation exchange and snowmelt in Pennsylvania. In *Proceedings of the 34th Annual Eastern Snow Conference*, Belleville, Ontario, Canada, pp. 105–17.

Eisel, L. M., Bradley, K. M., and Leaf, C. F. (1990). Estimated runoff from man-made snow. *Water Resour. Bull.*, **26**(3), 519–26.

Eisel, L. M., Mills, K. D., and Leaf, C. F. (1988). Estimated consumptive loss from man-made snow. *Water Resour. Bull.*, **24**(4), 815–20.

Federer, C. A., Pierce, R. S., and Hornbeck, J. W. (1972). Snow management seems unlikely in the Northeast. In *Proceedings Symposium Watersheds in Transition*,

American Water Resources Association, Ft. Collins, CO June 19–22, 1972, pp. 212–19.

Frederick, K. D. and Gleick, P. H. (1999). *Water and Global Climate Change: Potential Impacts on U. S. Water Resources*. Pew Center on Global Climate Change, www.pewclimate.org.

Gary, H. L. (1974). *Snow Accumulation and Melt Along Borders of a Strip Cut in New Mexico*, Res. Note RM-279. Rocky Mtn. For. Range Exp. Sta.: US Department of Agriculture, Forest Service.

Gary, H. L. (1980). Patch clearcuts to manage snow in lodgepole pine. In *Proceedings, Watershed Management Symposium*, vol. I, Irrig. and Drainage Div., American Society of Civil Engineers, Boise, ID July 21–23, 1980, pp. 335–46.

Gary, H. L. and Troendle, C. A. (1982). *Snow Accumulation and Melt Under Various Stand Densities in Lodgepole pine in Wyoming and Colorado*, Res. Note RM-417. Rocky Mtn. For. Range Exp. Sta.: US Department of Agriculture, Forest Service.

Golding, D. L. (1981). Hydrologic relationships in Interior West Watersheds., In *Proceedings, Symposium Interior West Watershed Management*, Spokane, WA April 8–10, 1980. Pullman, WA: Washington State University Cooperative Extension, pp. 107–16.

Golding, D. L. and Swanson, R. H. (1986). Snow distribution patterns in clearing and adjacent forest. *Water Resour. Res.*, **22**(13), 1931–40.

Heisler, G. M. and DeWalle, D. R. (1988). Chapter 2: Effects of windbreak structure on wind flow. In *Agriculture, Ecosystems and Environ.*, **vol. 22/23**, 41–69.

Ho, C. L. I. and C. Valeo (2004). Observations of urban snow properties in Calgary, Canada. *Hydrol. Processes*, **19**, 459–73.

Hoover, M. D. and Leaf, C. F. (1967). Processes and significance of interception in Colorado subalpine forest. In *Proceedings Seminar Forest Hydrology*, ed. W. E. Sopper and H. W. Lull, Univ. Park, PA Aug. 29–Sept. 10, 1965. Oxford: Pergamon Press, pp. 213–24.

Hornbeck, J. W. and Pierce, R. S. (1970). Changes in snowmelt runoff after forest clearing on a New England watershed. In *Proceedings of the Eastern Snow Conference*, 1969, pp. 104–12.

IPCC (2001). Chapter 2: Observed climate variability and change. In *Climate Change 2001: Working Group I: The Scientific Basis*, Intergovernmental Panel on Climate Change, www.grida.no/climate/ipcc_tar/wg1/index.htm.

Kongoli, C. E. and W. L. Bland. (2002). Influence of manure application on surface energy and snow cover. *J. Environ. Qual.*, **31**, 1166–1173.

Martinec, J. (1989). Hydrological consequences of snow avalanches. In *World Meteorological Organization*, Proc. Technical Conf., Geneva, Hydrology and Disasters, ed. O. Starosolszky. London: James and James, pp. 284–93.

McClung, D. and P. Schaerer. (2006). *The Avalanche Handbook*, 3rd Edition. Seattle, WA: The Mountaineers Books.

Meiman, J. R. (1987). Influence of forests on snowpack accumulation. In *Management of Subalpine Forests: Building Upon 50 Years of Research*, Proc. Tech. Conf., Gen. Tech. Report RM-149. US Department of Agriculture, pp. 61–7.

Perla, R. I. and Martinelli, Jr, M. (1978). *Avalanche Handbook*. In Agric. Handbook 489. Washington, D.C.: US Printing Office.

Rango, A., Gomez-Landesa, E., Bleiweiss, M., Havstad, K., and Tanksley, K. (2003). Improved satellite snow mapping, snowmelt runoff forecasting, and climate change simulations in the Upper Rio Grande Basin. *World Resource Rev.*, **15**(1), 25–41.

Schmidt, R. A. and Troendle, C. A. (1989). Snowfall into a forest and clearing. *J. Hydrol.*, **110**, 335–48.

Schmidt, R. A., Troendle, C. A. and Meiman, J. R. (1998). Sublimation of snowpacks in subalpine conifer forests. *Can. J. For. Res.*, **28**, 501–13.

Scholten, H. (1988). Chapter 20: Snow distribution on crop fields. In *Agriculture, Ecosystems and Environ.*, **vol. 22/23**, 363–80.

Semadeni-Davies, A., A. Lundberg, and L. Bengtsson. (2001). Radiation balance of urban snow: a water management perspective. *Cold Regions Sci. and Technol.*, **33**, 59–76.

Shaw, D. L. (1988). Chapter 19: The design and use of living snow fences in North America. In *Agriculture, Ecosystems and Environ.*, **vol. 22/23**, 351–62.

Slaughter, C. W. (1969). *Snow Albedo Modification: a Review of Literature*, Technical Rep. 217. Hanover, NH: US Army Cold Regions Research and Engineering Laboratory.

Sturges, D. L. (1992). Streamflow and sediment transport responses to snow fencing a rangeland watershed. *Water Resour. Res.*, **28**(5), 1347–56.

Sturges, D. L. and Tabler, R. D. (1981). Management of blowing snow on sagebrush rangelands. *J. Soil and Water Conserv.*, **36**(5), 287–92.

Tabler, R. D. (1980). Geometry and density of drifts formed by snow fences. *J. Glaciol.*, **26**(94), 405–19.

Tabler, R. D. (1991). *Snow Fence Guide*. Strategic Highway Research Program, SHRP-W/FR-91–106, National Res. Council, Washington, DC: National Academy of Sciences.

Tabler, R. D. (1994). *Design Guidelines for the Control of Blowing and Drifting Snow*. Strategic Highway Research Program, SHRP-H-91–381, Washington, DC: National Academy of Sciences.

Tabler, R. D. (2003). *Controlling Blowing and Drifting Snow with Snow Fences and Road Design*. Final Report, NCHRP Project 20–7(147).

Tabler, R. D., Pomeroy, J. W., and Santana, B. W. (1990). *Drifting Snow*. In *Cold Regions Hydrology and Hydraulics*, Tech. Council on Cold Regions Engin. Monograph, ed. W. L. Ryan and R. D. Crissman. New York: American Society of Civil Engineers, pp. 95–145.

Transportation Association of Canada (1999). *Road Salt and Snow and Ice Control*. (www.tac-atc.ca/)

Troendle, C. A. (1987). The Potential Effect of Partial Cutting and Thinning on Streamflow from the Subalpine Forest, Res. Paper RM-274. Rocky Mountain Forest and Range Exp. Sta.: US Department of Agriculture, Forest Service.

Troendle, C. A. and Kaufmann, M. R. (1987). Influence of forests on the hydrology of the subalpine forest. In *Management of Subalpine Forests: Building on 50 Years of Research*, Proc., Tech. Conf., Gen. Tech. Rept., RM-149. Rocky Mtn. Forest and Range Exp. Sta.: US Department of Agriculture, Forest Service, pp. 68–78.

Troendle, C. A. and King, R. M. (1985). The effect of timber harvest on the Fool Creek Watershed, 30 years later. *Water Resour. Res.*, **21**(12), 1915–22.

Troendle, C. A. and Leaf, C. F. (1981). Effects of timber harvest in the snow zone on volume and timing of water yield. In *Proceedings, Symp. Interior West Watershed Management*, Spokane, WA April 8–10, 1980. Pullman, WA: Washington State University Cooperative Extension, pp. 231–44.

Verge, R. W. and Williams, G. P. (1981). Chapter 16: Drift control. In *Handbook of Snow, Principles, Processes, Management and Use*, ed. D. M. Gray and D. H. Male. Toronto: Pergamon Press, pp. 630–47.

Warren, S. G. and W. J. Wiscombe. (1981). A model for the spectral albedo of snow. II: Snow containing atmospheric aerosols. *J. Atmos. Sci.*, **37**, 2734–45.

Weir, P. (2002). Snow Avalanche Management in Forested Terrain. *Land Management Handbook Series*, **55**, British Columbia, Canada, Ministry of Forests, Forest Science Program.

Wigley, T. M. L. (1999). *The Science of Climate Change: Global and US Perspectives.* Pew Center on Global Climate Change, www.pewclimate.org.

Appendix A

Physical constants

Table A.1 *Saturation vapor pressure variations with temperature (List, 1963)*[a]

Temperature (°C)	Over liquid water (Pa)[b]	Over ice (Pa)
−30	50.88	37.98
−25	80.70	63.23
−20	125.40	103.2
−15	191.18	165.2
−10	286.27	259.7
−5	421.48	401.5
0	610.78	610.7
5	871.92	
10	1227.2	
15	1704.4	
20	2337.3	
25	3167.1	
30	4243.0	

[a] List, R. J. (1963). Smithsonian Meteorological Tables, Sixth Revised Edn. Smithsonian Misc. Collections, Vol. 114, Washington, DC.
[b] 1 mb = 100 Pascals (Pa)

Table A.2 Density, specific heat, and thermal conductivity of air, ice and water (List, 1963; Monteith, 1973; US Navy, 2003)[a]

Temperature (°C)	AIR				Ice/Water		
	dry density[b] (kg m^{-3})	saturated density[b] (kg m^{-3})	specific heat[c] (J kg^{-1} K^{-1})	thermal conductivity (W m^{-1} K^{-1})	density (kg m^{-3})	specific heat (J kg^{-1} K^{-1})	thermal conductivity (W m^{-1} K^{-1})
−40	1.51	1.51	1005	0.023	924	1730	2.65
−30	1.45	1.45	1005	0.024	923	1825	2.55
−20	1.39	1.39	1005	0.024	922	1920	2.44
−15	1.37	1.37	1005	0.024	921	1970	2.39
−10	1.34	1.34	1005	0.025	921	2020	2.33
−5	1.32	1.31	1005	0.025	921	2060	2.28
0	1.29	1.29	1005	0.025	920/1000	2110/4220	2.230/0.567
5	1.27	1.26	1005	0.026	1000	4200	0.575
10	1.25	1.24	1005	0.026	1000	4190	0.582
15	1.22	1.22	1005	0.026	999	4190	0.590
20	1.2	1.19	1005	0.027	999	4180	0.597
30	1.16	1.15	1006	0.027	996	4170	0.612

[a] List (1963) See Table A.1.
Monteith, J. L. (1973). *Principles of Environmental Physics*. New York: American Elsevier Publishing Co., Inc.
US Navy, Coast Guard (1999). Chemical Hazards Response Information System, Selected Properties of Fresh Water, Sea Water, Ice and Air. (www.chrismanual.com/Intro/prop.htm)
[b] at standard pressure
[c] at constant pressure

Appendix B

Potential solar irradiation theory

Potential solar irradiation theory considers physical factors affecting receipt of shortwave radiation, but excludes effects of the atmosphere (see Lee, 1963; Frank and Lee, 1966). This theory includes the effects of times of year and day, latitude, slope inclination and aspect on incoming solar radiation. Once potential solar irradiation data are computed for a specific site and time, empirical adjustments can be added to approximate the effects of the atmosphere and the distribution of incoming solar radiation into diffuse and beam fractions as described in Chapters 6 and 7. This appendix provides a brief discussion of the theory with example calculations divided into two sections: one for horizontal surfaces and the second for sloping surfaces.

B.1 Potential solar irradiation on horizontal surfaces

Potential solar irradiation on a horizontal surface for any instant in time (I_s) can be computed as:

$$I_s = (I_0/e^2)\cos Z \tag{B.1}$$

where:

$I_0 =$ the solar constant
$e =$ the radius vector
$Z =$ zenith angle

The solar constant is the average flux density of radiation received outside the Earth's atmosphere perpendicular to the solar beam. The value of the solar constant has been given as 1360 to 1400 W m^{-2} (1.94 or 2.00 cal cm^{-2} min^{-1}) (Monteith, 1973) and 1360 W m^{-2} is used for calculations here. The radius vector adjusts the solar constant for the slight effects ($\pm3\%$) of variations in the Earth–Sun distance during the year. Radius vector values can be obtained from solar ephemeris

Table B.1 *Solar declination and radius vector for days during year (after Frank and Lee, 1966).*

Dates, approximate	Julian Dates (Jan 1 = 1)	Solar Declination δ, degrees	(Radius Vector)² e^2
6/22 summer solstice	173	+23.5	1.03297
6/1 and 7/12	152 and 193	+21.967	1.03090
5/18 and 7/27	138 and 208	+19.333	1.02728
5/3 and 8/10	123 and 222	+15.583	1.02190
4/19 and 8/25	109 and 237	+10.917	1.01528
4/4 and 9/9	94 and 252	+5.633	1.00739
3/21 and 9/23 equinoxes	80 and 266	0	0.99960
3/7 and 10/8	66 and 281	−5.633	0.99154
2/20 and 10/22	51 and 295	−10.917	0.98404
2/7 and 11/5	38 and 309	−15.583	0.97790
1/24 and 11/19	24 and 323	−19.333	0.97285
1/10 and 12/3	10 and 337	−21.967	0.96938
12/22 winter solstice	356	−23.5	0.96759

tables for a given date (see Table B.1). Zenith angle is the angle between the solar beam and a perpendicular to the horizontal surface, e.g. $Z = 0$ when the Sun is directly overhead and 90° at sunrise and sunset. The cosine law of illumination, represented in Equation (B.1), dictates that flux density of radiation received is reduced according to the cosine of the zenith angle; maximum irradiation occurs at normal incidence when $\cos Z = 1$. The complement of the zenith angle is the solar altitude or the elevation angle of the Sun above the horizon.

Zenith angle (Z) can be computed from three other angles: latitude, solar declination, and hour angle according to:

$$\cos Z = (\sin \theta \, \sin \delta + \cos \theta \, \cos \delta \, \cos \omega t) \qquad \text{(B.2)}$$

where:

θ = latitude
δ = solar declination
ωt = hour angle

Latitude indexes the position of a site in degrees north (+) or south (−) of the equator. Solar declination accounts for seasonal effects and ranges between $+23\frac{1}{2}°$ on the summer solstice and $−23\frac{1}{2}°$ on the winter solstice (see Table B.1). The hour

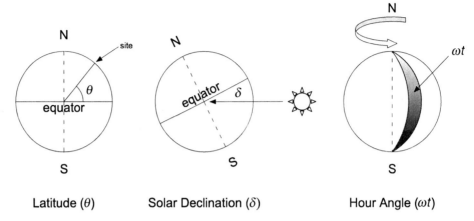

Figure B.1 Angles used to compute potential solar irradiation on horizontal surfaces.

angle is the angle the Earth must rotate so that the Sun is directly above the meridian of the designated site. The hour angle accounts for time of day and is the product of the angular velocity of the Earth's rotation ($\omega = 360°/24\ h = 15°\ h^{-1}$) and the true solar time (t) in hours before ($-$) or after ($+$) solar noon. These angles are illustrated in Figure B.1. True solar time can be computed from local standard time plus a correction for the difference in longitude between the meridian for the surface and the reference meridian for the appropriate standard time zone [(longitude of $RM°$ $-$ longitude of surface$°$)/15$°$ h^{-1}] and a small correction for the equation of time (-0.24 to $+0.27$ h) that also varies with time of year (see ephemeris tables such as List, 1968).

To illustrate the use of Equations (B.1) and (B.2) combined, I_s is computed for a site on February 7 ($\delta = -15.583°$ and $e^2 = 0.9779$ from Table B.1) at a time equal to two hours before solar noon [$\omega t = (-2\ h)\ (15°\ h^{-1}) = -30°$] and a latitude of $55°$ north of the equator ($\theta = 55°$) as:

$$I_s = (1360/0.9779)[(\sin 55°)(\sin -15.583°)+(\cos 55°)(\cos -15.583°)(\cos -30°)]$$
$$= (1390.74)[(0.8191)(-0.2686) + (0.5736)(0.9632)(0.8660)]$$
$$= (1390.74)(0.2584) = 359\ W\ m^{-2}$$

The flux density of incoming potential shortwave radiation is obviously reduced well below the solar constant of 1360 W m^{-2}. The zenith angle equivalent to this time of day can also be found from the term in brackets as $\cos Z = 0.2584$ and $Z =$ arcos $0.2584 = 75.02°$. Solar altitude ($90° - Z$) would thus be $14.98°$ for this time, date, and location.

Many applications require total potential solar radiation for an entire day or part of day rather than just an instant in time. By solving Equation (B.2) for sunrise and sunset times and integrating the combined Equations (B.1) and (B.2) over time, the daily total potential solar radiation can also be obtained. Since $Z = 90°$ and $\cos Z = 0$ at sunrise and sunset, Equation (B.2) can be set equal to zero and solved for times $(\pm t)$ of sunrise and sunset as:

$$\cos \omega t = -\tan \theta \, \tan \delta \qquad (B.3)$$

Note that daylength theoretically only depends upon latitude and solar declination. Also, times of sunrise and sunset are always symmetrical about solar noon for horizontal surfaces, but with opposite signs. Whenever the value of $-\tan \theta \tan \delta$ in Equation (B.3) is ≤ -1, the maximum daylength (maximum of 24 h) has been achieved and the Sun is at or above the horizon all day.

Once times of sunrise and sunset are obtained, the total daily potential solar radiation (I_q) can be computed by integrating Equations B.1 and B.2 from sunrise (t_1) to sunset (t_2) as (Frank and Lee, 1966):

$$I_q = (I_0/e^2)[(t_2 - t_1) \sin \theta \, \sin \delta + 1/\omega \, \cos \theta \, \cos \delta (\sin \omega t_2 - \sin \omega t_1)] \qquad (B.4)$$

In this integration with time in hours, the solar constant I_0 is converted to give $I_0 = 4.896$ MJ h^{-1} m^{-2}. The average value of the integral, obtained by dividing the term in brackets in Equation (B.4) by total daylength (N), represents the mean effective $\cos Z$ during the solar day and permits the derivation of the mean zenith angle during an entire solar day.

Mean effective $\cos Z$ has utility in computing beam and diffuse solar radiation on slopes as described in Chapter 7.

Continuing the example using Equation (B.3) for a horizontal surface at latitude $\theta = 55°$ and for February 7 with solar declination $\delta = -15.583°$, the times of sunrise and sunset can be computed as:

$$\cos \omega t = -\tan(55°) \tan(-15.583°) = 0.3982$$
$$\omega t = \arccos(0.3982) = 66.53°$$
$$t = \pm[66.53°/(15° \text{ h}^{-1})] = \pm 4.435 \text{ h}$$
$$\text{sunrise} = t_1 = -4.435 \text{ h and sunset} = t_2 = +4.435 \text{ h}$$

The total length of solar day or maximum possible duration of sunshine (N) is thus 2 times 4.435 or 8.87 hours on this day. The ratio of measured hours of actual sunshine to the maximum possible duration of sunshine (N) computed as above is often used to index cloud-cover effects on shortwave radiation as discussed in Chapter 6.

The total potential solar radiation on February 7 for a horizontal surface can be computed using Equation (B.4) with $1/\omega = 3.8197$ rad deg^{-1} as:

$$I_q = (4.896/0.9779)[(4.435 - (-4.435)) \sin 55° \sin -15.583° + 3.8197 \cos 55°$$
$$\times \cos -15.583° (\sin 66.53° - (\sin -66.53°))]$$
$$= (5.007)[1.919] = 9.609 \text{ MJ m}^{-2} \text{ from sunrise to sunset}$$

The total potential solar irradiation of 9.609 MJ m^{-2} can also be expressed as an average flux density of 111 W m^{-2} for a 24-hr daily period or an average flux density of 301 W m^{-2} for the 8.87-hr-long solar day. The average value of the integral, 1.919 in brackets above, divided by total daylength of 8.87 hours gives the cosine of the average effective zenith angle for the day, e.g. $\cos Z = 0.2163$ or $Z = 77.5°$ and an average effective solar altitude of $90 - Z = 12.5°$ on this day. Figure 6.2 in Chapter 6 shows the annual variations of total daily potential solar radiation in MJ m^{-2} received on a horizontal surface for various latitudes computed with Equation (B.4).

B.2 Potential Solar Irradiation on Sloping Surfaces

Computation of potential solar irradiation for sloping surfaces can also be based on theory given by Lee (1963) and Frank and Lee (1966) paralleling computations given in section B.1 above. In Chapter 7, the use of potential solar irradiation theory to determine the radiation balance of slopes is described. Correction of daily direct-beam irradiation for slope effects is based upon finding an "equivalent horizontal surface" that receives radiation at the same angle as the slope. For any instant in time the potential solar irradiation on a slope (I'_s) can be related to irradiation of the equivalent horizontal surface as:

$$I'_s = (I_0/e^2) \cos Z' = (I_0/e^2)[\sin \theta' \sin \delta + \cos \theta' \cos \delta \cos \omega t'] \qquad \text{(B.5)}$$

where:

Z' = angle between the solar beam and a perpendicular to the slope
θ' = latitude of the equivalent horizontal surface

$$= \arcsin [\sin k_s \cos h \cos \theta + \cos k_s \sin \theta] \qquad \text{(B.6)}$$

k_s = slope inclination angle
h = slope azimuth (degrees clockwise from north)
$\omega t'$ = hour angle for the equivalent horizontal surface

$$= \omega t + a \qquad \text{(B.7)}$$

a = difference in longitude between equivalent horizontal surface and slope

$$= \arctan[(\sin h \sin k_s)/(\cos k_s \cos \theta - \cos h \sin k_s \sin \theta)] \qquad \text{(B.8)}$$

In computations for a sloping surface the zenith angle is replaced by the angle of incidence between the solar beam and a perpendicular to the slope (see Figure B.2). Sunrise and sunset times for the equivalent horizontal surface are found in similar manner as in our previous example for a horizontal surface, but with the latitude of the equivalent horizontal surface rather the latitude of the slope as:

$$\cos \omega t' = -\tan \theta' \tan \delta \qquad (B.10)$$

and with the further stipulation that sunrise and sunset times for the slope can not occur before or after that, respectively, for a horizontal surface at the latitude and longitude of the slope. Times between the slope and the equivalent horizontal surface must be adjusted for the longitude shift.

Equation (B.5) must be integrated from sunrise to sunset times and angles for the slope to find a daily potential slope irradiation sum. The equation can be integrated from sunrise to sunset for the slope to obtain the total daily potential solar irradiation on the slope (I'_q) as:

$$I'_q = (I_o/e^2)[(t_2 - t_1) \sin \theta' \sin \delta + 1/\omega \cos \theta' \cos \delta (\sin \omega t'_2 - \sin \omega t'_1)] \qquad (B.11)$$

Again, the term in brackets on the right side of Equation (B.11) divided by the length of the solar day $(t_2 - t_1)$ gives an approximation of the mean cos Z' for the entire day for the slope. The equations appear daunting, but can be programmed and easily solved on the computer. A step-by-step computation is given below to illustrate the procedure.

B.2.1 Computing instantaneous potential irradiation on slope

The previous example in section B.1 above, where potential solar was computed for a horizontal surface at latitude 55° on Feb 7 ($\delta = -15.583$), will be continued. It will be assumed that slope inclination angle $k_s = 30°$ and that the slope is oriented to the southeast (SE aspect) equivalent to an azimuth angle (h) of 135° measured clockwise from north.

First, the latitude of the equivalent horizontal surface must be determined as:

$$\theta' = \arcsin[\sin 30 \cos 135 \cos 55 + \cos 30 \sin 55] = \arcin[0.5065] = 30.43°$$

which indicates that the latitude of the "equivalent horizontal surface" is south of the latitude of the actual slope (actual $\theta = 55°$). A negative θ' can occur and would indicate that the equivalent horizontal surface was in the southern hemisphere.

Next the longitude shift needed to find the "equivalent horizontal surface" must be evaluated as:

$$a = \arctan[(\sin 135 \ \sin 30)/(\cos 30 \ \cos 55 - \cos 135 \sin 30 \ \sin 55)]$$
$$= \arctan[0.4496] = 24.21°$$

Since the slope faces to the east, the equivalent horizontal surface is 24.21° of longitude to the east of the slope. For slopes facing toward the west, the longitude shift to the equivalent horizontal surface is towards the west. Equivalent horizontal surfaces for slopes facing due north or south have zero longitude shifts.

Given our earlier example, enough information is now available to compute the instantaneous potential irradiation on a slope with Equation (B.5). Computations assume an hour angle (ωt) equal to two hours before solar noon or ($\omega t = -30°$) on the slope for a comparison with computations for a horizontal surface at that same latitude and time. First the hour angle for the equivalent horizontal surface must be adjusted for the longitude shift using Equation (B.7), which gives:

$$\omega t' = -30 + 24.21 = -5.79^{o\prime}$$

and

$$I_s' = (1360/0.9779)[\sin 30.43° \sin -15.583° + \cos 30.43° \cos -15.583° \cos -5.79°]$$
$$= (1390.74)[0.6902] = 960 \text{ W m}^{-2}$$
$$Z' = \text{arcos}[0.6902] = 46.35°$$

The instantaneous flux density of potential solar irradiation for the slope of 960 W m^{-2} is much greater than that for a horizontal surface for the same date and time of 359 W m^{-2} computed in section B.1. This difference in receipt of radiation occurs because the angle of incidence Z' for radiation on the slope ($Z' = 46.35°$) is less than the zenith angle for the horizontal surface ($Z = 75.02°$ from section B.1) for the same time and date.

B.2.2 Computing daily potential irradiation on slope

Computation of the potential solar irradiation for an entire solar day begins with finding the sunrise and sunset times for the "equivalent horizontal surface" using Equations (B.7) and (B.10) as:

$$\cos \omega t' = -\tan 30.43° \tan -15.583° = 0.1638$$
$$\omega t' = \pm 80.57°$$
$$\text{sunrise } t_1' = -80.57°/15° \text{ h}^{-1} = -5.37 \text{ h}$$
$$\text{sunset } t_2' = +80.57°/15° \text{ h}^{-1} = +5.37 \text{ h}$$

These times represent sunrise and sunset for the equivalent horizontal surface at a different longitude. Again a correction is needed for the longitude shift to obtain the times for the slope and to compare them with the limiting case of sunrise and sunset times found previously for the horizontal surface at that same latitude and longitude; $\pm$ 4.435 h. The shift in longitude is equivalent to a time difference

between the equivalent horizontal surface and the slope of $a/\omega = 24.21°/15°\ h^{-1}$
$= 1.61\ h$, thus:

$$t_1' = t_1 + a$$
$$-5.37\ h = t_1 + 1.61\ h$$
$$t_1 = -6.98\ h$$

and

$$t_2' = t_2 + a$$
$$+5.37\ h = t_2 + 1.61\ h$$
$$t_2 = 3.76\ h$$

The indicated sunrise time of -6.98 hours before solar noon for the slope can not actually occur because it can not occur before sunrise for a horizontal surface for that latitude and day which is only $-4.435\ h$ before solar noon. Thus, the sunrise time for the horizontal surface ($t_1 = -4.435\ h$) becomes the limiting condition and its equivalent hour angle using Equation (B.7) [$\omega t_1' = (15°\ h^{-1})(-4.435) + 24.21°$ $= 42.32°$] must be substituted in Equation (B.11). The sunset time of $+3.76$ hours for the slope occurs before the sunset time of $+4.435$ for a horizontal surface for that day and no adjustment is needed; therefore for sunset $t_2 = +3.76\ h$ and $\omega t_2' = +80.57°$ in Equation (B.11).

Total solar daylength for the slope from sunrise to sunset is $8.195\ h$ ($4.435\ h + 3.76\ h$), which is less than daylength for the horizontal surface of $8.87\ h$ ($2 \times 4.435\ h$) found previously. Note that the sunrise and sunset times for a similarly inclined slope facing to the southwest would simply be the reverse of the southeast-facing slope case.

Using Equation (B.11) and the appropriate sunrise and sunset times and hour angles the potential solar irradiation on the slope for the day (I_q') can be computed as:

$$I_q' = (4.896/0.9779)[(3.76 - (-4.435))(\sin 30.43°)(\sin -15.583°)]$$
$$+ (3.8197)(\cos 30.43°)(\cos -15.583°)(\sin 80.57° - \sin(-42.32°))\ h]$$
$$= (5.007)[5.15353] = 25.8\ MJ\ m^{-2}$$

For comparison, the potential solar irradiation received on a horizontal surface for the same day and latitude was only $9.609\ MJ\ m^{-2}$ which illustrates the relatively large effect that slope orientation can have on receipt of radiation. The average value of the integral, 5.15353 in brackets above, divided by total daylength of 8.195 hours gives an effective $\cos Z' = 0.6289$ or an average effective angle of incidence of $Z' = 51°$ for the slope on this day. This angle of incidence becomes important to scaling of the fractions of diffuse and beam solar irradiation for slopes as described in Chapter 7.

B.3 References

Frank, E. C. and Lee, R. (1966). *Potential Solar Beam Irradiation on Slopes: Tables for 30° to 50° Latitude. Res. Paper RM-18*. Rocky Mtn. For. Range Exp. Sta.: US Department of Agriculture, Forest Service.

Lee, R. (1963). *Evaluation of Solar Beam Irradiation as a Climatic Parameter of Mountain Watersheds*, Hydrol. Paper. No. 2, Ft. Collin, CO: Colorado State University.

List, R. J. (1968). *Smithsonian Meteorological Tables*, 6th edn., Smithsonian Misc. Collection, Smithsonian Publ. 4014, vol. 114. Washington, DC: Smithsonian Institution Press.

Monteith, J. L. (1973). *Principles of Environmental Physics*. New York: American Elsevier Publishing Co., Inc.

Index

CPSIA information can be obtained at www.ICGtesting.com
Printed in the USA
BVOW071523140911

271177BV00003B/4/P